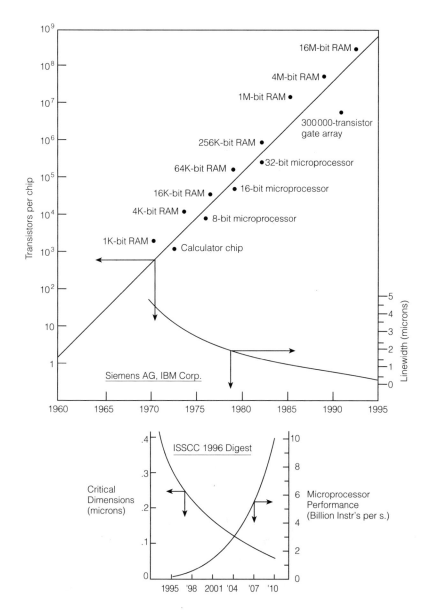

The increase of transistor density of the most pervasive IC types, memory and logic, has been steady and predictable: about every two years the density of active elements doubles on logic chips and quadruples on memory chips. Meanwhile, linewidths have steadily decreased, although recently the slope has flattened, reflecting the greater difficulty of scaling down by tenths of micrometers. *IEEE Spectrum, Sept. 1988.*

Join Us on the Internet

WWW: http://www.thomson.com
EMAIL: findit@kiosk.thomson.com

thomson.com is the on-line portal for the products, services and resources available from International Thomson Publishing (ITP).

This Internet kiosk gives users immediate access to more than 34 ITP publishers and over 20,000 products. Through *thomson.com* Internet users can search catalogs, examine subject-specific resource centers and subscribe to electronic discussion lists. You can purchase ITP products from your local bookseller, or directly through *thomson.com*.

Visit Chapman & Hall's Internet Resource Center for information on our new publications, links to useful sites on the World Wide Web and an opportunity to join our e-mail mailing list.
Point your browser to: **http://www.chaphall.com/chaphall.html**
or **http://www.thomson.com/chaphall/electeng.html** for Electrical Engineering

A service of

Single Event Phenomena

George C. Messenger
Nuclear Radiation Effects Consultant,
Messenger and Associate,
Las Vegas, Nevada

Milton S. Ash
Radiation Effects Consultant,
Santa Monica, California

CHAPMAN & HALL

 INTERNATIONAL THOMSON PUBLISHING

New York • Albany • Bonn • Boston • Cincinnati • Detroit • London
Madrid • Melbourne • Mexico City • Pacific Grove • Paris • San Francisco
Singapore • Tokyo • Toronto • Washington

Cover design: Curtis Tow Graphics

Copyright © 1997 by Chapman & Hall

Printed in the United States of America

Chapman & Hall
115 Fifth Avenue
New York, NY 10003

Chapman & Hall
2-6 Boundary Row
London SE1 8HN
England

Thomas Nelson Australia
102 Dodds Street
South Melbourne, 3205
Victoria, Australia

Chapman & Hall GmbH
Postfach 100 263
D-69442 Weinheim
Germany

International Thomson Editores
Campos Eliseos 385, Piso 7
Col. Polanco
11560 Mexico D.F
Mexico

International Thomson Publishing–Japan
Hirakawacho-cho Kyowa Building, 3F
1-2-1 Hirakawacho-cho
Chiyoda-ku, 102 Tokyo
Japan

International Thomson Publishing Asia
221 Henderson Road #05-10
Henderson Building
Singapore 0315

All rights reserved. No part of this book covered by the copyright hereon may be reproduced or used in any form or by any means—graphic, electronic, or mechanical, including photocopying, recording, taping, or information storage and retrieval systems—without the written permission of the publisher.

1 2 3 4 5 6 7 8 9 10 XXX 01 00 99 98 97

Library of Congress Cataloging-in-Publication Data

Messenger, George C.
   Single event phenomena / George Messenger, Milton Ash.
     p. cm.
   Includes bibliographical references and index.
   ISBN 0-412-09731-1 (alk. paper)
   1. Cosmic rays. I. Ash, Milton S. II. Title.
QC485.M47   1997
621.39'5 -- dc20   96-34533
                                         CIP

British Library Cataloguing in Publication Data available

"Single Event Phenomena" is intended to present technically accurate and authoritative information from highly regarded sources. The publisher, editors, authors, advisors, and contributors have made every reasonable effort to ensure the accuracy of the information, but cannot assume responsibility for the accuracy of all information, or for the consequences of its use.

To order this or any other Chapman & Hall book, please contact **International Thomson Publishing, 7625 Empire Drive, Florence, KY 41042.** Phone: (606) 525-6600 or 1-800-842-3636. Fax: (606) 525-7778. e-mail: order@chaphall.com.

For a complete listing of Chapman & Hall titles, send your request to **Chapman & Hall, Dept. BC, 115 Fifth Avenue, New York, NY 10003.**

To Our Wives
Priscilla Messenger

And

Shulamite Ash
And Our Children

Contents

	Page
Dedication	v
Preface	xi
Introduction	xvii

1 Preliminaries — 1

    1.1 Introduction — 1
    1.2 Fluence, Flux, and Current Density — 2
    1.3 Cross Sections — 5
    1.4 Chord Distribution Functions — 10
    Problems — 19
    References — 21

2 Extraterrestrial SEU-Inducing Particles — 23

    2.1 Introduction — 23
    2.2 Cosmic Rays — 24
    2.3 Other Cosmic Ray Particles — 34
    2.4 Alpha Particles — 38
    2.5 Solar Flares — 42
    2.6 Van Allen Radiation Belts — 48
    Problems — 57
    References — 58

3 Particle Penetration and Energy Deposition — 61

    3.1 Introduction — 61
    3.2 Particle Penetration in Materials — 62
    3.3 Range — 65

	3.4 Ionization Loss	67
	3.5 Bremsstrahlung Loss	68
	3.6 Pair Production, Cosmic Ray Showers	71
	3.7 LET Introduction	74
	3.8 LET Spectra (Heinrich) Curve	75
	3.9 Linear Energy Transfer (LET) Applicability	77
	Problems	85
	References	86
4	**Single Event Upset: Experimental**	**88**
	4.1 Introduction	88
	4.2 Heavy-Ion Accelerators	89
	4.3 Critical LET, Critical Charge	91
	4.4 Californum-252 Sources	100
	4.5 Lasers	112
	Problems	122
	References	124
5	**Single Event Upset Error Rates**	**127**
	5.1 Introduction	127
	5.2 Geosynchronous SEU	128
	5.3 Proton-Induced SEU	140
	5.3.1 Approximate Methods for Proton-Induced SEU	149
	5.4 Neutron-Induced SEU	155
	5.5 Alpha-Particle-Induced SEU	162
	5.5.1 Alpha-Particle-Induced Error Rate	167
	5.6 Ground-Level SEU	169
	Problems	174
	References	176
6	**Single Event Phenomena I**	**179**
	6.1 Introduction	179
	6.2 Funneling	180
	6.2.1 Funneling Models	182
	6.3 Track Transport	194
	6.4 SEU Cross-Section Morphology	199
	6.5 Device SEU Scaling	204
	6.6 Single Event Latchup	210
	6.7 South Atlantic Anomaly, Low Earth Orbits	216
	6.8 Single Event Gate Rupture, Device SEU Burnout	220
	Problems	227
	References	228

7	**Single Event Phenomena II**	**232**
	7.1 Introduction	232
	7.2 Multiple Event Upsets	233
	7.3 Ionizing Dose Effects	240
	7.4 Redundancy, Scrubbing	245
	7.5 Error Detection and Correction	249
	7.6 SEU Shielding	256
	7.7 Radiation Sensors and Detectors	263
	Problems	270
	References	272
8	**Single Event Upset Practice**	**275**
	8.1 Introduction	275
	8.2 Geosynchronous SEU Error Rate Computations	276
	8.3 Proton-Induced SEU II	284
	8.4 Neutron-Induced SEU II	293
	8.5 Single Event Latchup II	297
	8.6 Single Event Burnout II	300
	8.7 Guidelines	305
	8.7.1 How to Use Part Index Tables 8.13 Through 8.16	311
	Glossary for Tables 8.13-8.17	322
	Appendices 8.1–8.3	323
	Problems	325
	References	326

Appendix A - Answers to Problems   329

Index   347

Preface

This monograph is written for neophytes, students, and practitioners to aid in their understanding of single event phenomena. It attempts to collect the highlights as well as many of the more detailed aspects of this field into an entity that portrays the theoretical as well as the practical applications of this subject. Those who claim that "theory" is not for them can skip over the earlier chapters dealing with the fundamental and theoretical portions and find what they need in the way of hands-on guidelines and pertinent formulas in the later chapters. Perhaps, after a time they will return to peruse the earlier chapters for a more complete rendition and appreciation of the subject matter.

It is felt that the reader should have some acquaintance with the electronics of semiconductors and devices, some broad atomic physics introduction, as well as a respectable level of mathematics through calculus, including simple differential equations. A large part of the preceding can be obtained informally, through job experience, self-study, evening classes, as well as from a formal college curriculum.

This treatise attempts to put the subject matter in a proper perspective, of course as seen from the viewpoint of the authors, which is rather catholic in scope. Also, it is guided by our experience in the field and the multitude of subject journal articles. At least 90% of the latter appear, and will continue to appear, in one vehicle. It is the *IEEE Transactions on Nuclear Science*, from the middle 1970s to the present. Earlier issues contain little in the way of single event phenomena, especially as it is perceived today.

The problems at the end of each chapter should be attempted, as they provide supplementary information and drill to bolster intuition. They are constructed so as to be relatively easy for the serious reader, and all answers are provided at the end of the book.

Before describing the individual chapter details, it is well to provide a chronology of the Single Event Phenomenon. It is graciously made available by Dr E. Wolicki who played an instrumental role in the genesis of this subject. It is written in narrative form with interposed references and is excerpted verbatim. It is the only place in this treatise where names are mentioned in profusion, as we made valiant attempts to eschew them throughout the remainder of this work for reasons discussed later. The chronology follows.

The first paper that should be cited is one by Wallmark and Marcus, of RCA, that was published in 1962. These authors predicted that cosmic rays would start upsetting microcircuits when the feature size became small enough, both due to the heavily ionizing tracks and due to cosmic ray spallation reactions. They were, however, so far ahead of their time that the paper was entirely overlooked. They did turn out to be wrong in predicting that device feature sizes could not go much below about 10 microns. Questions of heat dissipation were part of the reason for estimating this size limit. Ref: J.T. Wallmark and S.M. Marcus, "Minimum Size and Maximum Packing Density of Non-redundant Semiconductor Devices," *IRE Proc.*, **50**, 286 (1962).

The next important paper was one by Binder, Smith, and Holman, of Hughes Corp., published in 1975, in which the authors reported on upsets in digital flip-flop circuits that were observed in certain space satellites. Partly because the number of upsets observed was small, their claim that the upsets were due to cosmic rays in the "iron group" was disputed by most radtiation effects specialists at the time. It was only after some difficulty that they managed to get their paper included in the 1975 IEEE Annual Conference on Nuclear and Space Radiation Effects (NSREC) and even then the paper was put in as the last paper on the last day of the conference. Their published paper was quite remarkable in that (a) it identified cosmic rays in the "iron group" as the likely cause of the upsets, (b) it already showed how to calculate the rate at which cosmic ray upsets could be expected to occur, and (c) their calculations were within a factor of 2 of the number of upsets they actually observed. Due to the aforementioned skepticism, this paper was also largely overlooked. Ref: D. Binder, E.C. Smith, and A.B. Holman, "Satellite Anomalies From Galactic Cosmic Rays," *IEEE Trans. Nucl. Sci.*, **NS-30**, 2675 (1975).

In 1978, two papers were published that would awaken the radiation effects community to the new phenomenon of, as it was then called, soft errors. The first was a paper by May and Woods in which the authors determined that the upsets which were being observed at sea level in dynamic RAMs were due to alpha particles emitted from trace amounts of uranium and thorium that were present in the materials from which the electronic devices were fabricated; these upsets were creating a serious problem from the standpoint of device reliability. Their experimental measurements were performed with polonium and americium alpha sources on 4K DRAMs and 16K CCDs. May and Woods introduced the concept of critical charge for upset and predicted that the soft error problem would continue to get worse as device feature size became ever smaller. Ref: Timothy C. May and Murray H. Woods, "Alpha Particle Induced Soft Errors in Dynamic Memories," *IEEE Electron. Dev.*, **ED-26**, 2 (1979).

The second was a paper by Pickel and Blandford in which a model was developed to predict the cosmic ray bit error rate in dynamic MOS RAMs. Specifically, this cosmic ray interaction model was used to estimate the bit error event rate in an operating satellite memory which consisted of 24 NMOS RAMs, each with 4096 memory bits. The upsets were explained as being due to energetic cosmic rays in the "iron and aluminum groups." In this paper, which was presented at the 1978 IEEE NSREC, the number of upsets was large enough to be convincing, and it was this paper that particularly alerted the space and nuclear radiatior effects community to the existence of a new and important radiation effect. The method used for calculating the number of upsets to be expected in a cosmic ray environment was similar to that of Binder et al., but used more detailed models both for the device geometry and for the cosmic ray environment. At this time, circuit upsets were typically referred to as soft errors or single-particle errors (for single cosmic ray particles). Ref: J.C. Pickel and J.T. Blandford, "Cosmic Ray Induced Errors in MOS Memory Cell," *IEEE Trans. Nucl. Sci.*, **NS-30**, 1166 (1978).

After hearing the Pickel and Blandford paper, and hearing about the work of May and Woods, Guenzer and Wolicki had the idea that if alpha particles from uranium and thorium could upset circuits, then alpha particles from nuclear reactions induced by energetic neutrons and protons might also be able to produce such upsets. In late 1978, they made measurements on 4K and 16K dynamic MOS RAMs, first in a 14-MeV (mean energy) neutron beam of a 75-MeV cyclotron at the Naval Research Laboratory, and then with 14-MeV neutrons from a D-T generator. First upsets were observed at neutron fluences of about 1E8 neutrons per square centimeter (10^8 cm^{-2}). The discovery that dynamic RAMs were vulnerable to such low neutron fluences was startling at the time because dynamic RAMs are majority carrier devices and were expected to be hard to neutron fluences three to four orders of magnitude larger than 1E8. Fluences for first upset, by 32-MeV protons, were also about 1E8 protons per cm^2. These results were presented at the 1978 IEEE NSREC Conference. This paper correctly identified neutron- and proton-induced nuclear reactions as the cause of the upsets, and because more than one particle was involved (unlike the case of single cosmic ray particles), the title of the paper was chosen to be "Single Event Upset of Dynamic RAMs by Neutrons and Protons." The field of SEUs (single event upsets) thus took its name from this paper. Ref: C.S. Guenzer, E.A. Wolicki, and R.F. Allas, "Single Event Upset of Dynamic RAMs by Neutrons and Protons," *IEEE Trans. Nucl. Sci.*, **NS-26**, 5048 (1979).

Also given at the 1979 NSREC was a paper by McNulty et al. in which proton-induced soft errors were reported. This paper also correctly identified proton-induced nuclear reactions as the cause of the errors. Ref: R.C. Wyatt, P.J. McNulty, P. Toumbas, P.L. Rothwell, and R.C. Filz, "Soft Errors Induced by Energetic Protons," *IEEE Trans. Nucl. Sci.*, **NS-26**, 4905 (1979).

Another important result presented at the 1979 NSREC was the discovery of heavy-ion-induced latchup in static RAMs by Kolasinski and Blake (Aerospace), Anthony (SAMSO), Price (JPL), and E.C. Smith (Hughes). These were the first upset experiments to be conducted with high-energy heavy ions from an accelerator and the first to observe cosmic-ray-induced latchup. Ref: W.A. Kolasinski, J.B. Blake, J.K. Anthony, W.E. Price, and E.C. Smith, "Simulation of Cosmic Ray Induced Soft Errors and Latchup in Integrated Circuit Computer Memories," *IEEE Trans. Nucl. Sci.*, **NS-26**, 5087 (1979).

In 1979, Ziegler (IBM) and Landford (SUNY, Albany) published a paper on the effects of sea-level cosmic rays on computer memories. Ref: J.F. Ziegler and W.A. Landford, "Effect of Cosmic Rays on Computer Memories," *Science*, **206**, 776 (1979).

Also published in 1979 were two papers by J.N. Bradford (RADC) entitled "A Distribution Function For Ion Track Lengths in Rectangular Volumes," *Jo. Appl. Phys.*, **50**, 3799 (1979) and "Cosmic Ray Effects in VLSI in Space Systems and Their Interaction With Earth's Space Environment," *Progr. Astrophys. Aerophys.*, **71**, 549 (1980).

The paper by Guenzer, Wolicki, and Allas came to the attention of Dr. Arden Bement, at that time Director of Research at Defense Advanced Research Programs Agency (DARPA), and he called NRL to ask if Dr. Wolicki could organize a workshop on the subject of soft errors in VLSI devices. DARPA has a major interest in computers and computer architecture and Dr. Bement wanted to know if this new phenomenon, which could perhaps be initiated by neutrons from a nuclear explosion and by trapped protons in space, might affect computers containing VLSI devices. The expectation at that time was that once the phenomenon was described to them, the DARPA computer experts could find a solution to any potential problem. In organizing this workshop, it was Dr. Wolicki's idea to involve DNA radiation effects experts in the workshop program. The workshop was held at NRL on September 5, 1979 and was the first official contact between DNA and DARPA on this topic. The result of the workshop, contrary to expectation, was that soft errors could not be solved easily, that the problem did have important implications for DOD systems using VLSI devices, and that the problem might get worse as devices were made still smaller and faster. An additional result was that DNA was to prepare a proposal for a joint DNA/DARPA program on this subject. Guenzer and Wolicki then prepared, for DNA, the technical portions of a proposal for a program that was to be entitled "Single Event Radiation Effects." Presentations were made to DARPA and a joint DNA/DARPA program was established in 1980. DNA project officers at the time were Maj. M. Kemp and LCDR W. Mohr; the DARPA project director was Dr. S. Roosild.

From those days to the present, papers have proliferated on this subject, examining its many facets, revealing that the SEU problem is still unsolved from any reasonably elegant or substantive viewpoint. Expensive brute-force solutions are on the edge of practicality. Hopefully, this treatise may aid a bit toward a discovery of the former.

The monograph begins with Chapter 1 where, following historical and introductory preliminaries, the fundamentals of particle transport such as flux, fluence and cross section, which are often foreign to the electronics engineer, are discussed. Also included is a discussion of chord distribution functions which, like cross sections, may be somewhat alien to many practitioners. It is felt to be sufficiently important as a basis in later derivations of SEU error rates that serious readers should peruse this portion meticulously "at least once in their lives."

Chapter 2 continues with a discussion of cosmic rays and other important particles within the aegis of single event phenomena, including alpha particles, gammas, mesons, protons, and neutrons. This is followed by the structural details of the Van Allen radiation belts.

In keeping with the fundamentals emphasis of the early chapters, Chapter 3 discusses the penetration of particles in matter including range–energy relations, Bragg curves, stopping power, linear energy transfer (LET), and Heinrich curves.

Chapter 4 describes, in detail, the relatively few types of SEU simulators available, and how the data are retrieved and transformed to provide wherewithwal to aid in building the subject matter edifice. The simulators include the cyclotron, synchrotron, Van de Graaf generator, californium-252 sources, and the laser.

Chapter 5 discusses and derives error rate computational formulas for SEU-inducing particles. They include geosynchronous heavy ions and their solar flare-enhanced counterparts: protons, neutrons, alphas, and mesons. Also discussed are the corresponding environments including ground-level SEU and the circumstances under which the SEU estimation formulas should be used.

Chapter 6, entitled "Single Event Phenomena I," begins by describing the phenomenon of funneling, a major adjunct for charge collection in addition to that collected from the upper regions near the device-sensitive nodes. Funneling details are given in the form of a number of different models couched in terms of device semiconductor parameters. This is followed by the dynamics of the incident ion-induced ionization tracks in the device, and how they transport charge to SEU-sensitive regions. The middle sections of this chapter take up important single event phenomena including SEU-induced latchup, MOSFET gate rupture, and SEU burnout. A further section describes the phenomenon of the South Atlantic Anomaly. Its origin is due to the tilt and displacement of the earth's magnetic field axis, which influences the SEU propensities of spacecraft in low earth orbit.

Chapter 7, entitled "Single Event Phenomena II," continues the discussion in Chapter 6 with a description of multiple event upsets, and the effect of SEU superimposed on a device already enduring ionization damage. The central sections discuss SEU amelioration methods, including redundancies, memory scrubbing, and error detection and correction (EDAC) methods. Finally, shielding possibilities against SEU, and effects on radiation detectors and sensors are discussed.

Chapter 8 together with Chapter 5 are written in part to provide practical methods for computing and coping with SEU in various radiation environments, including heavy ions, protons, and neutrons. Further details on SEU-induced latchup and device burnout are given. The final sections include a number of data tables, figures, and sets of guidelines from the chip level, through the circuit and system level, and to the mission level.

In our efforts to slight no one, all references are given by number with very few exceptions. We were tempted to refer to them by name to add a little dash to the flow as done in other treatises. However, we run the risk of omitting someone or, worse, attributing the work to the wrong authors, and also ultimately imputing a hierarchy of personages that perhaps too often skew actual contributions. It is true that mistakes can happen with either way of presentation, so we have made every effort to minimize such errors. In the end, all errors are ours.

We acknowledge the contributions of the many people who helped us prepare this book. We appreciate greatly the efforts of M.A. Shoga who read the manuscript and supplied salient part radiation data including that on SEU-induced latchup. Also, we extend heartfelt thanks to E.A. Wolicki for his guidance and providing the SEU chronology as part of the preface. Colleagues in the field and friends certainly know that we thank them profoundly for their many and varied contributions in this endeavor.

G.C. Messenger, Las Vegas, NV.
Milton S. Ash, Santa Monica, CA.

Introduction

In 1962, two perceptive researchers realized that the packing density of transistors in an integrated circuit (microcircuit) would eventually become so great that the individual transistors would suffer from damage due to cosmic rays, especially those used in spacecraft avionics; that is, these transistors would be so tiny that spurious charge induced by cosmic rays would upset the information in the storage cells of these circuits. In 1975, this was confirmed by the irrefutable establishment of cosmic-ray-induced soft upsets in a portion of the electronics of a spacecraft. Almost concurrently, single event upset (SEU) in DRAMs, due to alpha particles from the decay of heavy actinide nuclei in the earth's crust, in ceramic packages was observed. Also, concurrently, SEU in NMOS DRAMs in space as well as proton- and neutron-induced SEU was found. From then to now, activity in the area of single event upset (soft errors) and other SEU phenomena has been the subject of intense study by workers in this field. To date, many verifications of SEU in spacecraft, high-cruise-altitude aircraft, and in some ground-based electronic systems have been found. The problem of the nature of SEU and its effects on semiconductor devices is an ongoing one. Many questions regarding SEU phenomena have been answered, but many more remain.

With the specter of ever-decreasing microcircuit physical size, which means more and more transistors on a chip, as seen in the frontispiece, the SEU problem will become exasperated with time. Today, it takes hardly 0.1 pC ($\sim 10^6$ electrons) of charge to change the bit state of a nominal memory storage cell node. It may be that ultimately SEU will be the limiting factor in the successful production and operation of very small (submicron) devices. This is in contrast to other difficulties like quantum-electronic effects or, strictly, construction and manufacturing vagaries. The former include the resolution-limiting wavelengths of the beams used in integrated-circuit manufacturing processes.

1

Preliminaries

1.1. Introduction

Because this monograph attempts to reach a diversity of readers, from electrical/electronic engineers, to solid-state scientists, to nuclear physicists, it is at once a problem of how much to discuss in this chapter regarding preliminary aspects. The subject matter has a rather wide scope as well, drawing from the disciplines of electronics, solid state (condensed matter) and nuclear physics. Aspects of the appropriate physics concepts not usually familiar to engineers, as well as those familiar to engineers but not to physicists, was decided as the approach for this short chapter.

Section 1.2 discusses the main ideas of particle flux, fluence, and current density from a fundamental approach and using a minimum amount of mathematics to describe them. They are examples of concepts that are perhaps familiar in the physical sciences as well as in the engineering community.

Another important parameter of the preceding kind is cross section. Briefly, it is a semiempirical concept in the sense of it being a probability coefficient (yet with dimensions of area) that links the number of particle interactions of interest with corresponding incident flux or fluence levels. It, together with linear energy transfer, discussed in Sections 3.7–3.8 (Chapter 3), is one of the principal parameters employed to describe single event upset (SEU).

The chord distribution function is a rather unique part of SEU calculations. It is a product of the somewhat arcane area of geometric probabilistics. Its use herein places it equivalent to the cross section, but the chord distribution function instead of the cross section is used in important SEU formulas, for reasons discussed in Section 1.4.

In the Preface, it was alluded that the problem sets following each chapter would add another dimension to aid in the understanding of single event

2 / Single Event Phenomena

phenomena (SEP). This is strongly the case for this chapter because it discusses fundamentals. It is important for the readers, especially for those who initially find some of the ideas alien, to work the problems, even though the drill involved in a few might try their patience. The preceding applies to the mathematical aspects as well.

1.2. Fluence, Flux, and Current Density

Both fluence and flux are important concepts, together with current density, used in the computation of particle–SEU interactions with material media. As

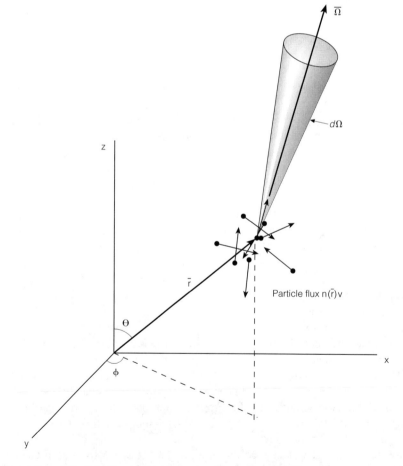

Figure 1.1a. Particle flux at position **r** in spherical coordinates.

will be seen, integrals of the product of fluence or flux with corresponding cross sections yield the number of interactions of interest (or their rates) per cubic centimeter or per gram of material. Particle–medium interactions, as opposed to particle–particle interactions, are of overwhelming interest. The latter interactions do occur, such as high energy gamma reactions that produce an electron and a positron. However, they have negligible bearing in the present context.

Let $n(\mathbf{r}, \mathbf{\Omega})d\Omega$, a scalar function of the radius position vector $\mathbf{r}$ and direction unit vector $\mathbf{\Omega}$, be the number density of particles per cubic centimeters at $\mathbf{r}$ traveling in the direction $\mathbf{\Omega}$. Their velocity vectors $v\mathbf{\Omega}$, where v is their speed, lie within a solid angle $d\Omega$ about $\mathbf{r}$. Then the particle density $n(\mathbf{r})$ in particles/cm^3 (Fig. 1.1a) is the scalar [1], with $d\Omega = \sin\theta d\theta d\phi$,

$$n(\mathbf{r}) = \int n(r, \mathbf{\Omega})\, d\Omega = \int_0^{2\pi} d\phi \int_0^{\pi} n(r, \mathbf{\Omega}) \sin\theta\, d\theta \qquad (1.1)$$

An interaction cross section σ, from its definition as discussed in Section 1.3 has the property that σv is the total volume of all the cylindrical segments that a single incident particle "cuts" out of the medium during its travel in 1 s, (Fig. 1.1b). This particle interacts with each and every medium (target)

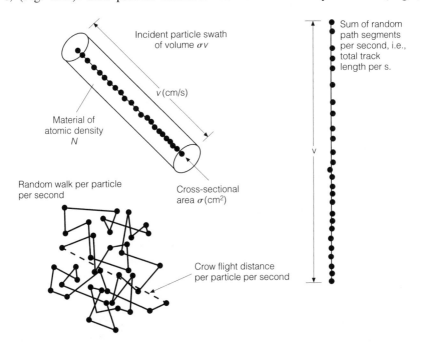

Figure 1.1b. Illustrations showing the number of collisions per second per incident particle in material of atomic number density N.

particle therein. So that $N\sigma v$ yields the number of interactions per incident particle per second that take place in the swath. N is the number of medium particles (targets)/cm^3. $N = \rho N_0/M$ or $\rho N_0/A$, where ρ is the medium density and N_0 is Avogadro's number. A is the medium mass number if the targets are nuclei, and M is the molecular (or atomic) weight if the targets are atoms or molecules. The atomic weight and the mass number are almost equal, as almost all of the mass of the atom is in it's nucleus. Then,

$$N\sigma v n(\mathbf{r}) = \frac{n(\mathbf{r})v}{\lambda} = \frac{\phi(\mathbf{r})}{\lambda} \qquad (1.2)$$

yields the number of interactions/cm^3 s that occur at $\mathbf{r}$, of a type characterized by the particular cross section σ. $(N\sigma)^{-1} = \lambda$ is the corresponding mean free path.

Now, a monoenergetic or one velocity (v) flux [1], with the usual limits on θ and ϕ in $d\Omega$,

$$\phi(\mathbf{r}) = \int n(\mathbf{r}, \mathbf{\Omega}) v \, d\Omega = n(\mathbf{r})v \qquad (1.3)$$

is defined as the total number of particles threading a sphere of unit cross-sectional area per second in the medium from all incident angles. An alternate definition of $\phi(\mathbf{r})$ is the total track length traveled by all these particles per unit volume in one second. To show this, consider a flux ϕ emanating from a sphere of unit cross-sectional area A_u per second, on which a height h is erected. This then forms a cylinder of volume $A_u h$. The total track length threading through the cylinder per second is seen to be $A_u h$. Hence, the expected total track length per second per unit volume is ϕ. The corresponding fluence is merely the time integral over ϕ, thereby eliminating the "per second" in the previous sentence, thus producing the definition of fluence. Unfortunately, all too often, the same symbol is used for both flux and fluence, requiring the context for clarification. This definition is useful in the determination of flux or fluence from track detector counting data [2].

If the particles are from a constant pulsed source t seconds long, then the corresponding monoenergetic fluence at $\mathbf{r}$ is given by nvt. Then the number of type i interactions/cm^3 during the pulse epoch is given by $\mu_i n t v$, where $\mu_i = N\sigma_i = 1/\lambda_i$. μ_i is the interaction probability/cm (Section 1.3) for the ith type of interaction process (scattering, SEU, ionization, etc.) also called the macroscopic cross section Σ_i in reactor circles. Generalizing to incident flux particles of energy E and cross section $\sigma_i(E)$, it is seen that

$$\int \mu_i(E) E\phi(\mathbf{r}, E) \, dE = E_T, \qquad \phi(\mathbf{r}, E) = \int n(\mathbf{r}, \mathbf{\Omega}, E) v(E) \, d\Omega \qquad (1.4)$$

where $\mu_i(E)\phi(\mathbf{r}, E)$ in the leftmost expression in (1.4) is the number of interactions/cm^3 s. Multiplied by the energy E of the individual incident particle, then the leftmost integral in (1.4), E_T, yields the energy absorbed cm^{-3} s^{-1} by the medium, as $\sigma_i(E)$ is defined here as the absorption cross section of the medium for these incident particles. The rightmost equation (1.4) is the flux generalization to all energies E of the monoenergetic flux given in (1.3). As a corollary, the incident beam of intensity I (particles cm^{-2} s^{-1}), which has the same dimensions as ϕ, discussed in Section 1.3, can be written as [1]

$$I(\mathbf{r}, \mathbf{\Omega}) = \int n(\mathbf{r}, \mathbf{\Omega}) v \cos\theta \, d\Omega = \int_0^{2\pi} d\phi \int_0^{\pi} n(\mathbf{r}, \mathbf{\Omega}) v \cos\theta \sin\theta \, d\theta \quad (1.5)$$

The concept of particle current density (not electric current density) is defined by the vector quantity,

$$\mathbf{J}(\mathbf{r}, \mathbf{v}) = \int n(\mathbf{r}, \mathbf{\Omega}) v \, d\Omega \quad (1.6)$$

where the limits on $d\Omega$ are the same as in (1.5). For a physical interpretation of $\mathbf{J}$, its x component is a scalar given as in (1.5), with the same limits on $d\Omega$,

$$J_x = \int d\phi \int n(\mathbf{r}, \mathbf{\Omega}) v \cos\theta \, d\Omega \quad (1.7)$$

where θ is the angle between v and the x axis. J_x yields the *net* number of particles cm^{-2} s^{-1} along the x axis. Note that for $\cos\theta < 0$ ($\pi/2 \leq \theta \leq 3\pi/2$), the particles coming from that direction subtract from those whose $\cos\theta > 0$ to yield the net particle number threading a cm^2 s^{-1} along the x axis. This is in contrast to the flux integrals which are over all angles, summing all particles regardless of their angles of incidence.

In general, it is seen that integrating flux over the duration time of pulsed sources yields the corresponding fluences. However for SEU cosmic ray particles, flux is the appropriate quantity multiplied by cross section in the aforementioned integrals to compute SEU error rates. Cosmic ray flux is ever present at nominal levels due to its extraterrestrial origins. Fluence is the correct quantity to use for computing energy absorption from radiation of finite duration. For example, gamma rays employed to simulate ionizing dose sources are expressed as gamma ray fluence, as the devices under test are exposed to this radiation for a finite interval only. This fluence is used to calculate the energy deposited in the device material due to the gamma rays absorbed therein. As is well known, the energy deposited can be expressed in ergs per gram of device material (i.e., rads) or joules per kilogram [i.e., grays (Gy)].

1.3. Cross Sections

In the important process of computing the interactions of particles of interest with pertinent materials, corresponding interaction probabilities must be avail-

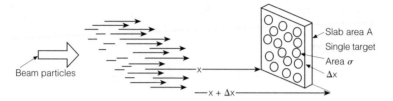

Figure 1.1c. Particle beam on thin target.

able. These are couched in what are called "cross sections." They are introduced in the following simple manner by considering a beam of particles of intensity I (particles/cm^2) at normal incidence on a thin slab of material of area A and thickness Δx, as in Fig. 1.1c.

The slab contains the particle beam targets in its material, which can be atoms, nuclei, complex organic molecules, or macroscopic matter, corresponding to whatever the particular interactions are being studied. The projected area (assumed circular) of each target normal to the beam direction is labeled σ (cm^2). The material slab is assumed thin, which implies that an interaction (e.g., collision) of a beam particle with a target scatters that particle out of the incident direction, and so it is lost to the beam intensity beyond the material slab; that is, it has an extremely low probability of being scattered back into the beam. Let $I(x)$ be the beam intensity incient at the slab face; then

$$I(x + \Delta x) = I(x) - I(x)\frac{A\Delta x N \sigma}{A} \tag{1.8}$$

Equation (1.8) simply asserts that the beam just beyond the slab is given by its intensity prior to interacting with the slab, reduced by beam–target interactions per square centimeter, the rightmost term in (1.8). The latter is merely the beam attenuation written in terms of a fraction of the incident beam intensity I. This fraction is given by the ratio of actual total target area to the slab material area A. The total target area is given by the number of targets in A multiplied by their individual projected areas σ. For N being the number of targets per cubic centimeter in the slab, $NA\Delta x$ is the total number in the slab. Also, in the limit of zero Δx, the slab is so thin that any shadowing of targets by each other is assumed to be negligible, so that all targets therein are facing the incident beam. Then $NA\Delta x\sigma$ is the total target area in A, and its ratio with area A yields the required fraction multiplying $I(x)$ in (1.8). Taking the limit as Δx approaches zero in (1.8) yields where $I(x + \Delta x) = I(x) + I'(x)\Delta x + $ higher-order terms in Δx

$$\frac{I'(x)}{I(x)} = -N\sigma, \qquad I(0) = I_0 \tag{1.9}$$

with an integral,

$$I(x) = I_0 \exp(-N\sigma x) \quad (1.10)$$

It is seen from (1.10) that $\exp(-N\sigma x)$ can be construed as the survival probability of a beam particle up to its interaction with the material target occurring at and after traveling a distance x. From (1.10), the difference in beam intensity due to the interaction is

$$\Delta I(x) \equiv I(x + \Delta x) - I(x) = [I_0 \exp(-N\sigma x)][\exp(-N\sigma \Delta x) - 1] \quad (1.11)$$

and for sufficiently small Δx,

$$\frac{dI(x)}{I_0} \cong -[\exp(-N\sigma x)](N\sigma dx) \quad (1.12)$$

The right-hand side of (1.12) is interpreted as the fractional reduction in beam intensity $|dI/I_0|$ due to the interactions and is given by the product of the probability of particles surviving to a distance x with their interaction probability $N\sigma dx$ per distance dx, about that x.

Introducing a mean free path λ, which is the mean distance a particle travels between interactions (collisions), by using the above survival probability to compute it from its expectation, that is,

$$\lambda = \frac{\int_0^\infty x \exp(-N\sigma x) dx}{\int_0^\infty \exp(-N\sigma x) dx} = -\frac{d}{d(N\sigma)}\left(\ln \int_0^\infty \exp(-N\sigma x) dx\right) = \frac{1}{N\sigma} \quad (1.13)$$

then the interaction probability per distance Δx is $N\sigma \Delta x = \Delta x / \lambda$, and $1/\lambda = N\sigma$ can be interpreted as the interaction probability per centimeter of particle travel. Note that the dimensions are $\sigma(\text{cm}^2)$, $N(\text{cm}^{-3})$, $\lambda(\text{cm})$, and $N\sigma(\text{cm}^{-1})$. Therefore, σ is the interaction probability per centimeter per target particle, even though its dimension is square centimeters. σ, the target projected area is defined as the interaction cross section for the collision type (e.g., scattering collisions). Much of the time, σ corresponds to the actual geometric cross section of the target particle. However, it can be smaller or a million times greater due to quantum mechanical collision phenomena.

Finer distinctions in the cross section can be made. For example, the cross section can be elastic (kinetic energy before and after the collision is conserved) or inelastic. They are designated by σ_{el} and σ_{inel}, respectively. The total attenuation is

$$-dI = -dI_{el} - dI_{inel} = -I\sigma_{el}N\,dx - I\sigma_{inel}N\,dx = -I(\sigma_{el} + \sigma_{inel})N\,dx,$$

$$I = I_0 \exp[-N(\sigma_{el} + \sigma_{inel})x] \qquad (1.14)$$

So that from (1.14), it is seen that the total cross section

$$\sigma_{tot} = \sigma_{el} + \sigma_{inel} + \cdots \qquad (1.15)$$

where $+\cdots$ implies that this process can be extended to include as many types of interaction processes as desired. σ is often called the microscopic cross section, whereas $\mu = N\sigma$ is also called the macroscopic cross section. Often μ_g (cm^2/g) $= \mu/\rho$ is employed. As mentioned, in nuclear reactor theory, Σ is used, where $\Sigma \equiv \mu$. Differential cross sections $d\sigma/d\Omega$ are defined as the cross section per incremental solid angle. The fraction of particles scattered into $d\Omega(\theta, \phi)$ can be written as

$$\frac{dI}{I(\theta, \phi)} = N\left(\frac{d\sigma_s(\theta, \phi)}{d\Omega}\right)d\Omega\,dx \qquad (1.16)$$

Then, integrating over the solid angle retrieves the cross sections previously discussed. Of further importance is that cross sections are frequently functions of energy of the incident particle [i.e., $\sigma = \sigma(E)$, as well].

A second and perhaps more pertinent example which compares the difference between the differential and integral cross section is that of certain empirical curve fits to the SEU cross section $\sigma_u(L)$ resulting from accelerator experimental data. Details of the latter are discussed in Section 1.4 and L is the linear energy transfer (LET) variable. $\sigma_u(L)$ as obtained in this manner, shown in Fig. 4.5a for example, is construed as the probability that a single incoming heavy ion of LET greater than the LET threshold L_{th} will cause an SEU. A frequently used empirical fit for such $\sigma_u(L)$ is the three-parameter Weibull probability distribution function. It is used in reliability hazard computations, with three fitting parameters resulting in a function which is analytically appealing. The Weibull cumulative probability distribution $F(x)$ is defined by [3]

$$F(x) = \begin{cases} 1 - \exp\left[-\left(\frac{x - x_0}{m}\right)^b\right], & x > x_0 \\ 0, & x \leq x_0 \end{cases} \qquad (1.17)$$

$F(x)$ yields the cumulative probability that a random variable x is greater than a lower limit x_0. The corresponding probability density function $f(x)$ [3] is

$$f(x)\,dx = dF(x) = \left(\frac{b}{m}\right)\left(\frac{x-x_0}{m}\right)^{b-1} \exp\left[-\left(\frac{x-x_0}{m}\right)^b\right] dx, \quad x \geq x_0 \quad (1.18)$$

As seen, the fitting parameters are m (scale), x_0 (position), and b (shape)[3]. $f(x)$ is the probability that x lies in the incremental region dx about x, so that $f(x)$ is frequently interpreted as the probability per unit x, which when multiplied by dx, yields the preceding definition. Also, $F(x) = \int_{x_0}^{x} f(x')\,dx'$, $x \geq x_0$

The Weibull probability distribution fit to $\sigma_u(L)$ and corresponding to the cumulative or integral distribution function is then [4]

$$\sigma_u(L) = \sigma_{\mathrm{asy}} \begin{cases} 1 - \exp\left[-\left(\dfrac{L-L_{\mathrm{th}}}{W}\right)^s\right] & L > L_{\mathrm{th}} \\ 0, & L \leq L_{\mathrm{th}} \end{cases} \quad (1.19)$$

σ_{asy} is the asymptotic or saturation cross section in the limit of large L. $\sigma_u(L)$ provides the probability distribution for an incoming particle whose LET $L > L_{\mathrm{th}}$ to cause an SEU in a device characterized by the three parameters L_{th} (threshold), W (LET scale), and s (cross-section shape). The corresponding probability density, or the SEU differential cross section $d\sigma_u(L)$, is within a constant

$$d\sigma_u(L) = \left(\frac{d\sigma_u(L)}{dL}\right) dL = \left(\frac{s}{W}\right)\left(\frac{L-L_{\mathrm{th}}}{W}\right)^{s-1} \exp\left[-\left(\frac{L-L_{\mathrm{th}}}{W}\right)^s\right] dL \quad (1.20)$$

which yields the probability that an incoming particle, whose LET $L \geq L_{\mathrm{th}}$, lies in the incremental region dL about L, will produce an SEU. Again, $d\sigma_u(L)/dL$ is the SEU probability per unit L, which when multiplied by dL, gives the differential cross section in (1.20). Often $d\sigma_u(L)/dL$ is also called the differential cross section.

Chapters 5 and 8 discuss how cross sections are used to determine SEU error rates. Suffice it to say here that the SEU error rate, E_r, is proportional to the integral of the product of the appropriate incident SEU-producing particle flux and the SEU cross section, either integral or differential; that is, if $\Phi(L)$ is the integral SEU flux of dimensions, say, the number of particles cm^{-2} da^{-1}, then

$$E_r \sim \int_{L_{\mathrm{min}}}^{L_{\mathrm{max}}} \Phi(L)\,d\sigma_u(L) = \int_{L_{\mathrm{min}}}^{L_{\mathrm{max}}} \Phi(L)\left(\frac{d\sigma_u(L)}{dL}\right) dL \quad \text{(errors per day)} \quad (1.21)$$

Integrating (1.21) by parts results in

$$E_r \sim \Phi(L)\sigma_u(L)\Big|_{L_{\min}}^{L_{\max}} - \int_{L_{\min}}^{L_{\max}} \sigma_u(L)\left(\frac{d\Phi(L)}{dL}\right)dL \qquad (1.22)$$

where the integrated term is shown to vanish, and the negative sign in front of the second term on the right-hand side of (1.22) vanishes as well. The differential flux $\Phi'(L) \equiv d\Phi(L)/dL$ has dimensions of particles cm^{-2} day$^{-1}\cdot$ LET^{-1}.

As an example, assume that $\Phi(L) = \Phi_0/L^2$ in appropriate units, so that $\Phi'(L) = -2\Phi_0/L^3$, where $|\Phi'(L)|$ is an analytical approximation to the LET abundance spectrum curve, as in Figure 5.1. It is used later to derive a "figure-of-merit" estimate of the geosynchronous SEU error rate as discussed in Section 5.1. Using $\sigma_u(L)$ from (1.19) yields, from (1.22),

$$E_r \sim \left(\frac{\Phi_0}{L^2}\right)\left(\sigma_{\text{asy}}\left\{1 - \exp\left[-\left(\frac{L-L_{\text{th}}}{W}\right)^s\right]\right\}\right)\Big|_{L_{\text{th}}}^{\infty}$$

$$+ (2\sigma_{\text{asy}}\Phi_0)\int_{L_{\text{th}}}^{\infty}\frac{dL(1 - \exp\{-[(L-L_{\text{th}})/W]^s\})}{L^3} \qquad (1.23)$$

where $L_{\max}$ has been extended to infinity for convience, because $\Phi'(L)$ decreases as $1/L^3$ beyond the sensible cosmic ray abundance "cutoff," so that the corresponding contribution to the integral is vanishingly small. The first term on the right-hand side of (1.23) is seen to vanish, and the second of the two integral terms is integrated by parts. This finally yields, where $\alpha = W/L_{\text{th}}$,

$$E_r \sim \left(\frac{\sigma_{\text{asy}}\Phi_0}{W^2}\right)F(\alpha, s), \qquad F(\alpha, s) = \alpha^2 \int_0^{\infty}\frac{\exp(-x)\,dx}{(1 + \alpha x^{1/s})^2}, \quad s > 0 \qquad (1.24)$$

$F(\alpha, s)$ is easily integrated numerically using a modern hand-held computer and is depicted in Fig. 1.2 which is a plot of F versus α^{-1} for various s values; or, written as a sum, $F(\alpha, s) = \sum_{n=0}^{\infty}(-1)^n(n+1)\Gamma(1+n/s)\alpha^{n+2}$, where the gamma function $\Gamma(1+n) = \int_0^{\infty}x^n \exp(-x)\,dx$.

1.4. Chord Distribution Functions

The chord distribution function is essentially the frequency distribution (i.e., the probability density of chord lengths threading a convex body), the latter being defined in the following. As will be seen, the chord distribution function plays an important part in the computational estimation of SEU or rate. It is also used, for example, in the calculation of escape probabilities of particles in nuclear reactor assemblies. Its importance stems as an outgrowth of the

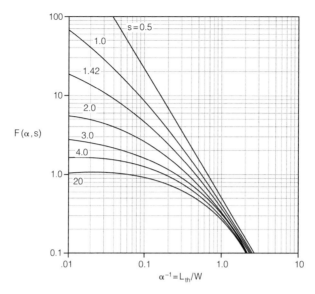

Figure 1.2. Single event upset error rate correction factor $F(\alpha, s)$ versus α^{-1} for various s for the Weibull SEU cross section fit.

prototypical manner by which SEU and other interaction rates are computed classically; that is, in general, the sought after number of interactions of type x in a material volume induced by a penetrating particle flux φ is given by $\int N\sigma_x \varphi\, dV$. σ_x is the cross section per flux particle per material "target" particle that characterizes the x interaction process. N is the number of material target particles per unit volume that will interact with φ in that way. In the case of SEU, there is a paucity of cross-section functions σ by which an estimate of SEU error rate could be computed using the above integral. However, the need for a cross section is bypassed in the above integral by employing geometrical probabilistic methods using chord distributions and other similar functions.

Of course, the SEU error rate can be computed today by employing the above integral with cross-section data obtained experimentally and fed into a computer integration program. For sensitivity studies to gain insight into the SEU process and its corollaries, this method would be rather cumbersome, considering the effort that goes into obtaining such cross-section data. The use of chord distributions to compute SEU error rate was introduced about 20 years ago concurrently with the lack of cross-sectional and general knowledge of SEU at that time and has well served the needs of practitioners in this field. Further comparisons to reconcile the classical cross section with the geometric probabilistic approach is given in Section 5.1, where it is shown that they are equivalent.

Consider a convex body of material of volume V, as in Fig. 1.3. A convex body physically means that if a particle leaves, it can return only if it suffers

12 / *Single Event Phenomena*

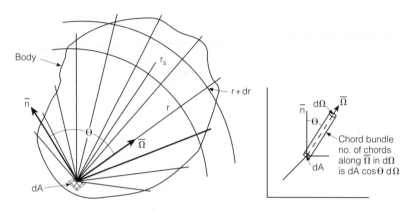

Figure 1.3. Chords emanating from an interior incremental surface area dA of volume V with an inner normal unit vector **n**.

a collision outside, redirecting it back toward the body [5]. It cannot leave and reenter while on the same vector. The bodies could be "L" shaped, but these are not convex. In Fig. 1.3, $\boldsymbol{\Omega}$ is a unit vector along a particular chord or chord bundle from an incremental surface area dA of the body. $\boldsymbol{\Omega} \cdot \mathbf{n} = \cos\theta$, where **n** is a unit vector inner normal from dA, and $d\Omega$ is an incremental solid angle subtended by dA at r, a particular chord length.

The number of chords in a bundle included in $d\Omega$ emanating from dA about the direction $\boldsymbol{\Omega}$ is given by $(\mathbf{n} \cdot \boldsymbol{\Omega})\, d\Omega\, dA = dA \cos\theta\, d\Omega$ (see Fig. 1.3, inset). This can be seen by noting that the number of particles that enter the volume V through dA in $d\Omega$ in the direction $\boldsymbol{\Omega}$ is $\Phi(d\Omega/4\pi)(\mathbf{n} \cdot \boldsymbol{\Omega})\, dA$; that is, as Φ is the incident isotropic homogeneous particle fluence (particles/cm^2) and $\Phi/4\pi$ is the fraction of particles per unit solid angle, so $(\Phi/4\pi)\, d\Omega$ is the fraction in $d\Omega$. Then multiplying by $(\mathbf{n} \cdot \boldsymbol{\Omega})\, dA = \cos\theta\, dA$ gives the projection of dA normal to the $\boldsymbol{\Omega}$ direction that these particles thread. Assuming that the fluence particle tracks produce one chord each, the corresponding number of chords in $d\Omega$, $\Phi(d\Omega/4\pi)(\mathbf{n} \cdot \boldsymbol{\Omega})\, dA$, must be given by the product of their fraction per unit solid angle $\Phi/4\pi$, and the sought after number that penetrate dA, becoming chords in a bundle within $d\Omega$ about $\boldsymbol{\Omega}$ viz. $d\Omega(\mathbf{n} \cdot \boldsymbol{\Omega})\, dA = dA \cos\theta\, d\Omega$.

The chord distribution function $f(r_s)$ (i.e., the probability density that a chord from dA is of length r_s between r and $r + dr$) is given by [5] (Fig. 1.3),

$$f(r_s) = \frac{\int dA \int (\mathbf{n} \cdot \boldsymbol{\Omega})\, d\Omega \Big|_{r=r_s}}{\int dA \int (\mathbf{n} \cdot \boldsymbol{\Omega})\, d\Omega} = \frac{\int dA \int \cos\theta\, d\Omega \Big|_{r=r_s}}{\int dA \int \cos\theta\, d\Omega}, \quad d\Omega = \sin\theta\, d\theta\, d\phi \quad (1.25)$$

where $0 \leq \phi \leq 2\pi$, $0 \leq \theta \leq \pi/2$, and the integral in the numerator is restricted to chord lengths r_s such that $r \le r_s \le r + dr$ and $\cos\theta = \mathbf{n}\cdot\mathbf{\Omega} > 0$ (i.e., $0 \le \theta \le \pi/2$ for this convex body). The numerator is seen to be equal to the total number of chords in the body of length r_s within dr, because the integral is over the solid angle and body surface area S. The denominator yields the total number of chords with no restriction on their length, or

$$\int dA \int \cos\theta\, d\Omega = \int dA \int_0^{2\pi} d\phi \int_0^{\pi/2} \cos\theta \sin\theta\, d\theta = \pi S \qquad (1.26)$$

Instead of using $f(r_s)$ directly to compute the mean chord length $\bar{r}$, it is easier to note that, because of the definition of $dA \cos\theta\, d\Omega$ with the same limits as in (1.26),

$$\bar{r} = \frac{\int dA \int r(\mathbf{n}\cdot\mathbf{\Omega})\, d\Omega}{\int dA \int (\mathbf{n}\cdot\mathbf{\Omega})\, d\Omega} = \frac{\int d\Omega \int r(\mathbf{n}\cdot\mathbf{\Omega})\, dA}{\pi S} \qquad (1.27)$$

It is seen that the incremental volume dV of the bundle of chords of length r in $d\Omega$ emanating from dA in the direction $\mathbf{\Omega}$ is $r(\mathbf{n}\cdot\mathbf{\Omega})\, dA$, or $dV = r\, dA \cos\theta$; that is, because $(\mathbf{n}\cdot\mathbf{\Omega})\, dA = \cos\theta\, dA$ is the projection of dA normal to $\mathbf{\Omega}$, then $r(\mathbf{n}\cdot\mathbf{\Omega})\, dA$ are the incremental volumes of cylinders with bases $(\mathbf{n}\cdot\mathbf{\Omega})\, dA = \cos\theta\, dA$ and heights r. Hence,

$$V = \int r(\mathbf{n}\cdot\mathbf{\Omega})\, dA = \int r \cos\theta\, dA \qquad (1.28)$$

Finally, integrating (1.27) over the solid angle of the convex body yields, from (1.25),

$$\bar{r} = \frac{\int d\Omega\, V}{\pi S} = \frac{4\pi V}{\pi S} = \frac{4V}{S} \qquad (1.29)$$

This is a general result holding essentially for all three-dimensional convex bodies, including rectangular parallelepipeds used to model transistor SEU-sensitive regions. It is straightforward to show that $f(r)$ in (1.25) is normalized to unity by removing the restriction of $r = r_s$.

An alternate definition of fluence Φ is the mean total particle track length per unit volume [11]. To appreciate this, an equivalent definition of Φ is the mean total number of particle traversals through a sphere of unit cross-sectional area. To reiterate from Section 1.2, where $A_u = 1$, by erecting a cylinder of height h over the unit cross section and with a homogeneous fluence parallel to the cylinder axis, it is seen that the total track length in the cylindrical volume is $h\Phi$. The mean number of traversals in the cylinder is still Φ, and that the mean total track length per unit volume is also seen to be Φ. This also can be shown to hold for an isotropic homogeneous fluence.

14 / *Single Event Phenomena*

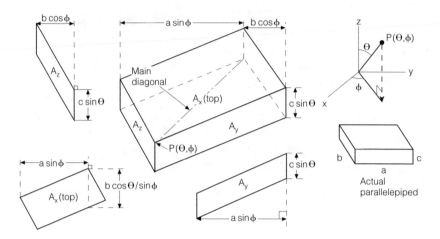

Figure 1.4. Center figure is parallelepiped projected onto plane $P_\perp$ normal to main diagonal through point $P(\theta, \phi)$. Peripheral figures are parallelogram projections of face areas A_x, A_y, A_z onto $P_\perp$ normal to main diagonal through $P(\theta, \phi)$. $P_\perp$ is in the plane of this page.

A_p, needed later, is the projected area of a convex body for a given orientation or aspect. For a parallelepiped, the area is projected onto a plane parallel to the eye as in Fig. 1.4. $\bar{A}_p$ is the mean projected area over all orientations. From Fig. 1.4, it is seen, for orientations that correspond to the projection of the three visible face areas A_x, A_y, A_z onto a plane normal to its main diagonal at a point $P(\theta, \phi)$, that the individual areas are given by

$$A_x = ab\cos\theta, \qquad A_y = ac\sin\theta\sin\phi, \qquad A_z = bc\sin\theta\cos\phi \qquad (1.30)$$

and

$$A_p = A_x + A_y + A_z = ab\cos\theta + c\sin\theta(b\cos\phi + a\sin\phi). \qquad (1.31)$$

Then in the chosen octant with A_p from (1.31),

$$\bar{A}_p = \frac{\int A_p\, d\Omega}{\int d\Omega} = \frac{\int_0^{\pi/2} d\phi \int_0^{\pi/2} A_p \sin\theta\, d\theta}{\int_0^{\pi/2} d\phi \int_0^{\pi/2} \sin\theta\, d\theta} = \tfrac{1}{2}(ab + bc + ac) = \tfrac{1}{4}S \qquad (1.32)$$

where S is the parallelepiped surface area.

From the preceding definition of fluence, it is appreciated that the mean number of chords $\bar{N}_{ch}$ threading through a sphere of radius r is given not by $(4/3)\pi r^3 \Phi$ but by

$$\bar{N}_{ch} = \pi r^2 \Phi = 4\pi r^2 \frac{\Phi}{4} = \frac{S\Phi}{4} \qquad (1.33)$$

where S is the sphere surface area. Equation (1.33) can also be shown to hold in general for any convex body. From (1.32), $\bar{A}_p = S/4$, then for a parallelepiped

$$\bar{N}_{ch} = \bar{A}_p \Phi \qquad (1.34)$$

As a corollary, note that from (1.29) and $\bar{A}_p = S/4$, the mean chord length $\bar{s}$ for a parallelepiped is given by

$$\bar{s} = \int_0^{s_{max}} sf(s)\,ds = \frac{4V}{S} = \frac{V}{\bar{A}_p} \;;\qquad \int_0^{s_{max}} f(s)\,ds = 1 \qquad (1.35)$$

where $V = abc$ is the parallelepiped volume, $f(s)$ is the corresponding chord distribution function [7] for chord lengths s therein, and $s_{max} = (a^2 + b^2 + c^2)^{1/2}$ is its main diagonal.

For a rectangular parallelepiped, the chord distribution function $f(s)$ is a complicated algebraic function of the parallelepiped dimensions a, b and c. It consists of close to a score of terms—far too many to describe here [7]. Figure 1.5 depicts the chord distribution function for a parallelepiped whose dimensions are $24 \times 18 \times 16\,\mu$m. It should be noted in that figure that $f(s)$ is relatively smooth except for peaks occurring at the parallelepiped edges. This can be appreciated by assuming that the eye pans across the parallelepiped, from the viewpoint of a homogeneous isotropic incident flux. At broadside, the chord lengths change somewhat slowly, whereas across an edge, the lengths vary rapidly and over a much larger magnitude. This is because the eye sees rapidly increasing and decreasing chord lengths that tend toward local maxima at each edge, corresponding to the peaks in the figure. The cumulative chord length distribution function [8]

$$C(s) = \int_s^{s_{max}} f(s')\,ds' \qquad (1.36)$$

which is of primary importance in the SEU error rate computations. $C(s)$ is the cumulative (integral) distribution of chords whose lengths lie between s and s_{max}, the main diagonal. Figure 1.6 shows $C(s)$ versus chord length for the $f(s)$ in Fig. 1.5. Also shown in Fig. 1.6 are approximations to $C(s)$, which will be discussed later. It is apparent that $C(s)$ is much smoother than its corresponding $f(s)$. This allows credible approximations of $C(s)$ to be made for

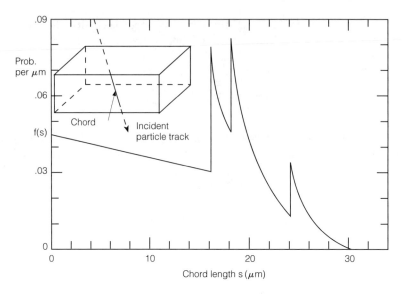

Figure 1.5. Distribution function $f(s)$ versus chord length s for rectangular parallelepiped of dimensions $24 \times 18 \times 16\ \mu m$, with $s_{max} = 34\ \mu m$. From Ref. 6.

certain parallelepiped dimensions of interest. This facilitates making SEU error rate integrals tractable. For example, Approximation No. 1 is described for the case where one parallelepiped dimension is much less than the other two (domino parallelepiped); that is, for dimensions a, b, and c, where $a, b \gg 3c$, then [8]

$$C(s) \cong \begin{cases} \frac{3}{4}(c/s)^{2.2}, & s \geq c \\ 1 - \frac{1}{4}(s/c), & s < c \end{cases} \quad (1.37)$$

In certain cases where the larger chord lengths predominate, a more complex approximation (Approximation No. 2) is described. For the parallelepiped main diagonal $s_{max} = (a^2 + b^2 + c^2)^{1/2}$ and again $a, b \gg 3c$, then [6]

$$C(s) \cong \begin{cases} \left(1 - \dfrac{2.37c}{1.80c + s_{max}}\right)\left(\dfrac{s_{max}^{3.8} - s^{3.8}}{s_{max}^{3.8} - c^{3.8}}\right)^3 \left(\dfrac{c}{s}\right)^2, & s \geq c \\ 1 - \dfrac{2.37c}{1.80c + s_{max}} \dfrac{s}{c}, & s < c \end{cases} \quad (1.38)$$

Figure 1.6 compares $C(s)$ using Approximations No. 1 and No. 2, with the numerical integration of $f(s)$ in (1.38), where $f(s)$ is the exact multiterm analytical expression. Chord length distributions $f(s)$ and $C(s)$ geometrically

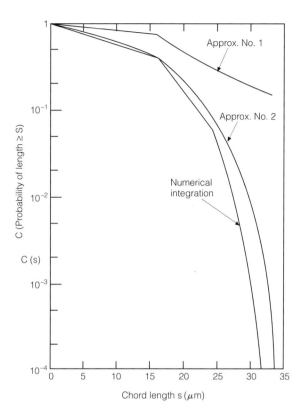

Figure 1.6. Cumulative chord distribution function $C(s)$ versus chord length s for rectangular parallelepiped $24 \times 18 \times 16\,\mu\text{m}$, with $s_{\max} = 34\,\mu\text{m}$. From Ref. 6.

model the incident heavy particle tracks very well. This is because when highly charged, heavy, high-energy cosmic ray ions penetrate the SEU-sensitive region, their principal energy loss in each atomic electron ionizing collision is very small. This is due to the electron mass and energy being extremely small compared to the cosmic ray particle. Hence, the incoming heavy ion hardly deviates from its chord path.

Expressions for SEU cross sections were not accessible in the past for use in computing SEU error rates for geosynchronous orbits. This meant that SEU error rate estimations involving integrals over products of cross sections with cosmic ray fluxes would be difficult to perform. The alternate method is to use chord length distributions in these integrals, where cross-section equivalence is implied. Chord length distribution functions for double-bit SEU (one incoming particle produces two SEUs) also have been derived [9] as well as multiple-bit distributions.

Generally, if the geometrical dimensions of the SEU-sensitive region as well as the chord distribution function (probability density) and cumulative distribution $C(s)$ are known, then the SEU error rate should be computed using expressions given in Section 5.2 that employ these functions. In too many situations, none of the above, except the incident flux, are known. Then the error rate should be calculated using integrals of the products of known flux with experimentally determined cross sections.

Some ion tracks that penetrate the material are not chords; that is, they do not travel completely through the body but are stopped within. These are termed random track segments. A relation connecting their mean length to the mean particle range and the mean chord length can be derived. It is used mainly to determine limits on the applicability of linear energy transfer (LET) as a sole relevant parameter in computing the energy deposited in a sensitive region [10]. This is discussed in Section 3.8.

For a body of volume V, the total track length within it is $V\Phi$ by an equivalent definition of fluence Φ as the total track length per unit volume. If the mean range of all particle tracks is $\bar{r}_t$, then the total number of *track end points*, P_e, plus *track start points*, P_s, is [11]

$$P_e + P_s = 2V\Phi/\bar{r}_t \tag{1.39}$$

$V\Phi/\bar{r}_t$ is the mean number of tracks, and is multiplied by the necessary factor of 2, as each has a *start point* and an *end point*. Now the mean number of *traversal points*, P_t, which are intersection points for those tracks that pierce the body surface area S, is given by

$$P_t = S\Phi/2 \tag{1.40}$$

To appreciate this, consider a sphere of radius r with a mean number of chords, $\bar{N}_{ch} = \pi r^2 \Phi$ threading it. Each chord intersects the sphere in two places; to keep from counting P_t twice, the number of intersections (i.e., *traversal points*/cm^2) of its surface is $\Phi/2$, and for the whole surface area, it is $S\Phi/2$.

Consider that each track segment has two terminal points which consist of (a) *track start points*, (b) *track end points*, and (c) *track traversal points*. So that the mean length $\bar{l}$ of a track segment in a volume V is given by the total track length $V\Phi$ divided by half the number of terminal points, as each track has two; that is,

$$\bar{l} = \frac{V\Phi}{\frac{1}{2}[(a)+(b)+(c)]} = \frac{V\Phi}{\frac{1}{2}(P_e + P_s + P_t)} \tag{1.41}$$

Using (1.29), (1.39), and (1.40) in (1.41) gives the desired expression [10] for the mean length of a track segment $\bar{l}$

$$\frac{1}{\bar{l}} = \frac{1}{\bar{r}_t} + \frac{1}{\bar{r}} \tag{1.42}$$

where, again, $\bar{r}_t$ is the mean range of an incident particle and $\bar{r}$ is the mean chord length.

Problems

1.1 Imagine that solar wind protons are entering the earth's neighborhood in only one direction, which is normal to the earth surface, at a rate of 10^5 protons s^{-1}, crossing a unit area at some extraterrestrial point. Also assume that galactic cosmic rays are deflecting about half of the solar wind protons strictly outward at that point, opposite to their incident direction. At that extraterrestrial point, compute (a) the solar proton flux, (b) the solar proton current, and (c) the solar proton fluence for the 5 years remaining in the sunspot cycle when their flux drops to an assumed negligible level.

1.2 For a point source from which is emanating S_o particles s^{-1} into a vacuum (i.e., no collisions with the medium), what is the expression for the corresponding flux?

1.3 In a slab made up of many very thin lamina, where each is of thickness Δx and N is a very large number of lamina, the probability of an incident particle being absorbed in a lamina is $p \cong \Delta x / \lambda$, where λ is the particle mean free path in the lamina material. In the limit of an infinite number of infinitesimally thin lamina, such that $\lim_{\substack{N \to \infty \\ \Delta x \to 0}} N \Delta x = x$, which is the slab thickness, show that the probability of a particle penetrating the slab is $\exp(-x/\lambda)$.

1.4 What is the mean free path for an accelerator ion in (a) silicon if its microscopic cross section $\sigma_{si} = 10\,b$? (1 barn $= 10^{-24}$ cm^2) and (b) in air assuming the same σ_{si}?

1.5 One fast neutron colliding with a silicon atom induces, in the mean, a cascade of about 1000 secondary (recoil) atoms in a volume of about 10^{-16} cm^3 of silicon material. What is the fluence level that would maintain the same cascade in one gram of silicon?

1.6 Modify the result in Section 1.3 for the particle beam current (viz. $I = I_0 \exp(-N\sigma x)$ for nonhomogeneous materials; that is, for

$\rho = \rho(x)$ to obtain $I = I_0 \exp[-\int_0^x (\rho/\rho_0)\,dx/\lambda_0]$. λ_0 is the mean free path immediately incident at the slab surface.

1.7 In the stratosphere, some cosmic-ray-produced neutrons descend to ground level to cause SEU in ground-based electronics. What is their flux at the ground after penetrating the atmosphere? An expression [12] for neutrons penetrating the atmosphere from altitude r (cm) is $\Phi_n(r) = (5 \times 10^{22} M/r^2) \exp\{-\int_0^r [\rho(x)\,dx/2.38 \times 10^4]\}$ neutrons/cm² s. At $r = R_0$, $M = M_0$ is the number of moles of neutrons produced per second (1 mole $= 6 \times 10^{23}$). A meteorological rule of thumb for atmospheric density $\rho(r)$ is that it decreases by a factor of 10 for every 10-mile increase in altitude. Show that the above neutron flux at the ground is 4.4×10^{-3}, where $R_0 = 1.5 \times 10^5$ ft, $M_0 = 200$, and $\rho_0 = 1.1$ g/liter.

1.8 For protons whose mean range is $\bar{R}$, assumed to be a factor of 10 larger than the mean chord length $\bar{r}$ in a homogeneous spherical sensitive volume, show (a) that the corresponding mean segment length $\bar{l} = 0.91\bar{r}$. (b) Calculate $\bar{l}$ in terms of the radius of the sphere r to obtain $\bar{l} = (40/33)\bar{r}$. (c) What interpretation can be made of the fact that $\bar{l} = 0.91\bar{r}$ in terms of the applicability of linear energy transfer as a single parameter by which to compute the energy deposited in a spherical sensitive region?

1.9 For a two-dimensional figure of surface area S_2 and circumference C, show that $\bar{r}_2$, its mean chord length, is given by $\bar{r}_2 = \pi S_2/C$.

1.10 This exercise provides further insight into the concept of the chord length probability distribution function and its consequences, such as the mean chord length. There is a probability of a chord length existing between l and $l + dl$ in a convex body, such as a memory cell in an integrated circuit. This probability function is called the probability density, whereas its integral over suitable limits is called the corresponding cumulative distribution, or cumulative probability distribution function. In principle, the probability density function $p(l)\,dl$ is obtained from the derivative of the cumulative distribution function, $P(a \leq l \leq b) = \int_a^b p(l')\,dl'$. $P(a \leq l \leq b)$ is read as the probability that l lies between a and b.

Consider a surface element, as in Fig. 1.3, with incidence angle θ which lies between the normal to the surface element and a particular direction of an incoming particle.

(a) What is the probability density distribution function $p(\theta)\,d\theta$ of the angle θ if the surface element is exposed to an isotropic flux of particles? In other words, how are the values of θ distributed in the probabilistic sense over all the angles between 0 and $\pi/2$ on the surface element?

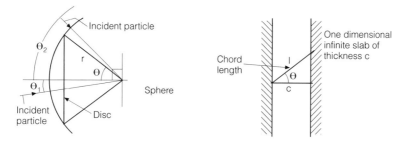

Figure 1.7. Calculation of the probability distribution of the incident angle θ (left) and chord length in a one-dimensional infinite slab of thickness c (right).

    (b) For a one-dimensional, infinitely long slab of thickness c, as in the rightmost Fig. 1.7, what is its chord length probability density function $p(l)\,dl$?

    (c) What is the value of the mean chord length $\bar{l}$ in the slab in (b)?

    (d) What is the value of the mean cosine of the angle of incidence $\overline{\cos\theta}$ in (a)?

    (e) What is the value of the mean angle of incidence $\bar{\theta}$ in (a)?

    (f) What is the value of the cosine of the mean angle $\bar{\theta}$ in (e)?

1.11 (a) Obtain the mean chord length directly from $\bar{r} = \int rf(r)\,dr$. **Hint:** Use the delta function $\delta(r - r_s)$ in the integration of the numerator of $f(r_s)$ in (1.25) instead of the $r = r_s$ symbology [i.e., use $\int f(r)\,\delta(r - r_s)\,dr = f(r_s)$, an important delta function property].

    (b) Show that with the above expression for $f(r)$, $\int f(r)\,dr = 1$, using the relation that $\int_a^b \delta(x - x_0)\,dx = 1$, $a \leq x_0 \leq b$, another delta function property.

1.12 Show that the mean chord length $\bar{s} = 4V/S$ from the fact that $\bar{s}$ is also given by the mean total track length in the convex body volume V divided by the mean number of chords threading the body.

References

1. J.R. Lamarsh, *Introduction to Nuclear Reactor Theory*, Addison-Wesley, Reading, MA, 1966, Chap. 5.

2. R.L. Fleisher, P.B. Price, and R.N. Walker, *Nuclear Tracks in Solids*, University of California Press, Berkeley, 1975, Chap. 9.

3. W.G. Ireson, *Reliability Handbook*, McGraw-Hill Book Co., New York, 1966, Sect. 26.
4. E.L. Petersen, J.C. Pickel, J.H. Adams, Jr., and E.C. Smith, "Rate Prediction For Single Events—A Critique," *IEEE Trans. Nucl. Sci.*, **NS-39** (6), 1577–1599 (1992).
5. K.M. Case, F. deHoffman, and G. Placzek, *Introduction to the Theory of Neutron Diffusion, Vol. I*, Los Alamos Scientific Laboratory, Los Alamos, NM, 1953, Sect. 10.1.
6. W.L. Bendel, "Length Distribution of Chords Through a Rectangular Volume," NRL Memo Report No. 5369, Naval Research Laboratory, Washington, DC, 1984.
7. M.D. Petroff, in J.C. Pickel, and J.T. Blandford, Jr., "Cosmic Ray Induced Errors in MOS Devices," *IEEE Trans. Nucl. Sci.*, **NS-27** (6), 1006–1015 (1980), Appendix I.
8. J.N. Bradford, "A Distribution Function For Ion Track Lengths in Rectangular Volumes," *J. Appl. Phys.*, **50**, 3799 (1979); "Cosmic Ray Effects in VLSI in Space Systems and Their Interaction with Earth's Space Environment," *Progr. Astronaut. Aeronaut.*, **71**, 549 (1980).
9. L.D. Edmonds, "A Distribution Function For Double Upsets," *IEEE Trans. Nucl. Sci.*, **NS-36** (2) 1344–1346 (1989).
10. A.M. Kellerer and D. Chmelevsky, "Criteria for the Applicability of LET," *Radiat, Res.*, **63** (1975).
11. A.M. Kellerer, "Considerations on the Random Traversal of Convex Bodies and Solutions for General Cylinders," *Radiat. Res.*, **47**, 359–376 (1971).
12. H.L. Brode, "Close-in Weapon Phenomena," *Annu. Rev. Nucl. Sci.*, **8** 153–202 (1969).

2

Extraterrestrial SEU-Inducing Particles

2.1. Introduction

The major causes of single event upset (SEU) are due to particles originating from deep space, and a main concern is their impact on spacecraft avionics. These particles include heavy cosmic ray ions of galactic origin. Other cosmic ray particles include the pion and muon groups of mesons. They are secondary particles produced by the interaction of incident cosmic rays with the earth's atmosphere. Generally, the meson flux is greater than that of cosmic rays near the ground.

Another source of SEU-inducing particles is gamma rays from interstellar space. Analogous to secondary SEU-inducing particles from primary proton reactions in device material, galactic gamma rays can also produce secondary SEU-causing particles such as alphas, protons, and neutrons from (γ, α), (γ, p), and (γ, n) reactions, respectively. Although the total flux from these gammas is greater than that from cosmic ray heavy ions, the corresponding gamma ray ionizing-dose damage is under study. SEU rates are still to be investigated. Gamma rays can cause direct ionizing radiation damage, as used in ionizing-dose parts testing. The competition between the SEU impact of the secondaries of gamma ray origin and the gamma ray ionizing-dose is also under study.

Alpha particles can cause SEU directly and are produced by (n, α) and (p, α) reactions in impurities in the chip. This is in addition to alphas from the chip package entering the chip itself to cause SEU within (Section 2.4).

Solar flares, which are caused by eruptions on the solar surface, are described. They temporarily destabilize most of the SEU-inducing particles fluxes in the earth's vicinity and greatly increase their levels. Formulas for error rates (Section 5.2) from galactic cosmic rays at geosynchronous altitudes can be used for estimation under flare conditions by using a premultiplier of up to 5000.

Finally, details of the Van Allen belts are discussed with respect to protons therein inducing SEU in spacecraft transiting through the belts. Their SEU effects principally impact spacecraft at altitudes much less than geosynchronous. Also, low earth spacecraft orbit altitudes ($\leqslant 800$ miles) are "under" the belts, so their SEU difficulties are mainly confined to those associated with the South Atlantic Anomaly (Section 6.7).

2.2. Cosmic Rays

Cosmic rays are of concern because it is well known that these heavy ions are certainly incident on integrated circuits (microcircuits) in spacecraft and aircraft electronics. They supply charge by producing ionization tracks within these microcircuits. At a minimum, this charge can temporarily upset microcircuit function by its ultrarapid migration (picoseconds) to memory-chip-sensitive regions to change the information stored in them.

Cosmic rays have been observed since about the turn of this century. Experimenters at that time found that their "ion chambers" (leaf electroscopes) never completely discharged, even in the perceived absence of radiation. Thick lead did not suppress the effect entirely. This radiation did not diminish when these chambers were borne aloft, as it would if it came from a terrestrial source. Actually, the radiation intensity increased with height above ground, as was shown in early balloon experiments. The intensity was just as great by night as by day, so it probably did not come from the sun. The radiation appeared to come from interstellar space. Cosmic rays were found to have very high energy and be highly ionized, as became apparent by their detection at up to 1650 ft under lake water [1]. Later, cloud chamber studies showed by measurement of their track radii that cosmic rays possessed very high energies up to more than 10^{20} eV. Particles with energies greater than about 10^{18} eV are presumed to be of extragalactic origin. What is observed at ground level is not only the original cosmic radiation; it includes a variety of secondary particles due to cosmic ray interactions with interstellar gases and the earth's atmosphere on their way to ground level. The intensities at ground level are relatively low, being harmless to many but not all integrated circuits.

It has been shown that cosmic rays and debris from nova and supernova detonations have endured in galactic space for 5–10 million years before reaching the space around earth. Cosmic ray relative abundances represent nearly all elements in the periodic table plus high-energy electrons. Figure 2.1 depicts the relative abundance of cosmic ray elements versus their atomic number [2]. As can be seen from the figure, elements of $Z \gtrsim 25$ (Z at and

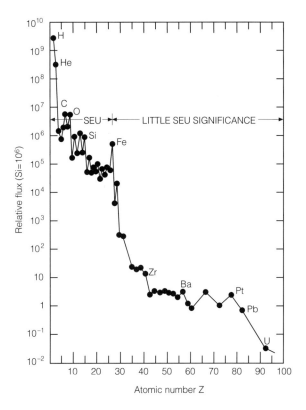

Figure 2.1. Cosmic ray relative abundance versus nuclei atomic number. From Ref. 2.

greater than iron) have intensities orders of magnitude less than those for $Z \lesssim 25$. Therefore, $Z > 25$ cosmic rays are of little concern because of their extremely low intensity. The reason that the abundance curves appear "choppy" is that even-Z elements, compared to their immediately adjacent odd-Z elements, have relatively greater nucleon-binding energy, as indicated by the nuclear shell model. Thus, even-Z elements are more stable and, therefore, more abundant cosmologically. Actually, the abundance curve is not a curve at all, because noninteger Z is meaningless, but it is actually a sequence of points indicating individual relative abundances.

The principal reason for the drop in abundances for elements whose $Z \gtrsim 25$ has to do with stellar nucleosyntheses processes, especially in stars that become novas and supernovas. Briefly, such studies claim that within the star, the element nuclei are built up, starting with hydrogen, through accretion of neutrons and alpha particles. Starting when the star is born, this process then proceeds over the millions of years of star life to $Z \sim 20\text{--}30$. However, each successive nuclear reaction in that progression is less and less exothermic.

Thus, each reaction provides less initiating energy for those that follow. The star then reaches the state, at about a Z corresponding to iron nuclei, where the nuclear-derived energy does not balance its gravitational potential energy. With the subsequent collapse on itself, certain stars detonate to form novas or supernovas. In this cosmological instant (~ 1 week), the detonation shock wave supplies the required initiation energy to form the elements whose $Z \geqslant 25$, using neutrons and alphas as before. As this process is extremely short compared to that prior to collapse, the $Z \geqslant 25$ element yield is very much less than that for $Z \leqslant 25$.

The most abundant cosmic ray element is hydrogen (protons), making up nearly 94% of the total. Helium is next with about 5%, with everything else summing to about 1%. The most abundant cosmic ray elements are hydrogen, helium, carbon, oxygen, neon, magnesium, silicon, sulfur, calcium and, iron, in that order. As mentioned, high-energy electrons of cosmic ray origin are also present. These electrons can cause some SEU directly. However, except for charge-coupled device (CCDs), these electrons cannot produce sufficient charge to impact Class S (spacecraft grade) integrated circuits (ICs) of current feature size. High-energy electrons can cause SEU indirectly by producing bremsstrahlung photons from charged-particle accelerations in matter. Typical photon reactions which follow in silicon are $Si(\gamma, n)$, $Si(\gamma, p)$, or $Si(\gamma, \alpha)$, where the secondaries (n, p, α) may go on to cause SEU. $Si(\gamma, x)$ means an incident photon (γ) interacts with a silicon nucleus to produce x. In the case of electrons, the flux must be enormously larger than, for example, the proton flux in the same vicinity to compete in producing SEU.

The sun, through the solar wind, contributes to the cosmic rays incident in our magnetosphere. These are mainly protons, but there is a heavier component as well. An example of the latter is depicted in Fig. 2.2 for iron.

Figure 2.2 depicts the cosmic ray spectrum of iron. The ordinate is the differential flux of iron particles in units of particles per square meter, steradian, second and MeV per atomic mass unit. The abscissa is their energy distribution in units of MeV per atomic mass unit. Curves like these are the results of the work of many people, as exemplified by the different graphical point symbols, whose identities are given in Ref. 4. The upper solid curve corresponds to the situation during a sunspot minimum, whereas the lower solid curve is that for a sunspot maximum. The reasons for this difference between the two curves is given in the ensuing discussion.

Another important characteristic of particle-element spectral curves is that most are made up of two main components. The major component, and most of the curve, is due to galactic cosmic rays with their typical peak. The second but much smaller component is that due to anomolous sources, mainly singly ionized penetrating particles [3]. This also includes trapped heavy ions in the Van Allen belts, as well as those from the sun. Both of these curves are indicated by dashed line extensions on the left side of Fig. 2.2. This part of

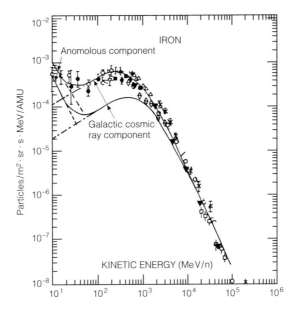

Figure 2.2. The cosmic ray iron spectrum curves for solar max (lower) and solar min (upper). From Ref. 4.

the curve is usually "faired –in," giving a smooth but obfuscating minimum. It exists because of the above two components and has no intrinsic physical meaning.

All of these contributions are periodic, varying with the solar cycle (11-year period). This "solar modulation" anticorrelates with the galactic cosmic ray flux; that is, the latter competes with the solar wind plasma which carries an embedded magnetic field with it that diverts cosmic rays from the magnetosphere. Hence, during a solar (sunspot) minimum, the solar wind is relatively weak, allowing more cosmic ray flux through the magnetosphere and for it to achieve a relative maximum. Solar flares, occurring at random intervals, affect the strength and direction of the solar wind. They also influence SEU error rates considerably.

The magnetosphere is the region surrounding the earth's neighborhood in which cosmic rays can be magnetically trapped by the earth's magnetic field. The magnetosphere boundary is called the magnetopause, beyond which particles are not trapped, although the geomagnetic field extends beyond the magnetopause. The magnetopause lies about 10 earth radii ($10R_e$) from the earth's center on the sun side. Its shape is initially ogival on the sun side, but elongated on the night side, extending into a cylinder of revolution out to the order of $40R_e$, with a diameter of more than $35R_e$. The magnetopause is formed

28 / *Single Event Phenomena*

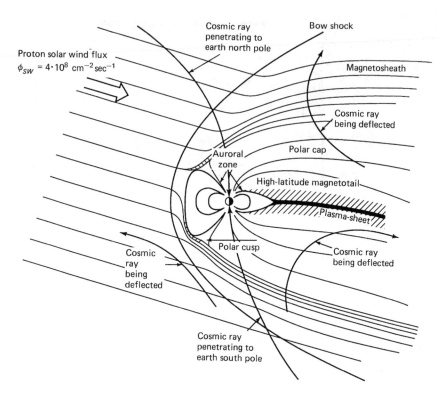

Figure 2.3. Regions of the magnetosphere in noon–midnight meridian plane showing cosmic ray trajectories and the solar wind. From Ref. 5.

by the confining action of the solar wind, plasma moving magnetic field, interacting with the stationary geomagnetic field. This action delineates the magnetopause to form a "cavity" about the earth which contains the magnetosphere. The solar wind also forms a detached bow hydromagnetic shock wave about $20R_e$ in front of the earth on the sun side. Its shape is similar to the magnetopause and somewhat concentric with it, especially as they both "flare," normal to the solar direction out to $60R_e$ (moon distance). This is all depicted in Fig. 2.3.

Rigidity and geomagnetic cutoff are two parameters that characterize the influence of the earth magnetic field on cosmic ray flux. An incident cosmic ray ion in the geomagnetic field is acted upon by the magnetic field force [viz. $q(\mathbf{v} \times \mathbf{H})$ and the centrifugal force $M(\mathbf{v})^2/r$. M is its mass and $q = Ze$ is its ionization charge state. The familiar prototype relationship is $Mv^2/r = Hev$. For the sake of simplification, first assume that the magnetic field is constant (DC field) and the cosmic ray velocity $\mathbf{v}$ is normal to the magnetic field $\mathbf{H}$. The ion orbit is then normal to the $\mathbf{E} \times \mathbf{H}$ plane. Because all the parameters are constant in magnitude, a dynamic equilibrium ensues between the two forces along the cosmic ray orbit; that is,

$$\frac{M|\mathbf{v}|^2}{r} = |q(\mathbf{v} \times \mathbf{H})| \quad \text{or} \quad \frac{Mv^2}{r} = ZHev. \qquad (2.1)$$

The threshold rigidity Hr is then given by

$$Hr = \frac{Mv}{Ze} = \frac{p}{Ze} \qquad (2.2)$$

where $p = Mv$ is the scalar momentum of the cosmic ray. Because r is now constant as well, the ion will travel in a circle, which is the locus of constant radius r, always normal to the rotating $\mathbf{v} \times \mathbf{H}$ plane.

The earth's magnetic field, as seen from the incoming ion trajectory, has a positive gradient so that Hr is not constant. The rigidity becomes a function of the geomagnetic coordinates as well as relativistic corrections for the heavy mass of the high-velocity cosmic ray ions. This implies that the heavy-ion trajectory arc radii decrease as the ion spirals into the magnetosphere. Its radius can be written as

$$r = \frac{Mv}{ZeH} = \frac{Av}{N_0 ZeH} \qquad (2.3)$$

where A, Z, and N_0 are the cosmic ray ion mass number, atomic number, and Avogadro's number, respectively, and $M = A/N_0$ (g/mole)/(particles/mole).

Geomagnetic rigidity herein implies threshold rigidity; that is, for each position and direction in the magnetosphere, there is a threshold rigidity vector $\mathbf{R}_T(\mathbf{p}) = \mathbf{u}R(p)$, whose magnitude $R(p)$ is called the magnetic rigidity cutoff. $\mathbf{u} = \mathbf{v}/v$ is a unit velocity vector giving the direction of the incoming heavy ion with respect to the magnetic field $\mathbf{H}$. If the rigidity of any charged particle (ion) as characterized by its ratio of momentum to charge p/q is less than its threshold $\mathbf{R}_T$, then the particle cannot reach the above position with direction $\mathbf{u}$ [6]; that is, it will be magnetically deflected, as shown in Fig. 2.3. A charged particle whose rigidity is equal to or greater than the cutoff value can reach that position and proceed in the direction $\mathbf{u}$.

Spacecraft coordinates are given in terms of the coordinates of the magnitude of the earth's magnetic field (viz. $\mathbf{B}$ and its parameters [5]). For a spacecraft at an altitude R_s over a point on the earth's surface measured from the geomagnetic center, $\mathbf{B}(R_s)/B_0$ is the value of the geomagnetic field intensity at the spacecraft normalized to its value B_0 at the equatorial altitude R_0, also measured from the geomagnetic center.

30 / Single Event Phenomena

A McIlwain parameter L can be defined as $L = R_0/R_e \simeq R_s/R_e$, for L not too large, where R_e is the earth's radius. For the assumed symmetric geomagnetic dipole field, one connective relation between these parameters is

$$\frac{B}{B_0} = \frac{(R_0/L)^3 (1 + 3\sin^2\lambda)^{1/2}}{\cos^6\lambda} \qquad (2.4)$$

where λ is the magnetic latitude (north or south). L shells are simply shell-like surfaces formed by the myriad geomagnetic field lines of constant distance L (not constant B). L is the distance, measured in earth radii, from the center of the magnetic dipole (approximation) to where the field lines of this particular L shell cross the equator at altitude $L = R_0/R_e$. This is shown in Fig. 2.4. If the dipole approximation is not used, then the L values are difficult to compute [5]. Another property of L shells is that they are surfaces of field lines, each of which is a guiding center for an incoming cosmic ray as it spirals around that line in a cycloidal path during its sojourn in the magnetosphere (Fig. 2.4).

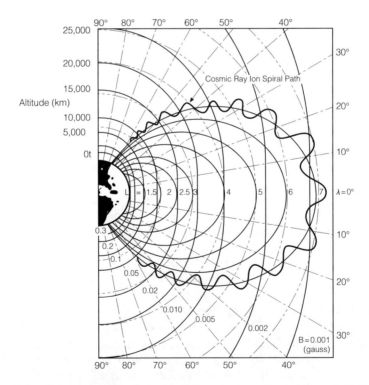

Figure 2.4. Cross section of constant L shell magnetic surfaces and constant B surfaces. From Ref. 5.

Transformation relations between spherical coordinates and B, L coordinates are given by

$$\frac{B}{B_0} = \left(\frac{R_0}{r}\right)^3 \left(4 - \frac{3r}{L}\right)^{1/2}, \quad r = L\cos^2 \lambda = R_0 \sin^2 \theta, \quad \theta = \frac{\pi}{2} - \lambda \quad (2.5)$$

The approximation of a symmetric geomagnetic dipole field implies spherical symmetry about the spherical coordinate azimuthal angle ϕ, so that it does not appear in these expressions. For the actual nonsymmetric geomagnetic field, the B, L coordinates are more complex [5]. In terms of B, L coordinates, the geomagnetic rigidity cutoff at the earth's surface using the dipole field approximation is given by

$$R = \left(\frac{60}{r^2}\right)\left(\frac{(1 - (1 - \cos\theta_t \cos^3 \lambda)^{1/2})^2}{\cos\theta_t \cos \lambda}\right)^2 \quad (2.6)$$

where θ_t is the angle that a cosmic ray arc makes with the magnetic west as it strikes the ground. Figure 2.5a depicts the rigidity versus the cosmic-ray-allowed solid angles [6]. The incoming cosmic ray ions have extremely high energies, implying very high velocities that begin to approach c, the speed of light. Then taking relativistic mass changes into account by eliminating v between $E_T = mc^2 = m_0 c^2 (1 - v^2/c^2)^{-1/2}$ and $p = mv = m_0 v (1 - v^2/c^2)^{-1/2}$, where p

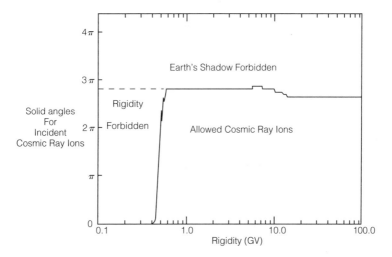

Figure 2.5a. The geomagnetic cutoff behavior at $-50°$ latitude (Southern Hemisphere), 120° longitude, 400 km altitude; near the location of the lowest cutoff experienced by a spacecraft in a 50° inclination orbit. From Ref. 6.

is the ion momentum, E_T its total energy, and m_0 its rest mass, yields (for any particle) the well-known relation

$$E_T^2 = p^2c^2 + m_0^2c^4 \tag{2.7}$$

and where the rest energy $E_0 = m_0c^2$. Because the ion is assumed to be a free particle in that its energy consists of kinetic and rest mass energy only, $E_T = E + E_0$. E, the kinetic energy, is usually called the particle energy. Inserting E_T into (2.7) yields

$$E = (E_0^2 + p^2c^2)^{1/2} - E_0 \tag{2.8}$$

From (2.2), the rigidity $Hr \equiv R = p/Ze$, when inserted into (2.8), results in

$$E = [E_0^2 + (ZeRc)^2]^{1/2} - E_0 \tag{2.9}$$

which is a sought after relation between the ion energy and the rigidity of the magnetic field. Then substituting R for all ion orbits allowed or not (rigidity cutoff) provides allowed or forbidden ion energies for incoming cosmic ray heavy ions. If (2.9) is divided by the ion mass number A, the result is [4]

$$E = [E_0^2 + (ZR/A)^2]^{1/2} - E_0 \tag{2.10}$$

where now E and E_0 are in units of energy per nucleon (E/A, E_0/A) and R has the dimensions of energy per nucleon. The rest energy of a nucleon (neutron or proton) $E_{on} = M_n c^2 = (A_n/N_0) c^2/(1.6 \times 10^{-6}) \cong [1/6 \times 10^{23}] (9 \times 10^{20})/(1.6 \times 10^{-6}) \cong 938$ MeV, as $A_n = 1$ and there are 1.6×10^{-6} ergs/MeV. N_0 is Avogadro's number. Figure 2.5(b) depicts (2.10) for various E/A shells (i.e.,

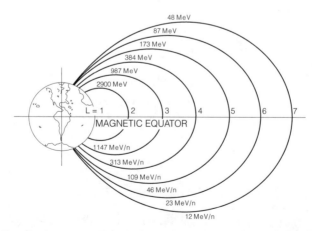

Figure 2.5b. Total energy required to penetrate magnetosphere for various L shells. From Ref. 7.

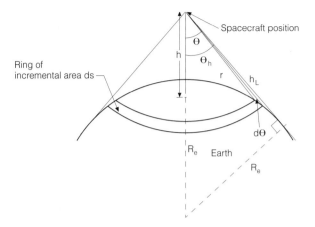

Figure 2.6. Space earth-shadowed from cosmic ray flux in the magnetosphere.

L shells). Extracting the rigidity R from (2.10) gives, for energies in MeV per nucleon,

$$R = (A/Z)(E^2 + 2E_0 E)^{1/2} \cong 2(E^2 + 1876E)^{1/2} \tag{2.11}$$

since $A/Z \simeq 2$ for most electron-stripped heavy ions of interest, and $2E_0 = 1876$ MeV per nucleon. However if the ion is, say, only singly ionized, then $Z = 1$ and $A/Z = A$.

Another aspect of cosmic ray incidence on spacecraft is that the earth shadows (occults) the spacecraft from a fraction of the cosmos, and so from the corresponding cosmic ray component. This is especially the case for near-earth orbits. This fraction is obtained by calculating the solid angle subtended by the earth at the spacecraft altitude h. Figure 2.6 shows the spacecraft at a given instant and the earth shadowing it from a portion of the celestial sphere. The incremental solid angle $d\Omega$ subtended by a ring of incremental area ds on the earth's surface as in Figure 2.6 is obtained using

$$ds/r^2 = d\Omega = 2\pi \sin\theta \, d\theta \tag{2.12}$$

where r and θ are depicted in the figure. Then

$$\Omega = 2\pi \int_0^{\theta_h} \sin\theta \, d\theta = 2\pi(1 - \cos\theta_h) \tag{2.13}$$

θ_h is the angle between the spacecraft altitude h and its distance to the horizon h_L, so that

$$\cos\theta_h = \frac{h_L}{R_e + h} = \frac{(h^2 + 2hR_e)^{1/2}}{R_e + h} \tag{2.14}$$

The fraction of the cosmos occulted by the earth at the spacecraft altitude, $\Omega/4\pi$, is obtained by inserting $\cos\theta_h$ from (2.14) into (2.13). This gives

$$\frac{\Omega}{4\pi} = \frac{1}{2}\left(1 - \frac{(1 + 2R_e/h)^{1/2}}{1 + R_e/h}\right) \qquad (2.15)$$

For near-earth orbits where $h/R_e \ll 1$, (2.15) yields

$$\frac{\Omega}{4\pi} \cong \frac{1}{2}\left[1 - \left(\frac{2h}{R_e}\right)^{1/2}\right] \qquad (2.16)$$

Equations (2.15) and (2.16) are first approximations, in that for low magnetic rigidities, the effective shadow edges are distorted by the field. The edges are skewed in an easterly direction, in the sense that cosmic rays can arrive below the optical shadow region from westerly directions. Also, $\Omega/4\pi$ decreases more rapidly with decreasing h than that given by (2.15), as the shadow density is highly variable in terms of penetrating cosmic ray flux [4].

2.3. Other Cosmic Ray Particles

Besides the major cosmic ray particles (high-energy heavy ions) that are detrimental to avionics and electronics systems, there are other particles whose energies and fluences are much less in magnitude, which can also contribute to SEU in these systems. As mentioned, cosmic rays incident in the earth's neighborhood atmosphere contain a secondary particle component. These include those particles produced mainly from interactions between the primary rays, such as the heavy ions, and extraterrestrial matter, as well as the atmosphere.

The principal secondary particles are mesons. They are closely related to electrons but have masses of 150–1500 times m_e, the electron mass. There are many kinds of mesons, but there are two main families of interest. They are the pions (273 m_e) and the muons (207 m_e). The members have different charge signs, such as π^+, π^-, π^0 (no charge), and μ^+ and μ^- (no μ^0 exists). Pion and muon fluences produced by cosmic ray ion interactions in solid material (e.g., silicon) often exceed the cosmic ray proton and neutron component external to the solid. This is especially the case at lower altitudes in the atmosphere. It is also claimed that neutrons and muons are the primary cause of ground-level SEU [9].

Pions of either charge ($\pi^\pm$) decay spontaneously into corresponding muons of either charge ($\mu^\pm$), plus other reaction products, with a mean life of about 20 ns. Neutral pions (π^0) decay into gamma rays ($\pi^0 \rightleftarrows 2\gamma$) with a very much shorter mean life of about 0.2 femtosecond (fs). However, for high-speed mesons whose velocity approaches the speed of light, their mean life increases

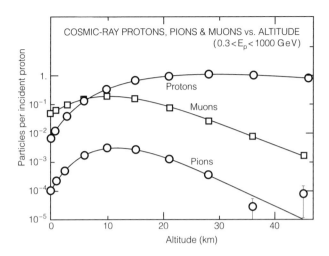

Figure 2.7. Calculated proton, muon, and pion numbers versus altitude produced from typical distribution of protons incident upon the earth's atmosphere at 50 km. From Ref. 8.

due to relativistic time dilation, and so their reaction probabilities increase as well [15]. The SEU cross section for accelerator-produced charged pions in a 4K CMOS static random access memory (SRAM) is about 10^{-7} cm^2 [8]. This cross section is comparable to that from low-energy protons or neutrons in similar microcircuits [10].

Charged muons ($\mu^{\pm}$) decay into electrons, plus other reaction products, with a mean life of about 2 μs. Using accelerator-produced muons, it is found that the SEU cross section for moderate-energy muons in 4K CMOS microcircuits is vanishingly small [8]. However, it can be shown analytically that high-energy muons can produce SEU in microcircuit memories of smaller feature size than 4K SRAMs. This includes integrated circuits whose critical charge is less than or equal to 20 fC [8].

Figure 2.7 depicts the relative number of pions and muons normalized to atmospheric protons that produced them at an altitude of 50 km [8]. Typical pion and muon production reactions are $p + p \to \pi^+ + d$, $n + p \to \pi^0 + d$, $n + p \to \pi^- + 2p$, and $\pi^{\pm} \xrightarrow{20 \text{ ns}} \mu^{\pm} + \nu$. p is the proton, n is the neutron, d is the deuteron, and ν is the neutrino. These particles must have the required energy thresholds in a corresponding nuclear environment, such as the upper atmosphere, for these reactions to proceed and result in the given reaction products. One complementary muon (decay) mode is $\mu^{\pm} \xrightarrow{2.15 \, \mu s} e^{\pm} + 2\nu$ (i.e., electron production). Also, when a μ^- enters a material nucleus, $\mu^- + p \to n + \nu$ can occur.

Another source of cosmic ray particles, discovered relatively recently, are gamma ray "bursters." The energy spectra of these gamma ray bursts range from about 10 keV up to 100 MeV [11]. They were discovered about 25 years ago by member spacecraft of the VELA program. A portion of the VELA spacecraft was used to monitor mainly U.S.S.R. compliance with the treaty barring nuclear testing in space. After many gamma ray bursts were detected by the VELA satellites, it was deduced that they did not come from banned nuclear detonations, but were a new source of extraterrestrial radiation [12]. The gamma burst pulse widths vary from about 10 ms to 1000 s, with rise times as short as 100 μs. The exact nature of their sources has so far not been identified with certainty. Evidence points to galactic pulsars and binary neutron stars [16] with and without accretion disc companions of some sort. One unique feature of the bursts is their apparent periodic behavior. In some cases, the period lasts from 1 to 20 s, especially after an unusually strong initial burst pulse. This behavior is reminiscent of a "ringing" phenomenon following a strong impulse event occurring in many branches of scientific investigation. Figures 2.8a and 2.8b depict this behavior [11].

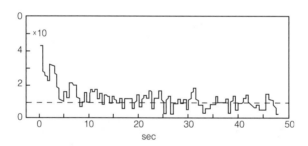

Figure 2.8a. Time profile of an ordinary burst; dashed line is detector background. From Ref. 11.

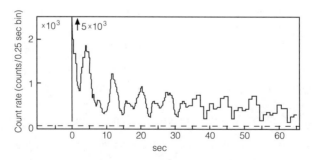

Figure 2.8b. Time profile of a repeater burst; dashed line is detector backgrounds. From Ref. 11.

The bursts have been separated into two main classifications: (a) "classical bursts" are those with an energy spectra greater than about 0.15 MeV and seem to appear from space isotropically, with a repetition rate of about one per day; (b) "soft repeaters" are bursts whose energy spectra are less than 0.15 MeV, but each apparently comes from almost the same celestial coordinates with a repetition rate of about one every 20–25 weeks.

Of the thousands of gamma bursts observed over the past 25 years, few have been repeaters. Otherwise, they seem to be distributed isotropically throughout our galaxy, with no two coming from the same sky coordinates. The mean gamma burst energy flux incident in the earth's neighborhood is about 6.25 GeV cm^{-2} s^{-1} for $E > 0.1$ MeV [11]. Table 2.1 compares gamma ray

Table 2.1. Comparison of Gamma Ray Bursts with Nuclear Detonations and Cosmic Rays

		Gamma ray bursts	Nuclear detonation	Cosmic rays
(1)	Particle Type	X- and gamma rays	X- and gamma rays	High–energy heavy ions
(2)	Origin	Pulsar (Neutron star)	Weapon detonation	Debris from nova and Supernova detonation
(3)	Energy flux (Earth) (ergs/cm^2 s)	10^{-2}	10^8–10^{12}	5×10^{-3}
(4)	Energy Spectrum range (MeV)	0.01–100	0.01–10	1–10^8
(5)	Simultaneous optical burst	No	Yes	—
(6)	Pulse duration (s)	10^{-2}–10^{-3}	$(20–50) \times 10^{-9}$	CW
(7)	Pulse rise time (s)	$\geq 10^{-4}$	10^{-9}	—
(8)	Occurrence frequency	~1 Per day	—	CW
(9)	Temporal Structure	Multiple peaks (periodic)	Double peak—endo[a] Single peak—exo[a,b]	Anomolously large solar flare peaks ~2 ea. per 10 years
(10)	Spatial Distribution	Isotropic	Anisotropic—Endo[a] Isotropic—Exo[a]	Isotropic
(11)	Earth magnetic field Effects	None	Minimal (EMP)	Deflects ions below energy threshold
(12)	Years since discovery	25	55	85

[a] Atmospheric.

[b] Single-stage device.

burst parameters with those of cosmic ray heavy ions and nuclear detonations [14]. It is seen from Table 2.1 that the gamma ray bursts share certain parameters with nuclear detonations as well as cosmic ray heavy ions.

Even though gamma rays cannot cause SEU directly due to their miniscule stopping power, they can cause such upsets indirectly. There are a number of indirect modes including (γ, α), (γ, n) and (γ, p) reactions, whereby the alphas, neutrons, and protons thus produced will cause SEU indirectly. For example, accelerator-induced bremsstrahlung gamma rays interacting with silicon can produce both protons and alpha particles. However, as in the case of gamma rays, the stopping power of these protons, [e.g., those from $^{28}Si(\gamma, p)\,^{27}Al$] cannot deposit sufficient energy in the microscopic device volumes to cause SEU. Protons can, in turn, produce SEU indirectly (Section 5.3). However, gamma-induced alpha particles from $^{28}Si(\gamma, \alpha)\,^{24}Mg$ and $^{24}Mg(\gamma, \alpha)\,^{20}Ne$ reactions producing energies of 4–10 MeV also cause upsets [13].

Gamma ray bursts also can cause SEU indirectly by way of SEU-inducing neutrons from (γ, n) photoneutron reactions in deuterium, lithium, beryllium, carbon, and other heavier materials [14] that can make up device components. Of course, gamma rays at much greater fluence levels easily do permanent damage to microcircuits. Gamma ray ionizing-dose testing is standard procedure today as part of component radiation test programs. Presently, gamma ray bursts are a component of cosmic rays. How their unique properties, such as their periodic behavior, affect SEU of microcircuits is still to be investigated.

Other cosmic ray derivative particles include spallation or fragmentation products. Fragmentation is a descriptor for both spallation and fission and other similar disintegration processes. Spallation is the disintegration of, say, a cosmic ray nucleus into a large number of free nucleons and/or very small fragments such as alpha particles, deuterons, and ^3He particles, as opposed to fission which results in the disintegration of the nucleus into mainly two fragments. Fission can be induced by many particles including neutrons such as in a nuclear reactor. Sufficiently high-energy particles can induce fission in virtually any nucleus, such as Cu. Spontaneous fission can also occur, usually in very heavy, high-mass-number nuclei, such as ^{252}Cf (Section 4.4), whereas spallation is usually induced through interactions of the incident particle with other particle nuclei, such as silicon in the device.

2.4. Alpha Particles

In the first 40 years of this century, alpha particles were the only "heavy" particles, in contrast to electrons, known to be the decay product of other radioactive processes. Figure 2.9 shows half-lives of decay alphas for a few heavy radioactive elements [17]. This was the situation until fission was

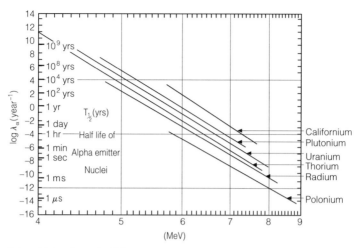

Figure 2.9. Alpha decay half-lives of heavy nuclei versus alpha energy. From Ref. 17.

discovered in about 1939. It was known that alphas are actually helium ions (He^{2+}). Alphas also participate in the radioactive decay chain of actinide elements, such as the decay of uranium through successive daughter elements, ultimately decaying to lead. One estimate of the age of the earth uses uranium to lead ratios in these computations.

Alphas further play a role in the relatively short history of SEU in that they were among the first particles to be identified in 1979 as causing SEU in microcircuits [18]. The other particles are, of course, heavy cosmic ray ions, which were first attributed to the production of SEU in a particular spacecraft in 1975 [19].

Prior to the turn of this century it was found that uranium and other actinide elements emitted a hard (penetrating) component called beta rays, which turned out to be energetic electrons. The other, a soft component, is now called an alpha particle. Alpha particles of a few megaelectron-volts in energy can be stopped by the 2–3 mil thickness of cigarette paper. Some of the actinides are trace elements in the earth's crust; their decay by alpha particles provides alphas in the chip silicon which can cause SEU. The addition of a passivation layer about the chip is sufficient to prevent alphas from entering the chip proper. However, many more alpha particles than from the earth's crust can be produced within the chip by penetrating particles, such as neutrons and protons, that interact with the silicon as well as with the boron p-dopant of the individual transistors. To exasperbate the situation, not only are alphas produced but in neutron-inducing reactions much heavier ions are also created to cause SEU. For example, in an (n, α) reaction ($^1_0n + {}^{28}_{14}Si \rightarrow {}^{25}_{12}Mg + {}^4_2\alpha$), and in an (n, p) reaction ($^1_0n + {}^{28}_{14}Si \rightarrow {}^{28}_{13}Al + {}^1_1p$), it is seen that protons, magnesium, and aluminum are formed. The SEU is predominantly caused by ionization from the protons, magnesium and aluminum moving as charged particles through the lattice. As for the boron, thermal (low energy) neutrons incident can interact with it to form a number of deleterious products. Specifically, $^{1}_{0}n + ^{10}_{5}B \rightarrow ^{7}_{3}Li^* + ^{4}_{2}\alpha$. Besides the alpha particle, excited lithium (Li*) deexcites to give off gamma rays, [i.e., $^{7}_{3}Li^* \rightarrow ^{7}_{3}Li + \gamma(0.478 \text{ MeV})$]. Lithium can combine with a neutron as well; that is, $^{7}_{3}Li + ^{1}_{0}n \rightarrow ^{8}_{3}Li^* \rightarrow ^{8}_{4}Be^* + ^{0}_{-1}e$, where the $^{0}_{-1}e$ electron produces bremsstrahlung, and beryllium can split into two alphas as $^{8}_{4}Be^* \rightarrow 2(^{4}_{2}\alpha)$.

Another important aspect of alpha particle behavior is that when they are used to bombard nuclei, they ultimately exited the nucleus, but with a time delay between their incident absorption and exit from the nucleus, which is very long compared to the classical transit time through the nucleus. This poses a quandary from the classical point of view. If the nuclear potential barrier is not of sufficient strength to hold the alpha particle within it, then the alpha should pass directly through in an extremely short but measurable time. This time is simply given by nominal nucleus diameter/alpha speed = $d_{\text{nucl}} (2E_\alpha/m_\alpha)^{-1/2}$. On the other hand, if the nuclear potential barrier is of sufficient strength to hold the alpha particle within the nucleus, it should almost never leave. In 1929, quantum mechanics provided the answer to the time delay by discovering that the time it takes for particles to tunnel through a potential barrier corresponds to the decay half-life of that nucleus. This is a bizarre concept classically, in that the nuclear potential barrier is somehow semitransparent. Quantum mechanical calculations showed that there is a small but finite probability for an alpha particle to tunnel through the barrier, which was validated by experiment. Today, the tunneling phenomenon is a mainstay of the detailed function of a number of active devices such as zener diodes, MOSFET interface dynamics, and even the operation of the carbon granule microphone.

Theoretical and corroborating experimental results verified that the quantum mechanical tunneling probability depends very sensitively on the alpha particle energy it has when it leaves the nucleus. This is seen in Fig. 2.9 where halving the alpha energy from 8 MeV to 4 MeV corresponds to an increase in the decay alpha half-life from about 30 μs to about 10^8 years, a factor of 12 orders of magnitude. The theory of the tunneling phenomenon also provides an explanation of the empirical Geiger–Nuttall relation (1911). This relation connects the alpha particle range R with the mean life λ^{-1} of the alpha emitter nucleus as given by

$$\ln R = A \ln \lambda + B \qquad (2.17)$$

A is the slope of the curves of $\ln R$ as a function of $\ln \lambda$, as depicted in Fig. 2.10; which shows that A is constant for the two radioactive series depicted. B is an arbitrary constant determined experimentally. The Geiger–Nuttall rela-

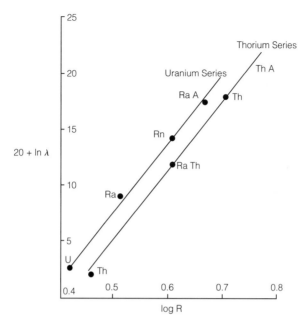

Figure 2.10. The Geiger–Nuttall law. From Ref. 20.

tion of (2.17) asserts that, with $\lambda T_{1/2} = \ln 2$, a fundamental relation of exponential radioactive decay relating half-life to reciprocal mean life λ [i.e., for $n/n_0 = \exp(-\lambda t)$; when $n/n_0 = \frac{1}{2} = \exp(-\lambda T_{1/2})$], then

$$R = C_2 \lambda^A = C_3 T_{1/2}^{-A} \tag{2.18}$$

where C_1 through C_4 are constants to be determined. The tunnel phenomenon verification curves given in Fig. 2.9 can be expressed as

$$T_{1/2} = C_1 E^{-m} \tag{2.19}$$

Algebraically eliminating $T_{1/2}$ between (2.18) and (2.19) yields

$$R = C_4 E^{mA} \tag{2.20}$$

Geiger's empirical formula [20] (1912) between range and the speed v of the alpha particles is, since $E \sim v^2$,

$$R = av^3 = bE^{3/2} \tag{2.21}$$

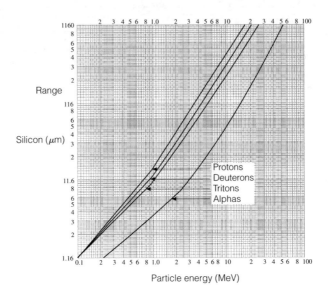

Figure 2.11. Range of alphas, tritons, deuterons, and protons in silicon. From Ref. 21.

which is the same relation obtained in (2.20) when the identification of the constants $C_4 = b$ and $mA = 3/2$ is made. This is one procedure for showing that the tunneling relationship results in the Geiger–Nuttall range relation.

Actually, alpha particle ranges are relatively small in solids including materials of interest such as silicon. This is simply because the alpha particle, although at the light end of the heavy particle "spectrum," is of sufficient mass and charge to easily ionize these materials. It can be seen from Fig. 2.11 that the range of a typical alpha particle of, say, 3 MeV is about 2.7 mg/cm^2 or about 12 μm in silicon. For such a small range, which is, however, comparable to microcircuit dimensions, alphas can ionize the material atoms to provide spurious charge to cause SEU in the individual transistors in microcircuits.

2.5. Solar Flares

A solar flare is a giant eruption of solar gases and plasma on the surface of the sun, whose arcs can extend to altitudes of more than 100,000 miles above the sun. They are also characterized by photon bursts including x-rays, optical line spectra, and microwave emissions up to hundreds of megahertz. This produces magnetic storms on the earth due to the resulting solar plasma shock waves incident on the earth's magnetosphere. These events cause extreme distortion of the geomagnetic field lines so that at high latitudes, the magnetic rigidity cutoff can drop to negligible values. This means that the energy spectrum, and so the fluxes, of allowed cosmic ray ions is greatly increased. This causes increases in error rates, often by orders of magnitude (factors of 2000 are not uncommon).

Solar flares, including anomolous (very large) flares, occur at random with a dispersion of about one per week to about one per 100 weeks. The individual flare phenomena, described above, lasts from about 2 h to 250 h in the earth's neighborhood. This corresponds to the observation that about 98% of the time, the galactic cosmic rays (enhanced at low energies by small flares) are the dominant particles in the magnetosphere. For the remaining 2% of the time, the particle environment consists of solar flare particles of mainly low- and moderate-energy protons compared to energies of cosmic rays. These low energies allow the use of passive shielding materials—certainly for ground-based electronics.

The statistics of flare occurrences have been investigated. However, they suffer the same small sample (too few flares) difficulties associated with active electronic component radiation testing (too few parts) statistics. One aid is to lump the populations of past flare occurrences with those of predicted future occurrences, as will be seen. As flare occurrences are assumed to be independent and random events, the binomial probability distribution describes them. Suppose that a mean of $\bar{N}$ flares have occurred in the past over a span of T years. Given this past record, what is the probability of n flares occurring in a span of the next t years? A maximum likelihood estimate of the probability of a single flare occurrence is taken as $p = t/(t+T)$. Then the probability P of the occurrence of n flares in an upcoming t-year interval is given by the well-known binomial distribution function as [6]

$$P(n, t; \bar{N}, T) = \left(\frac{(\bar{N}+n)!}{\bar{N}!\, n!}\right)\left(\frac{t}{t+T}\right)^n \left[1 - \left(\frac{t}{t+T}\right)\right]^{\bar{N}} \quad (2.22)$$

or

$$P(n, t; \bar{N}, T) = \binom{\bar{N}+n}{n} \frac{(t/T)^n}{(1 + t/T)^{\bar{N}+n}} \quad (2.23)$$

where the binomial coefficient $\binom{x}{y} = x!/(x-y)!y!$. For example, during a $T = 7$-year epoch of the 20th solar cycle, $\bar{N} = 24$ flares occurred. The probability of exactly $n = 1$ flare, no more no less, occurring in the next $t = 5$ years is, from (2.23),

$$P(1, 5; 24, 7) = \binom{25}{1} \frac{5/7}{(1 + 5/7)^{25}} = 2.5 \times 10^{-5} \quad (2.24)$$

The very small value of P is due to the precision of the query. P is used as a basis to compute more meaningful binomial statistics of interest.

As an example, the probability that *at least* one flare occurs in the next year, P_1^1, given the past flare record, is computed from $P_1^1 = 1 - P_0^1$, where P_0^1 is the probability that no flare occurs in the next year; that is,

$$P_1^1 = 1 - P_0^1 = 1 - P(0, 1; 25, 7) = 1 - (1 + 1/7)^{-24} = 0.96 \qquad (2.25)$$

The binomial distribution expectation (i.e., the average number of flares occurring in, say, the next 5 years) is

$$E(P) = np = 5(5/12) = 2.08 \qquad (2.26)$$

with a standard deviation

$$\sigma(P) = [np(1-p)]^{1/2} = [5(5/12)(7/12)]^{1/2} = 1.10 \qquad (2.27)$$

The large value of $\sigma(P)$ compared to $E(P)$ is due to the small sample size.

It is also claimed that flare occurrences can be considered rare events, the previous discussion notwithstanding. As such, the Poisson distribution would be an apt descriptor of statistical flare activity. The Poisson distribution can be considered as a limiting form of the binomial distribution for rare events, (i.e., for events of very small individual probability p). The Poisson distribution function for the occurrence of k events in a time interval t is,

$$P_P(k, \lambda, t) = [(\lambda t)^k / k!] \exp(-\lambda t) \qquad (2.28)$$

where λ is the mean rate of flare occurrences. The corresponding expectation and standard deviation are $E(P_P) = \lambda$ and $\sigma(P_P) = (\lambda)^{1/2}$, respectively. As an example, using the parameters from the binomial case, assume that 24 flares occur in a time interval of 7 years. Then let $\lambda = 24/7 = 3.43$, the mean number of flares per year. Then the probability of exactly one flare in 5 years is given by

$$P_P = (1, 3.43, 5) = (3.43)(5) \exp[-(3.43)(5)] = 6.11 \times 10^{-7} \qquad (2.29)$$

where $E(P_P) = 3.43$ and $\sigma(P_P) = 1.85$. The chance of at least one (one or more) flares occurring in the next year is

$$P_{P1}^1 = 1 - P_{P0}^1 = 1 - P_P(0, 3.43, 1) = 1 - \exp(-3.43) = 0.97 \qquad (2.30)$$

which compares well with the binomial result of (2.25).

In the 20th solar cycle, an anomalously large solar flare occurred (August 1972), with solar fluxes orders of magnitude above moderate flare fluxes. This flare should not be included in the population of the preceding flare events, because of its size [16], and is called an outlier in statistical parlance. However, when such are included in the mission radiation time profile, they will dominate all other radiation events experienced during the whole mission duration. If the mission is of such length that to include the anomously large flare would pose an unacceptable risk, then this event must be considered part of the mission radiation environment and appropriate mitigating measures taken.

It is found that the solar flare proton integral flux spectra apparently can be fitted by a log-normal distribution in energy E. Log-normal means that the log of their energies fits a normal distribution. The mean log of the flare fluence is used to provide a fit for a solar flare proton differential spectra $\varphi_s(E)$ for $E > 10$ MeV, which is given by [4]

$$\varphi_s(E) = 10^5 [3.3 \exp(-0.05E) + 1013 \exp(-0.33E)] \text{ (protons cm}^{-2}\text{sr}^{-1}\text{MeV}^{-1}) \tag{2.31}$$

Equation (2.31) gives the solar flare particle spectrum (mainly protons) that are incident in the earth's neighborhood, that have been integrated over the flare duration, to yield the differential fluence. Figure 2.12 depicts this fluence, as well as a worst-case 90% confidence-level ordinary flare fluence (larger only 10% of the time) and an anomalously large flare fluence [4].

Figure 2.13 depicts the galactic cosmic ray solar minimum and solar maximum fluxes. Also included in the figure are the anomalously large solar flare, worst-case ordinary flare, and the typical ordinary flare fluxes. All of these are computed from spacecraft measurements at 10,700 nautical miles (nmi.) altitude [16].

Upon investigation of flare frequency distributions during the 9.2-year period from 1973 to 1982, it has been found that the number of flares, except for the anomalously large flares, is distributed log-normally [22], as seen in Fig. 2.14. This figure shows the plots of the cumulative number of flares $N(\Phi)$ of fluence Φ and greater, versus the heavy-ion fluence Φ integrated over the flare duration. It depicts two types of flares (a) those whose ions have a range of greater than 100 mils Al and (b) those whose energy is greater than 100 MeV per nucleon. For both (a) and (b), it can be appreciated that their number $N(\Phi)$ is distributed log-normally; that is,

$$N(\Phi) = (N_0/\sqrt{2\pi})\sigma \int_\Phi^\infty \exp(-[(\ln\phi - \mu)^2/2\sigma^2]) \frac{d\phi}{\phi}$$

$$= \frac{1}{2} N_0 \operatorname{erfc}\left(\frac{\ln\phi - \mu}{\sqrt{2}\sigma}\right) \tag{2.32}$$

46 / Single Event Phenomena

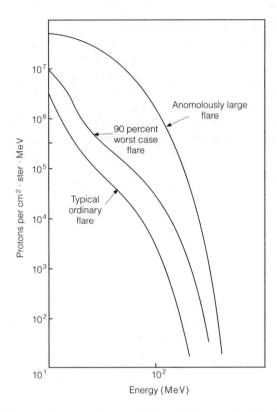

Figure 2.12. The event-integrated proton differential fluence for (a) a typical ordinary flare, (b) a 90% Worst-Case Flare, (c) an anomalously large flare. The 90% worst-case means that the fluence is greater only 10% of the time. From Ref. 4.

where the complementary error function $\text{erfc}(x) = (2/\sqrt{\pi}) \int_x^\infty \exp(-y^2)\,dy$ $N_0 = 23$ for type (a) flares as seen in Fig. 2.14. μ is the mean $\ln(\phi)$ [i.e., $(1/N) \sum_{i=1}^{N} \ln \phi_i$ for a group of N discrete $\ln \phi_i$ measurements]. σ is the standard deviation, and the variance is $\sigma^2 = \overline{(\ln \phi)^2} - \mu^2$. For a log-normal distribution, the mean fluence is given by

$$\bar{\Phi} = \frac{\int_0^\infty \Phi N(\Phi)\,d\Phi}{\int_0^\infty N(\Phi)\,d\Phi} = \exp(\mu + \tfrac{1}{2}\sigma^2) \qquad (2.33)$$

For the type (a) fluence, as in the figure, $\mu = 4.11$ and $\sigma = 1.15$ [22] yielding $\bar{\Phi} \cong 113$ cm^{-2} sr^{-1}. The corresponding number of flares $N(\bar{\Phi}) \cong 6.6$ over a 9-year period. It is claimed that the log-normal distribution of flare numbers, in the

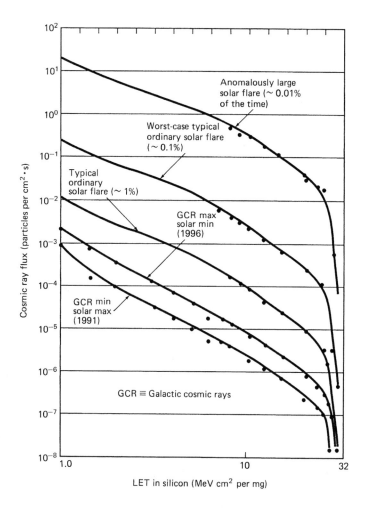

Figure 2.13. Solar flare cosmic ray flux at 10700 nautical miles (nmi) altitude through 1.23 gm/cm^{-2} aluminum shielding versus LET in silicon, including no-flare flux during sunspot extrema. From Ref. 16.

aforementioned period (21st sunspot cycle), is typical of virtually any cycle [22].

Recently claimed improvements [23] on earlier solar flare models [24] have resulted in refined results for SEU error rates during flare occurrences. A new model [25] (JPL-92), reevalutes the three major model components: (1) solar event proton flux, (2) solar helium flux, and (3) heavy-ion composition flux. Simultaneously, a refinement of the 11-year sunspot cycle with respect to solar

48 / Single Event Phenomena

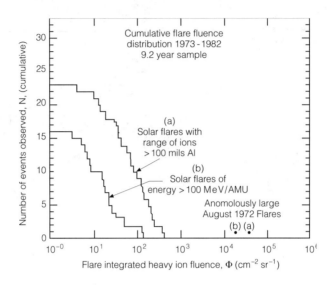

Figure 2.14. Cumulative solar flare fluence distribution. From Ref. 22.

cosmic rays (mainly protons) was obtained. Each cycle now contains four inactive and seven active years where the annual fluence is greater than 5×10^7 protons/cm^2 for solar protons whose energy is greater than 10 MeV. The active period spans the sunspot maximum from 2.5 years before to 4.5 year after its occurrence. This holds even if no major flare occurs in any of the active years. The fluences of the inactive years are relatively negligible and thus are included in the model only by omission. The new model asserts that both ordinary and anomalously large flares are in the same population and can be included in the same log-normal distribution (cf. Fig. 2.14). One result of the new model is adjustments in SEU error rates versus linear energy transfer (LET) threshold (L_{th}) for various environments as shown in Fig. 2.15. The M and Jo options of Fig. 2.15 correspond to the CREME and JPL-92 Models [25].

Although solar flares, especially the anomalously large flares, loom menacingly to greatly increase SEU error rates in spacecraft, shielding can mitigate this difficulty. Figure 2.16 shows the integral LET spectra dependence on aluminum shielding of flare flux. This implies that shielding is a practical means for attenuating solar flares, in contrast to high-energy cosmic heavy ions requiring ridiculously massive amounts of shielding.

2.6. Van Allen Radiation Belts

The Van Allen radiation belts constitute the major SEU threat for spacecraft orbits in these regions. About 90 years ago, the ionosphere, formerly called

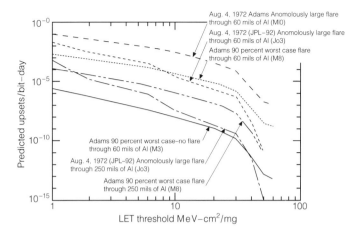

Figure 2.15. SEU rates for JPL and adams environments for two shield thicknesses for a device charge collection depth, d = 1 μm, and saturation cross section of 10^{-8} cm^2 per bit. From Ref. 22.

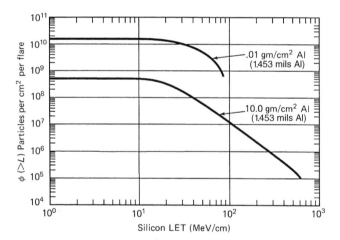

Figure 2.16. Integral LET spectra for solar flare proton fluence undiverted by earth's magnetic field (behind spherical Al Shields). From Ref. 7.

the Kennelly–Heaviside layer, was discovered. Its structure was later measured using vertically directed pulsed radio beams. The ionosphere plays a vital role in the propagation of high-frequency (HF) long-haul communications as is well known. It extends from above the atmosphere to several hundred miles and consists of several layers, each with similar but different ionospheric reflection properties. In 1957, J. Van Allen claimed discovery of the inner Van Allen

50 / *Single Event Phenomena*

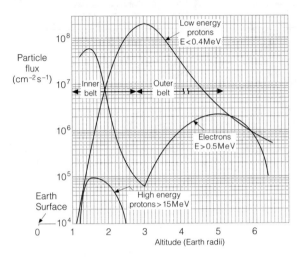

Figure 2.17. Van Allen belt equatorial trapped particle fluxes versus altitude in earth radii. From Ref. 7.

belt, as depicted in Fig. 2.17. The belts are usually two in number—the inner and the outer and extend from above the atmosphere out to about 40,000 miles. The inner belt consists principally of protons and also electrons plus trace numbers of heavy particles. The outer belt consists mainly of electrons. The ionosphere also contains protons and electrons and includes light ions such as O_2^- and N_2^-, constituents of the upper atmosphere. It is now known that the ionosphere is essentially the lower edge of the inner Van Allen belt, as shown in Fig. 2.18.

The inner (proton) belt is centered about an altitude of approximately 6000 miles, and was discovered first. It contains mainly protons plus some electrons and heavy particle ions whose energies are greater than 30 MeV [22]. The protons, especially the high-energy component, are produced mainly by the decay of high energy neutrons (mean life about 12 min) leaking up out of the atmosphere. The nuclear reaction is $n \to p + e^- + \nu$ where ν is the neutrino, an evanescent, chargeless, nearly massless particle with an infinitesimally small cross section for all matter, and which is still not fully understood. The source of these neutrons is the result of collisions of cosmic ray ions with the oxygen and nitrogen nuclei of the atmosphere. The low-energy protons are produced principally in the polar atmosphere by the same reaction. However, here the source neutrons are produced by solar proton interaction with upper atmosphere O_2 and N_2, analogous to the cosmic ray neutron producers for the high-energy neutrons [26].

Because the cosmic ray flux in the earth's neighborhood has been secularly constant over eons (except for solar flares), their production of neutrons to decay into protons and electrons, as previously described, is relatively stable.

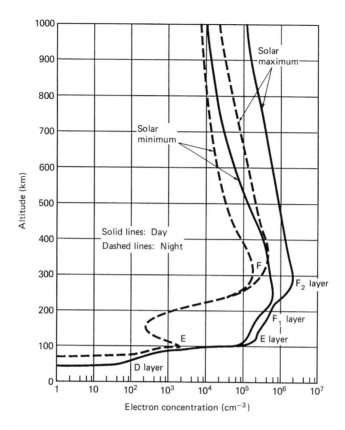

Figure 2.18. Ionosphere layers of electron density merging into the lower region of the inner Van Allen belts. From Ref. 5.

Thus, the inner belt proton and electron densities, with loss mechanisms that are principally collisional in nature, provide a relative stability to the inner belt particle densities.

However, in the outer Van Allen belt, centered at about 20,000 miles altitude, electrons are the major particles, and their levels fluctuate greatly with time, especially with solar-wind-induced geomagnetic storms. The electron flux seems to be in synchrony with magnetic storm activity, reflecting the influence of the storms on the outer belt electrons. Also, the mean lifetime for electrons in the outer belt is about 2–3 days, whereas their corresponding lifetime in the inner belts is about 400 days. Within about 600 miles above the earth, particles are depleted from the inner belt by collisions with the constituents of the upper atmosphere, losing energy until they become part of the ionosphere. At higher altitudes in the inner belt, the particle losses are mainly due to interactions with the ambient electromagnetic fields, ultimately suffering the same fate as those in the lower part of the belt. As an aside, the boundary between the inner and the outer belts is somewhat arbitrary, as shown in Fig. 2.17. It is well known that the source of the outer belt electrons is not from the decay of neutrons as in the inner belt. It is generally claimed that electrons are injected into the outer belt from the ambient electron density in and around the solar wind shock wave outside the magnetopause. This is especially the case during a magnetic storm. As mentioned, the latter causes the magnetosphere fields to fluctuate greatly, thereby producing corresponding electric fields (Faraday's law), which accelerate the electrons from beyond the magnetopause into the magnetosphere across its magnetic field lines. That the outer belt electron density fluctuates greatly during magnetic storms lends credence to this description of the outer belt electron sources. The electron energy spectrum (up to 5 MeV) is such that the electrons can be considered as fast (relativistic) particles (i.e., $v/c \lesssim 1$). They represent a serious radiation hazard for spacecraft orbits in these regions and may constrain the deployment of new semiconductor technologies in spacecraft.

The electron and magnetic field fluctuations in the belts are an important factor in the determination of spacecraft charging effects. In terms of the spacecraft avionics (electronics), the inner belt electrons result in a cumulative ionization damage to electrical components over the mission duration. With respect to protons in the outer belt, their effect on spacecraft electronics at geosynchronous orbit is negligible due to their relatively low fluxes and energies. For example, they can be stopped by about 2 mils of aluminum. Typical spacecraft skin thickness is about 40 mils of aluminum. However in the inner belt, high-energy protons can penetrate several centimeters of aluminum structure and cause radiation damage to electronic components in the spacecraft avionics. On the positive side, families of radiation-hard active electrical components are beginning to appear, so that future spacecraft orbits within the inner belt will become more feasible. In the outer belt regions, the particles are stopped much more easily, confining radiation damage to the very susceptible components such as CCDs and solar cells. In the 1962 high-altitude nuclear detonation series, certain spacecraft telemetry down-link channels ceased to function because of power loss due to solar cell radiation damage.

To verify experimentally that electrons could be injected into the magnetosphere in substantial numbers, three high-altitude nuclear devices (Argus I–III) were detonated in 1958 [5]. They were part of a series of high-altitude detonations, as shown in Table 2.2. The electron source is fast electrons (beta particles) which are part of the fission product menagerie produced during nuclear fission in the detonations. To maximize the electron injection efficiency, the detonations took place over the Atlantic Ocean, where the surface magnetic field is relatively low. It is well known that electron injection efficiency varies inversely with surface magnetic field strength [5]. All of these

Table 2.2. High-Altitude Nuclear Detonations

Event	Altitude (km)	Date	Latitude	Longitude	Approximate L-value of detonation	Yield[a]	Electron decay time
Teak	76.8	1 Aug. 1958	17°N	169°W	1.12	MT range	~few days
Orange	42.97	12 Aug. 1958	17°N	169°W	1.12	MT range	~1 day
Argus 1	~200	27 Aug. 1958	38°S	12°W	1.7	1–2 KT	0–20 days
Argus 2	~250	30 Aug. 1958	50°S	8°W	2.1	1–2 KT	10–20 days
Argus 3	~500	6 Sept. 1958	50°S	10°W	2.0	1–2 KT	10–20 days
Starfish	400	9 July. 1962	16.7°N	190.5°E	1.12	1.4 MT	1–2 years
USSR 1	—	22 Oct. 1962	—	—	~1.8	—	~30 days
USSR 2	—	28 Oct. 1962	—	—	~1.8	—	~30 days
USSR 3	—	1 Nov. 1962	—	—	1.75	—	~30 days

[a] KT: kiloton; MT: megaton.

Source: Ref. 5.

high-altitude tests showed that electrons could be injected into the magnetosphere from within and without the magnetopause. This then lends credence to the model that asserts that electrons are accelerated across the field lines of the magnetosphere by fluctuating magnetically induced electric fields. Experimental evidence shows that radiation belts on other planets exist if their magnetic moments are sufficiently large. This is the case for the larger planets, such as Jupiter, Saturn, Uranus, and Neptune, which have sufficiently strong magnetic fields. However, Mercury, Venus, and Mars have magnetic moments too small to support radiation belts. An interesting aspect of Jupiter's intense radiation belts pertains to Io, one of its moons. Io is volcanically active, ejecting ions of sulphur, oxygen, and sodium. These ions are subsequently captured by the Jovian radiation belts. This poses an additional difficulty for the Galileo spacecraft whose mission includes the exploration of the Jovian atmosphere and its radiation belts.

Both the North and South Pole auroras are produced by the belt electrons and protons ionizing their rarified atmospheres. The geomagnetic field lines form guiding centers which guide the particles in their spiral paths toward the poles and the corresponding atmospheres, as shown in Fig. 2.4. These particles begin to encounter a magnetic force gradient due to the increasing density of field lines, which lessens their spiral pitch angle. The force gradient increases as the particle descends, until the pitch angle changes sign as it passes through zero (mirror point). The magnetic field now propels the charged particle in its spiral path back up the field line in the direction opposite from its descent. It then oscillates between the magnetic poles. For an electron, this period is about 0.1–1.0 s. The electron single spiral gyro period is about 1 μs. This produces the well-known synchrotron radiation. The third periodic motion of the trapped belt electrons is the precession of the whole pole-to-pole oscillation orbit. It is about 10^4 s or about 2.8 h once around the earth.

As a foretaste of what is to come, the SEU import of the Van Allen belts is displayed in Fig. 2.19. It is seen in the figure that the altitude that maximizes the SEU rate for a moderately sensitive microcircuit occurs near the center of the inner (proton) belt [27]. This SEU rate is substantially greater than that at geosynchronous altitudes, as can be inferred from the figure.

The CRRES (combined release and radiation effects) spacecraft provided experimental SEU space data, including the Van Allen belts. As shown in Fig. 2.20, the preponderance of (proton) SEU occurs in the belt regions $\sim 1.5 R_E$, where R_E is the radius of the earth. This contrasts with SEU occurring above the belts in the vicinity of the geosynchronous altitudes (i.e., $R_E \sim 6$). In Fig. 2.20. the single- and multiple-bit-induced SEU data points are "faired-in" with curves.

The relative magnitudes of geosynchronous and Van Allen belt SEU levels depend on the spacecraft orbital parameters, as well as on the part types on board subject to SEU from the external environments. The CRRES orbit was

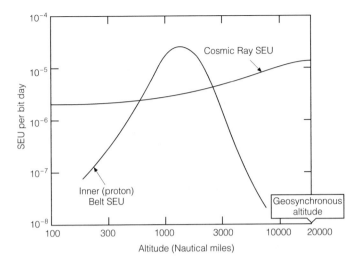

Figure 2.19. Single event upset rate versus altitude for 60° circular orbit passing through the inner Van Allen belt. From Ref. 27.

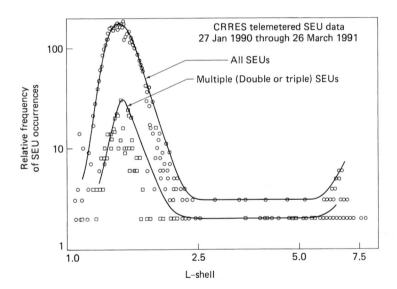

Figure 2.20. Single and multiple SEUs versus L Shell for seven static RAMs. From Ref. 28.

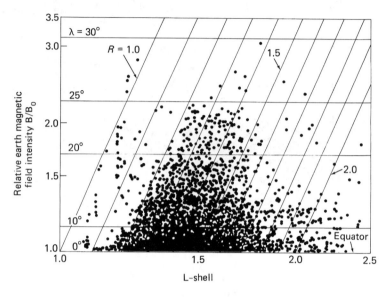

Figure 2.21. SEUs in the Van Allen belt region shown in B,L coordinates. From Ref. 28.

a highly elliptical one, at an inclination angle of 18.2° with respect to the equator. Its apogee was 33,532 km, whereas the perigee was only 348 km, and the orbit period was 9.87 h. Not unlike a Molniya orbit, the spacecraft passed through geosynchronous and Van Allen belt altitudes twice during each orbit. The SEU data were telemetered from averages taken over seven representative SRAMs [28, 29].

Figure 2.21 Van Allen belt presents SEUs in McIlwain B,L coordinates, which are discussed in Section 2.2. Suffice it to say here that L-shell values correspond approximately to R_E values in the neighborhood of the earth. A given L-shell labels the apogee distance of a particular horizontal toroid of earth magnetic field lines from the geocenter, as in Fig. 2.4.

Single event upsets occurring at values of the L parameter less than 2 are roughly those attributed to Van Allen belt protons, whereas those whose L parameter is greater than 2 are similarly attributed to heavy cosmic ray ions [28, 29]. Processing of CRRES data reveals the capability of distinguishing between proton-induced and cosmic ray ion-induced SEUs. An enhanced SEU was experienced by devices on board the CRRES spacecraft due to the large solar flare of March 23, 1991, that included CRRES orbit number 536. The flare redistributed the SEUs with respect to altitude above the earth, which accounted for their increase for $L \geq 5$. This also resulted in the creation of a temporary second and third peak in the Van Allen belt regions [30].

Problems

2.1 The extraterrestrial magnetic field H is estimated at about 10^{-7} Wb/m^2, spread over a region of diameter 1.6×10^{12} m. For a galactic cosmic ray ion whose $Z = 10$ (neon), what is the maximum energy this ion can possess and still be captured in orbit in this region if (a) the ion is completely stripped of electrons and (b) if it is singly ionized?

2.2 Spacecraft in low earth orbit (<1000 miles) are provided partial shadowing in appreciable amounts from the cosmos, and so from galactic cosmic ray ions by the solid angle Ω, subtended by the earth in their orbit. (a) At what altitude h is the spacecraft such that the earth's subtended solid angle at the spacecraft is 2π? (b) What is the minimum altitude h for which the solid angle subtended vanishes? (c) What solid angle does the earth subtend at the moon? (d) Moon at the earth? R_e is the radius of the earth.

2.3 Muons can cause SEU in charge-coupled devices (CCDs). They are produced in the upper atmosphere by cosmic rays. Once produced, those that penetrate the atmosphere take on the order of 10^{-4} s to reach the ground. Muons produced in particle accelerators have a well-measured lifetime of only 10^{-6} s. How do muons produced by cosmic rays, whose lifetime is but 10^{-6} s, endure more than 10 times their lifetime in their passage through the atmosphere, in order to cause SEUs in CCDs in ground-based equipment?

2.4 In Table 2.1, it is seen that the total energy flux of gamma ray bursts is about twice that of cosmic ray heavy ions. Why then do not the former pose at least as much of an SEU hazard as cosmic ray heavy ions?

2.5 Uranium-233 has a half-life of 1.6×10^5 years, decaying by alpha particle emission, as part of the heavy-element actinide chain reaching ultimately to lead.

(a) What is the value of the decay constant?

(b) What is the corresponding number of disintegrations per second in a 1-mg sample?

(c) What is the energy of its decay product alpha particle?

(d) What is the range of this alpha particle in silicon?

2.6 The mean rate of solar flare occurrences is λ_f, with a standard deviation $\sigma_f = (\lambda_f)^{1/2}$, for their assumed Poisson distribution (rare events). Suppose 24 flares occur in a period 7 years. What is the mean flare rate per annum and the corresponding standard deviation?

2.7 The inner and outer Van Allen belts extend from a few hundred miles altitude to more than 40,000 miles above the earth. If specific spacecraft SEU measures have been installed aboard the avionics, to combat galactic cosmic ray ions only, what is the minimum orbit altitude that can be allowed for these vehicles in an assumed nearly circular orbit in order not to intercept Van Allen belt SEU-inducing radiation?

2.8 Why do most of the cosmic ray ion spectra curves, such as that for iron in Fig. 2.2 have a pronounced dip in their lower energy ranges?

2.9 One of the very important statements regarding the energy of the heavy ion is that its total energy (kinetic plus rest mass energy), $E = mc^2$, is related to its rest energy, $E_0 = m_0 c^2$ by $E^2 = p^2 c^2 + E_0^2$. m_0 is its rest mass, $p = mv$ its momentum, $m = m_0 (1 - v^2/c^2)^{-1/2}$, and v is its speed. (a) Derive this relation (i.e., $E^2 = p^2 c^2 + m_0^2 c^4$). (b) Show that for the heavy ion at the end of its track inside the part that its energy $E = E_0 = m_0 c^2$.

2.10 The Galileo spacecraft in orbiting the planet Jupiter is experiencing sulphur and sodium ion impacts in addition to the other components present in the Jovian "Van Allen" radiation belts. What is the origin of these sulphur and sodium ions, as there are virtually no such ion species in the earth's Van Allen belts?

2.11 In Section 2.2, the computation of the rest energy E_0 involved Avogadro's number N_0 [i.e., $E_0 = M_0 c^2 = (A/N_0) c^2$]. Why include N_0 in the calculation?

References

1. Sir J.J. Thomson, and G.P. Thomson, *Conduction of Electricity Through Gases*, 3rd ed., Cambridge University Press, Cambridge, 1928.

2. J. Feynman, and S. Gabriel, "NASA Conf. Interplanetary Particle Environments," JPL Publ. 88-28, 1988; "High Energy Charged Particles at 1 AU," *IEEE Trans. Nucl. Sci.*, **NS-43** (2), 344 (1996).

3. J.R. Jokipii, and F.B. McDaniel," Quest for the Limits of the Heliosphere," *Sci. Am.*, 58–63 (1995).

4. J.H. Adams, R. Silberberg, C.H. Tsao, "Cosmic Ray Effects on Microelectronics: Part I. "The Near Earth Environment," NRL Memo Rept. No. 4506, Chap. 2, 1981.

5. J.B. Cladis, G.T. Davidson, and L.L. Newkirk, "The Trapped Radiation Handbook," DNA Report No. 2524H, Lockheed Missile Space Division, Palo Alto, CA, 1971.

6. J.H. Adams, J.R. Letaw, and D.F. Smart, "Cosmic Ray Effects on Microelectronics: Part II, "The Geomagnetic Cutoff Effects," NRL Memo Rept. No. 5099, Sec. 1, 1983.
7. E.G. Stassinopoulos, and J.P. Raymond, "The Space Radiation Environment for Electronics," *Proc. IEEE*, **76**, (11) 114–139 (1983).
8. J.F. Dicello. M.E. Schillaci, C.W. McCabe, J.D. Doss, M. Paciotti, and P. Berardo, "Meson Interactions in NMOS, CMOS Static RAMs," *IEEE Trans. Nucl. Sci.* **NS-32** (6), 4201–4205 (1935).
9. J.F. Ziegler, and W.A. Landford, "Effects of Cosmic Rays on Computer Memories," *Science*, **206**, 776–788 (1979).
10. C.S. Guenzer, R.C. Allas, A.B. Campbell, J.M. Kidd, E.L. Petersen, N. Seeman, and E.A. Wolicki, "Single Event Upsets in RAMs Induced by Protons at 4.2 GeV and Protons and Neutrons Below 100 MeV," *IEEE Trans. Nucl. Sci.*, **NS-27** (6), 1485–1489 (1980).
11. J.C. Higdon, and R.E. Lingenfelter, "Gamma Ray Bursts," Annu. Rev. Astron. Astrophys., **28**, 401–436 (1990).
12. B.E. Schaefer, "Gamma Ray Bursts," *Sci. Am.*, 52–58 (1985).
13. A.B. Campbell, and E.A. Wolicki, "Soft Upsets in 16K Dynamic RAMs Induced by Single High Energy Photons," *IEEE Trans. Nucl. Sci.*, **NS-27** (6), 1509–1515 (1980).
14. E. Segre, *Nuclei and Particles*, W.A. Benjamin Inc., New York, 1964, Table. 12.2.
15. G.C. Messenger, and M.S. Ash, *The Effects of Radiation on Electronic Systems*, 2nd ed., Van Nostrand Reinhold, New York, 1992, Sec. 8.10, Prob. No. 2.
16. E.L. Petersen, P. Shapiro, J.H. Adams, and E.A. Burke, "Calculations of Cosmic Ray Induced Soft Upsets and Scaling in VLSI Devices," *IEEE Trans. Nucl. Sci.*, **NS-29** (6), 2055–2063, (1982).
17. C.J. Gallagher, and J.D. Rasmussen, *J. Organic Chem.*, **3**, 333 (1957).
18. T.C. May, and M.H. Woods, "Alpha Particle Induced Soft Errors in Dynamic Memories," *IEEE Trans. Electron Dev.*, **ED-26** (1) (1979).
19. D. Binder, E.C. Smith, and A.B. Holman, "Satellite Anomalies From Galactic Cosmic Rays," *IEEE Trans. Nucl. Sci.*, **NS-22** (6), 2675–2680 (1975).
20. H. Semat, *Introduction to Atomic Physics*, Farrar and Rinehart, New York, 1939, p. 270.
21. *American Institute of Physics Handbook*, McGraw-Hill Book Co., New York, 1957, pp 8–32.
22. D.L. Chenette, and W.F. Dietrich, "The Solar Flare Heavy Ion Environment for Single Event Upsets: A Summary of Observations over the Last Solar Cycle, 1973–1983," *IEEE Trans. Nucl. Sci.*, **NS-31** (6), 12171-1222 (1984).
23. J. Feynman, T.P. Armstrong, L. Dao-Gibner, and S. Silverman, "New Interplanetary Proton Fluence Model," *J. Spacecraft,* **27** (4), 403-410 (1990).
24. J.H. Adams, Jr. and A. Gelman, "The Effects of Solar Flares on Single Event Upset Rate," *IEEE Trans. Nucl. Sci.*, **NS-31** (6), 1212–1216 (1984).

25. P.L. McKerracher, J.D. Kinnison, and R.H. Maurer, "Applying New Solar Particle Event Models to Interplanetary Satellite Programs," *IEEE Trans. Nucl. Sci.*, **NS-41** (6), 2368–2375 (1994).
26. NASA, "Significant Achievments in Particles and Fields, 1958–1964," NASA Report No. SP-97, Washington, DC. 1966.
27. E.L. Petersen, and P. Marshall, "Single Event Phenomena in the Space and SDI Areas," *J. Radiat Effects* Res. Eng. (1989).
28. A.B. Campbell, "SEU Flight Data from the CRRES MEP," *IEEE Trans. Nucl. Sci.*, **NS-33** (6), 1991.
29. A.B. Campbell, P.M. McDonald, K. Ray, and H.R. Schwartz, "Single Event Upset Rates in Space," *IEEE Trans. Nucl. Sci.*, **NS-39** (6), 1823–1835 (1992).
30. M.S. Gussenhoven, G.C. Mullen, M. Sperry, K.J Kerns, and J.B. Blake, "The Effect of the March 1991 Storm on Accumulated Dose for Selected Satellite Orbits: CRRES Dose Models," *IEEE Trans. Nucl. Sci.*, **NS-39** (6), 1765–1772 (1992).

3

Particle Penetration and Energy Deposition

3.1. Introduction

The penetration of particles and radiation into material is of prime importance in the determination of single event upset (SEU) and single event phenomena (SEP). This is because of the incident particle energy transfer to the device material to ionize it and thus release extraneous charge, which can ultimately be collected by memory nodes to produce SEU and other SEP.

Also discussed is stopping power, which is defined as the amount of incident particle energy deposited per unit track length ($-dE/dx$) in the material of interest. This parameter has become synonymous with linear energy transfer (LET), as used in SEU investigations. However, it is actually not quite the same. This aspect and how it affects single event phenomena is discussed in Section 3.9.

The major energy loss mechanisms from charged particles are ionization and bremsstrahlung (Section 3.5), where the former is much more predominant. Other energy loss mechanisms include pair production and annihilation (Section 3.6) and cosmic ray showers. They are cascades formed by cosmic rays as they enter the earth's neighborhood to influence SEU induced by the corresponding particles and their secondaries. They include neutrons, protons, and the various members of the meson family, which latter affect ground-level SEU.

The last three sections discuss aspects of the all-important LET parameter. This includes the idea of the LET spectra (Heinrich curves) by which the SEU error rate due to galactic cosmic rays can be derived. Finally, the applicability of the LET parameter in regard to the limitations of SEU error rate computations and how it conjoins the stopping power parameter is discussed.

The study of the penetration of particles and their energy deposition in materials has been investigated since before the turn of this century. It is one of the mainstays of the determination of particle behavior as extracted from their collisions with materials. Further, it is one of the pillars of the study of shielding and its requirements in fission and fusion reactors, high-energy accelerators, and other systems.

3.2. Particle Penetration in Materials

Because cosmic ray heavy ions are the principal particles that cause SEU in electronics systems, it is important to examine some of the processes involved in their penetration through materials that compose these systems. The passage of other (lighter) particles such as alphas through matter will also be investigated, as they play a somewhat lesser though important role in SEU.

One of the major characteristics of SEU-producing particles is their stopping power. The generally agreed-upon definition of stopping power of a particular particle incident on a material is its energy loss per unit path length of penetration into the material, or its corresponding incremental counterpart; $-dE/ds$ is the particle incremental energy loss per incremental distance ds traversed in the material. The stopping power is not only a characteristic of the particle parameters, such as its mass, charge, and energy, but it is also a function of the parameters of the material in which it deposits its energy; these include density, atomic weight (mass number A), and atomic number Z.

Figure 3.1 shows curves of stopping power for alpha particles and protons in silicon. In the figure, the abscissa used indicates the distance remaining to

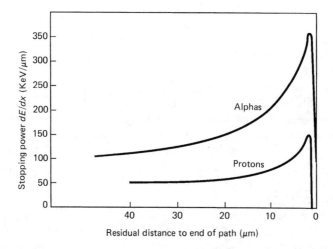

Figure 3.1. Bragg curves of stopping power versus residual distance to end of track for alpha particles and protons in silicon.

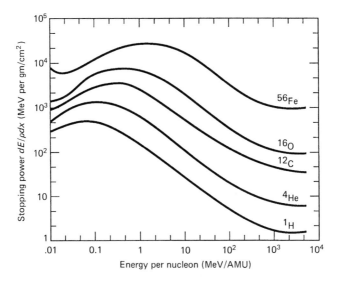

Figure 3.2. Stopping power for proton, alpha, oxygen, carbon, and iron nuclei in silicon versus particle energy per nucleon.

the end of the particle path (residual range), instead of indicating the penetration distance from a reference (e.g., material incident boundary); the curves are called Bragg curves. The end of the particle path usually coincides with its position when essentially all of its energy is lost to the material. Stopping power curves can also be given in terms of incident particle energy, as in Figs. 3.2 and 3.3 for various particles. In general, most stopping-power-curve data

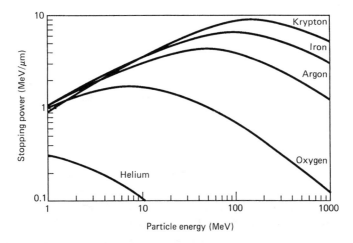

Figure 3.3. Heavy-ion stopping power in silicon versus particle energy. From Ref. 3.

is obtained by experiment, and then fitted empirically. However, for the case of heavy particles losing energy in material by ionization, there are well-known theoretical expressions available for stopping power. One such for nonrelativistic energies [1] is, with a number of qualifiers [2],

$$-\frac{dE_p}{dx} = \left[\left(\frac{2\pi e^2 (Z_p e)^2}{(m_0/M_p) E_p}\right) \ln\left(\frac{4(m_0/M_p) E_p}{I}\right)\right] N_m Z_m \quad (3.1)$$

where

m_0 = electron mass

$Z_p e$ = incident particle charge

E_p = incident particle energy

M_p = incident particle mass

Z_p = incident particle atomic number

N_0 = Avogadro's number

N_m = number density of material atoms/cm^3 = $\rho N_0/A_m$

Z_m = atomic number of material

A_m = mass number of material (atomic weight)

x = particle penetration distance

I = geometric mean ionization potential (amount of energy needed for a single ionization of atomic electron)

In (3.1), it is seen that the first factor (within the square brackets) contains only particle parameters with the exception of ln I, whereas the second factor, $N_m Z_m$, is the product of material parameters. It is apparent that $N_m Z_m$ is the concentration of material electrons per cubic centimeter, as N_m is the material atomic density and Z_m is the number of electrons per atom. Because $N_m = \rho_m N_0/A_m$, $N_m Z_m = \rho_m N_0 (Z_m/A_m)$, so $N_m Z_m/\rho_m = N_0(Z_m/A_m)$ is almost constant because $Z_m/A_m \simeq \frac{1}{2}$ for essentially all elements except hydrogen (for hydrogen $Z/A = 1$). Let $\xi = \rho x$ be the material mass areal density (g/cm^2), and use $d\xi = \rho dx$ in (3.1); $-dE_p/\rho dx$ obtains. With the areal density as the independent variable, the stopping power or ionization losses, $-dE_p/d\xi$, become approximately independent of the material in which the incident particle deposits its energy. This is one reason penetration distance is often spoken of in terms of grams per square centimeter of material; in other words, mass per unit area.

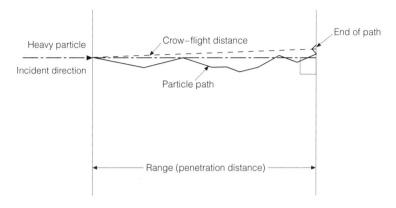

Figure 3.4. Particle path and range of heavy energetic ions.

3.3. Range

A quantity closely associated with particle stopping power is its range. The range $R(E_0)$ is essentially the projection of the "crow-flight" distance of the incident particle into the material in the direction of incidence, as in Fig. 3.4; that is, the range is roughly the penetration distance.

For energetic incident heavy ions, the connection between range and stopping power is seen from the approximation relation between them, where E_0 is the incident particle initial energy and the integral is a line integral along the particle path:

$$R(E_0) = \int_0^{R(E_0)} ds = \int_{E_0}^{0} \frac{dE}{(dE/ds)} = \int_0^{E_0} \frac{dE}{(-dE/ds)} \quad (3.2)$$

Implied in (3.2) is that the range is not very different from the corresponding random walk-distance, because the integral gives the total path distance equal to the range R. This is usually the case for heavy energetic particles of interest here. $|dE/ds|$ is often called the linear energy transfer (LET); that is, $|dE/ds|$ = LET. $|dE/d\xi|$ is frequently called the mass energy transfer (per unit areal density). However, the distinction usually dissolves so that both are often called linear energy transfer or LET (L). There are other subtle differences between stopping power and LET, including energy transfer (i.e., ΔE limitations on $\Delta E/\Delta s \simeq dE/ds$ discussed in Section 3.9).

Figure 3.5 depicts range versus particle energy for heavy ions in silicon. As an aside, "heavy ion" means heavy-ion mass with respect to an electron; therefore, even a proton is considered a heavy ion in this context. For these

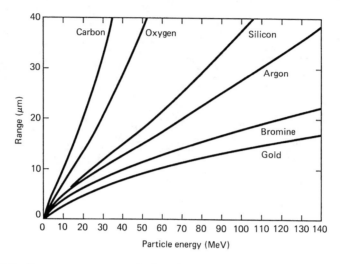

Figure 3.5. Range–energy curves for ions in silicon. From Ref. 3.

particles, an empirical range–energy relation is available for losing energy in silicon through ionization; it is [4]

$$E_p(R) = C_0 R^n, \quad n \cong 1.6 \tag{3.3}$$

where C_0 is a proportionality constant. If E_p is eliminated in (3.1) by using (3.3), the result yields the residual range, $R_s = R - x$ in terms of stopping power,

$$\frac{dE_p}{dR_s} = \left(\frac{C_1}{C_0}\right) R_s^{-n} \ln(C_0 C_2 R_s^n), \quad 0 \le R_s \le R \tag{3.4}$$

and

$$C_1 = \frac{2\pi e^2 (Z_p e)^2 N_m Z_m}{m_0/M_p} \quad \text{and} \quad C_2 = \frac{4 m_0}{M_p I}.$$

It is seen that (3.4) has the form $x^{-n} \ln(ax^n)$, which has a definite maximum. Equation (3.4) can be used to plot the stopping curves under discussion, delineating the Bragg peaks as in Fig. 3.1. The stopping power increases to a maximum near the end of the range, where most of the particle energy is lost, especially for low-Z particles. It is claimed that the behavior of stopping power with respect to residual range is due the following: (a) dE/dx increases initially because the incident heavy particle energy is sufficiently high that its coulomb field contracts for relativistic reasons. This allows it to approach the material atoms more closely, and thus transfer more collisional energy into the ionization process. This effect continues to increase at a somewhat slower rate until the Bragg peak is approached, where (b) the energy of the incident particle is now so low that its velocity is comparable with that of the material atomic electron environment. The heavy ion then begins to capture some of the electrons for part of the time, and therefore its charge is partially neutralized. This results in a rapid decrease in ionization near the end of the range; thus, the sharp drop following the peak ensues [5].

Besides concerns with heavy particles that produce SEU, fast relativistic electrons can also produce SEU in some charge-coupled devices (CCDs). Relativistic means that the speed of the particle is closely approaching the speed of light. Using (2.8) and substituting for the momentum $p = m_0 v/(1 - \beta^2)^{1/2}$ and $\beta = v/c$ results in $\beta = [1 - (1 + E/E_0)^{-2}]^{1/2}$. For a 5-MeV electron ($E_0 = 0.5$ MeV), $\beta = [1 - (1 + 5/0.5)^{-2}]^{1/2} = 0.996$, which is deemed relativistic. As well as being a component of cosmic rays, these electrons comprise the majority of particles of the outer Van Allen belt, where the energy spectrum exceeds 10 MeV. Therefore, for spacecraft that may traverse the outer belt, (e.g., in a high eccentricity orbit), these electrons must be considered. In addition to SEU, they also represent a serious ionizing radiation damage problem that could constrain the deployment of new device technology for space use. Cosmic ray electrons along with neutrons, protons, and muons can reach the ground and cause ground-level SEU. This is in contrast to heavy charged energetic cosmic ray ions, almost all of which interact with the atmosphere long before reaching the ground.

3.4. Ionization Loss

For relativistic electrons losing energy by ionization, their stopping power is given by [6]

$$-\frac{dE_{el}}{dx} = \left(\frac{4\pi e^4}{m_{0e}c^2}\right) N \left[\ln\left(\frac{2m_{0e}c^2}{I}\right) - \ln\left(\frac{(1-\beta^2)^{3/4}}{2^{3/2}}\right) + \frac{1}{16}\right] \quad (3.5)$$

where m_{0e} is the electron rest mass. For protons, the stopping power is

$$-\frac{dE_{pr}}{dx} = \left(\frac{4\pi e^4}{m_{0p}c^2 \beta^2}\right) N \left[\ln\left(\frac{2m_{0p}c^2 \beta^2}{I}\right) - \ln(1-\beta^2) - \beta^2\right] \quad (3.6)$$

where m_{0p} is the proton rest mass ($E_{0p} \cong 1000$ MeV). Comparing (3.5) and (3.6), it is seen that the stopping power of relativistic electrons and protons are quite similar in form. For equal values of β, they differ by less than 10% up to proton energies of 10^{10} eV [23]. An important difference between heavy ions and electrons is that, because the electron mass is much smaller (1/2000 that of a proton), they are more susceptible to all types of electromagnetic fields, including externally applied, atomic, and nuclear fields. Hence, the electron path is more likely to resemble random-walk behavior. This implies that the concept of range can correspondingly lose its validity. This is not the case for fast electrons: They arise in fission products, spallation reactions, and in the outer Van Allen belt. Briefly, spallation refers to the process of fissioning of nonactinide nuclei by an impinging particle of sufficiently high energy. Essentially, any nucleus can be fissioned if the incident particle is sufficiently energetic. For example, fast neutrons and protons produced in the Van Allen belts can interact with device silicon to cause a spallation reaction. In addition to fast electrons, this reaction can produce recoil ions from the silicon such as ^{25}Mg, which, in turn, can cause SEU.

3.5. Bremsstrahlung Loss

In addition to fast-electron energy loss by ionization, they also lose energy by radiation, called bremsstrahlung (braking radiation). It is well known that classically, any accelerating or decelerating free charged particle must lose energy through radiation. For fast electrons, the radiated energy is usually a continuous spectrum (as opposed to a discrete line spectrum) in the X-ray and/or gamma ray regions. This is the manner by which most commercial X-ray machines operate; that is, electrons hit a target and are deviated by the target nuclei, thereby accelerating about these nuclei to radiate X-rays. The electron radiation intensity (power density) is proportional to the acceleration squared. The acceleration about the electric field of a nucleus of charge $Z_n e$ of a particle of charge $Z_p e$ and mass m is proportional to $Z_p Z_n e^2/m$. The radiated power is then proportional to $(Z_p Z_n)^2 (e^2/m)^2$. Hence, protons and alpha particles, not to mention heavier particles, of mass 2000 and 8000 times that of an electron, respectively, will only produce about 10^{-6} the bremsstrahlung radiation of an electron because of their greatly increased mass. Practically, bremsstrahlung is considered of negligible importance for all fast particles except electrons. Another form of bremsstrahlung is the synchrotron radiation of electrons in the Van Allen belts spiraling (accelerating) around the earth's magnetic field lines as guiding centers, as discussed earlier.

One unfortunate aspect of the importance of bremsstrahlung radiation loss by fast electrons is that attempts to shield electronic and other systems is often thwarted. Increasing shield thickness to further attenuate bremsstrahlung produced X-rays is often futile. This is because the X-ray electron secondaries, in turn, produce further bremsstrahlung photons, whose range is very long compared to the electrons. This results in a rapidly decreasing marginal improvement for additional shielding. All this can be described by using the stopping power expression for bremsstrahlung [6], namely

$$-\frac{dE_{br}}{dx} = \left(\frac{8\pi e^2 Z^2}{hc}\right) NE_{br}\, \sigma_T \ln\left(\frac{180}{Z^{1/3}}\right) \qquad (3.7)$$

where h is Planck's constant, c is the speed of light, $\sigma_T = (8\pi/3)r_e^2$ is the Thomson electron scattering cross section, and $r_e = 2.82 \times 10^{-13}$ cm, the classical electronic raduis. By defining a bremsstrahlung radiation length $\lambda_R^{-1} = [8\pi e^2 Z^2\, \sigma_T N \ln(180 Z^{-1/3})]/hc$, which is roughly proportional to ρZ, the density and atomic number product of the penetrated material, (3.7) can be written as [6]

$$\frac{dE_{br}}{dx} = -\frac{E_{br}}{\lambda_R} \qquad (3.8)$$

Table 3.1 provides radiation lengths for some materials of interest.

As the incident (primary) electron density $n_e(x)$ is being attenuated as it transported through the material, it produces a (primary) bremsstrahlung photon density $E_{br}(x)$. Then adding a bremsstrahlung source term to (3.8), where now λ_R is assumed to be a mean free path for bremsstrahlung photons, yields

$$\frac{dE_{br}}{dx} = -\frac{E_{br}}{\lambda_R} + \alpha_e n_e, \quad E_{br}(0) = 0 \qquad (3.9)$$

In this simple model, the source term is $\alpha_e n_e$, where α_e is a suitable coefficient representing the conversion of the primary electron density to bremsstrahlung photons. The bremsstrahlung as well as the electrons deposit energy in the material through ionization and radiative processes, as represented by the loss terms $-E_{br}/\lambda_R$ and $-n_e/\lambda_e$, respectively. The bremsstrahlung photons can produce secondary (compton) electrons, which, in turn, can produce secondary bremsstrahlung, and so on. Such secondaries are much reduced relative to their primaries and, thus, are ignored in this model. The incident electron

Table 3.1. Bremsstrahlung Radiation Lengths. From Ref. 6

Material	Radiation lengths λ_R (cm)
Air (STP)	3.6×10^4
Silicon	10.
Lead	0.5

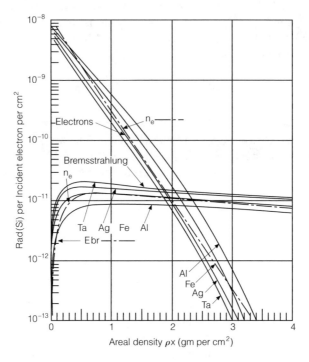

Figure 3.6. Fission product electron dose including bremsstrahlung production versus areal density for a family of shielding materials. From Ref. 7.

density $n_e(x)$ is attenuated in the material as mentioned and loses energy to it, as given by

$$\frac{dn_e}{dx} = -\frac{n_e}{\lambda_e}, \quad n_e(0) = n_{e0} \tag{3.10}$$

Equations (3.9) and (3.10) are a simultaneous pair, that, when integrated, yield respectively

$$E_{\text{br}}(x) = \left(\frac{\lambda_e \lambda_R \alpha_e n_{e0}}{\lambda_R - \lambda_e}\right)\left(\exp\left(-\frac{x}{\lambda_R}\right) - \exp\left(-\frac{x}{\lambda_e}\right)\right) \tag{3.11}$$

$$n_e(x) = n_{e0} \exp\left(-\frac{x}{\lambda_e}\right) \tag{3.12}$$

Using the set of parameters obtained from Table 3.1 and Fig. 3.6 viz. $\lambda_R = 10$, $\lambda_e = 0.309$, $\alpha_e = 0.0059$, $n_{e0} = 1.16 \times 10^{-8}$, fits the electron density n_e and bremsstrahlung photon density from (3.11) and (3.12), as shown in Fig. (3.6), in terms of energy deposited. In the figure, the abscissa is ρx in this context.

It is well known that the ratio of bremsstrahlung energy loss to electron energy ionization loss $(dB_{br}/dx)(dE_{ion}/dx)^{-1} \cong ZE/700$, where Z is the atomic number of the material and E is the energy of the incident fast electron in megaelectron-volts. Hence, bremsstrahlung is a much less formidable energy loss mechanism than ionization loss. However, it emerges when the other losses are diminished in the depths of materials, as seen in Fig. 3.6.

3.6. Pair Production, Cosmic Ray Showers

Very high-energy photons such as gamma rays produced by bremsstrahlung can, in the neighborhood of an atomic nucleus, be annihilated with the immediate appearance of an electron (e^-) and a positron (e^+) pair. The nucleus provides the third particle by which momentum in the reaction is conserved. The initiating photon of this pair production must have an energy of at least 1.022 MeV to provide the rest mass energy of the emerging electron and positron, each of whose rest mass energy is equivalent to 0.511 MeV. The cross section for pair production is [6]

$$\sigma_{pp} \cong \left(\frac{6\pi Z^2 e^2 \sigma_T}{hc}\right) \ln\left(\frac{180}{Z^{1/3}}\right) \tag{3.13}$$

If a bremsstrahlung mean free path for pair production is defined as $\lambda_{pp} = (N\sigma_{pp})^{-1}$, then the intensity of a beam of bremsstrahlung photons n_{br} will attenuate as

$$\frac{dn_{br}}{dx} = -\frac{n_{br}}{\lambda_{pp}} \tag{3.14}$$

or

$$n_{br} = n_{br0} \exp\left(-\frac{x}{\lambda_{pp}}\right) \tag{3.15}$$

This is analogous to bremsstrahlung loss by ionization of electrons as in (3.8), In actuality, $\lambda_{pp} = (9/7)\lambda_R$.

One reason for the preceding discussion is that high-energy bremsstrahlung and pair production combine to produce what are called cosmic ray showers. These are avalanches or cascades of bremsstrahlung photons, electrons, and positrons, each providing a source for the other until their energies are dissipated and the shower ceases. The high-energy particles in this process can deleteriously affect spacecraft avionics. Cosmic ray electrons of energy in the BeV range will radiate very energetic bremsstrahlung gamma rays. These produce electron–positron pairs. The latter then radiate this energy in bremsstrahlung photons, which, in turn, produce electron–positron pairs, and so forth. This multiplicative process divides the energy E_0 of the initiating electron between the many shower particles. When the energy decreases to a critical energy E_c, the shower ceases because the bremsstrahlung photon energies are too low to cause further pair production. The conservation of momentum provides an axis for the shower in the direction of the momentum of the initiating particle. The shower propagates longitudinally with little lateral spread. In the mean, a bremsstrahlung electron radiates about three gamma rays per radiation length λ_R. They disappear to produce about three electron–positron pairs in one λ_{pp} in the shower. It is plain that the total number of particles in the shower, $N(x)$, will grow exponentially along the assumed x direction; that is,

$$N(x) \cong \exp(mx) \qquad (3.16)$$

The exponent m can be found in a rudimentary manner. In a total mean free path distance d of one radiation length plus one pair production mean free path (i.e., $d = \lambda_R + \lambda_{pp}$)

$$N(d) = \exp(md) = 6 \qquad (3.17)$$

In the mean, one radiation length produces three gamma rays, which disappear to leave three electron–positron pairs in one mean free path from pair production, for a net of six particles. Because $\lambda_{pp} = (9/7)\lambda_R$, $d = [1 + (9/7)]\lambda_R = (16/7)\lambda_R$. Hence, $m = \ln(6)(16\lambda_R/7)^{-1} = 0.784/\lambda_R$. As mentioned, the shower stops when the particle energies are reduced to the critical energy E_c. At the instant the shower ends, a good approximation is that each particle type—electron, positron, and photons—has the same amount of energy E_c. Also, it is known that at this juncture before the photons disappear into various processes, they outnumber the electrons and positrons by two to one. Let (N_-, E_-), (N_+, E_+), and (N_p, E_p) represent the particle type numbers and corresponding energies for electrons, positrons, and photons, respectively. Then at the instant the shower ceases,

$$E_0 = N_- E_- + N_+ E_+ + N_p E_p \qquad (3.18)$$

where E_0 is the energy provided by the initiating particle. As each particle type has the same energy E_c, $E_- = E_+ = E_p = E_c$,

$$E_0 = (N_- + N_+ + N_p) E_c \qquad (3.19)$$

Before the photons disappear, the number of photons $N_p = 2(N_- + N_+)$. Inserting this in (3.19) gives

$$E_0 = 3E_c(N_- + N_+) \tag{3.20}$$

After an instant, the photons are gone and the total number of particles N are $N = N_+ + N_-$; or from (3.20),

$$N = N_{max} = \frac{E_0}{3E_c} \tag{3.21}$$

The particles reach their their maximum number N_{max} just prior to the shower demise, at which time the shower has spanned the distance λ_s. From (3.16),

$$N = N_{max} = \frac{E_0}{3E_c} = \exp\left(\frac{0.784}{\lambda_R}\right)\lambda_s \tag{3.22}$$

or

$$\lambda_s = \left(\frac{\lambda_R}{0.784}\right)\ln\left(\frac{E_0}{3E_c}\right) \tag{3.23}$$

This estimate neglects energy losses due to ionization, rest mass energy of the individual pairs, and so forth. These actually reduce the maximum number of shower particles to $N_{max} \cong E_0/6E_c$. The results of more refined calculations are shown in Fig. 3.7. It should be appreciated that the cosmic ray bremsstrah-

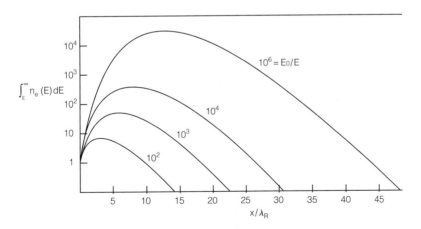

Figure 3.7. The average of shower electrons greater than energy E, begun by an electron of energy E_0, versus penetration depth x/λ_R. From Ref. 8.

lung–pair production shower is analogous to the bremsstrahlung–electron ionization process discussed earlier.

Again, an overriding reason for discussing these particles and their ramifications is that they can not only cause SEU but can also participate in other radiation-produced damage to system electronics and avionics. Recall that SEU is the mildest form of damage mechanism in the hierarchy of radiation damage; that is, from least to most, the hierarchy is SEU, hard errors (latchup), electronic parameter degradation, physical burnout of components and systems, and catastrophic physical disintegration.

3.7. LET Introduction

Linear energy transfer (LET), having been discussed previously as essentially the stopping power, with certain restrictions, in a particular material for a particular penetrating species of particles, has an obvious impact on the SEU probability of, for example, a microcircuit memory-cell-sensitive region (sensitive node). This can be appreciated by the fact that a threshold minimum or critical LET must obtain for the incident particle to produce sufficient ionization-derived charge to cause an SEU within a sensitive node of that cell. This minimum charge is labeled as the critical charge. It is a function of the physical and electrical characteristics of the cell. The electrical characteristics include the cell circuitry details, its bias voltage network, circuit physical layout architecture, and special SEU mitigation measures, such as cross-coupled resistances in the memory-cell-storage component and error detection and correction (EDAC) circuitry if present [9].

The parallelepiped model of the sensitive region of the cell with respect to its exposure to ions is shown in Fig. 3.8. Piercing it is a heavy ion at a polar

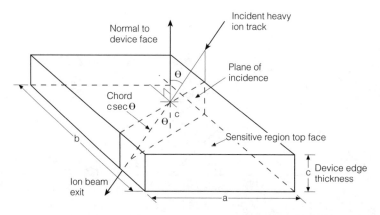

Figure 3.8. Parallelepiped model of the sensitive region with piercing heavy-ion-track-forming plane of incidence with the normal to the sensitive region face.

incident angle θ with respect to the sensitive region face normal. The plane of incidence as shown in the figure is formed by the ion track and that normal. The heavy-ion track remains essentially straight, as the ion-material electron ionization interaction leaves the track essentially undeviated. This is because of the very high energy and momentum of the incident ion, whose state vector is hardly altered in the interaction.

It is assumed that all incident ion tracks within the sensitive region are chords of length $c \sec \theta$ in Fig. 3.8, (i.e., the ions have sufficient range to penetrate the sensitive region). This is in contrast to tracks that happen to end within the sensitive region which are called track segments as discussed in Section 1.4.

3.8. LET Spectra (Heinrich) Curve

The linear energy transfer (LET) cosmic ray abundance spectra curves are the keystones of the SEU error rate computation for the particular cosmic ray environment in which the microcircuit of interest is used. These curves (e.g., the differential or integral LET spectra) are typically a plot of the number of cosmic ray particles per square centimeter per LET, or cosmic ray particles per square centimeter, respectively, that are incident in the earth's neighborhood as a function of their LET, where the latter is usually expressed in dimensions of MeV cm^2/mg i.e., MeV/mg/cm^2 of material absorbed.

For heavy cosmic ray ions, which are the main particles of interest, experimentally determined energy spectra of such ions as carbon, oxygen and iron were obtained from spacecraft measurements [10]. An interesting and experimentally verifiable phenomenon is that the shape of the individual energy spectra of most cosmic ray ions are essentially the same [10]. This allows the corresponding energy spectrum of ions of interest to be obtained merely by normalizing their amplitudes according to their relative abundances, from the above known energy spectra.

For cosmic ray ions (primaries) that have penetrated various material thicknesses, diffusion theory is used to determine their so-altered differential energy spectra. This included the production of secondary particles due to fragmentation processes undergone by the primary ions in the material [10]. The LET (dE/ds) spectra for pertinent ions are then computed from their corresponding differential energy spectra, for atomic numbers, Z, up to 26 (iron) and normalized by their known relative abundances where necessary. Then the overall LET abundance curve (Heinrich curves) were obtained by summing the LET spectra of the individual ions. It is shown in Fig. 3.9 for three environments, and for various altitudes and orbit inclination angles in Figs. 3.10 and 3.11, respectively [10, 14, 15].

If ion stopping power data are available, the LET spectrum also can be

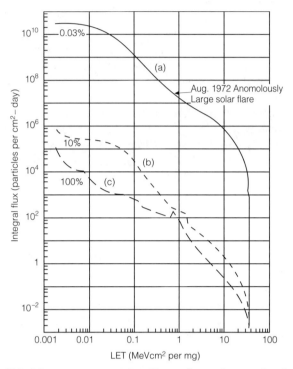

Figure 3.9. Heinrich spectra curves in silicon of cosmic ray abundance for three environments: (a) worse only 0.03% of the time—an anomolously large solar flare; (b) worse only 10% of the time—Adam's de facto 90% standard environment; (c) worse 100% of the time. From Refs. 10 and 14.

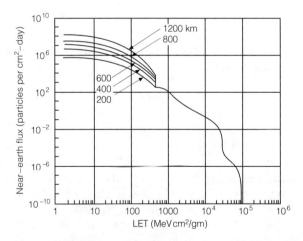

Figure 3.10. Integral LET spectra (Heinrich curve) through 25-mil of aluminum wall of spacecraft at 60° inclination measured from equator for various altitudes. From Ref. 15.

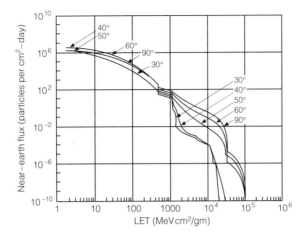

Figure 3.11. Integral LET spectra (Heinrich curve) through 25-mil of aluminum wall of spacecraft at 460 km altitude for various inclination angles measured from equator. From Ref. 15.

constructed from the individual LET spectra of the cosmic ray constituent heavy ions. The latter are obtained by cross-plotting between energy spectra and corresponding known stopping power curves of the individual heavy ions. The LET spectra of the individual ions are normalized by their known abundances and then summed to obtained the overall LET spectrum, as above.

3.9. Linear Energy Transfer (LET) Applicability

Linear energy transfer, a term derived from the health physics discipline, plays an important part in the computation of SEU error rates for electronic system components. One of its main uses is to determine the amount of ionization energy deposited in a microcircuit SEU-sensitive volume from the track of an incident ionizing particle, However, the employment of LET as a single parameter to accomplish this can suffer from its limitations.

At least two difficulties encountered include the following: (a) LET actually yields the energy lost by the incident ionizing particle, which is not necessarily the same as the energy deposited in the SEU sensitive volume; (b) LET is defined as an average quantity that does not allow for dispersion in the amount of ionization energy deposited (energy loss straggling [12]).

It has also been shown that SEU cross sections as obtained from accelerator beam ion simulators depend on the type of ion for a given LET. Further, it has been observed that high-energy beams induce more charge collected in CMOS/SOS MOSFETs than lower-energy beam ions, where both high- and low-energy ions have the same LET [11]. Hence, there is more involved in the description of the SEU process than attempting to characterize it by a single parameter, namely LET.

Also, there is a difference between LET and the usual definition of stopping power, even though both are defined in terms of energy lost per unit track length distance ($-dE/ds$). Frequently, a distinction is made between stopping power and LET, in that the former is the mean energy lost per unit path length, whereas the latter is the mean energy deposited per unit path length. The two are not always identical due to the occurrence of intervening processes, such as energetic particles leaking from the SEU sensitive region instead of depositing their energy there.

The classical definition of LET, dating back more than a half century, is the average energy lost by the incident ion per unit path length (L_Δ) in the material, due to ionizing collisions with energy transfers, but such transfers are equal to or less than a maximum energy transfer cutoff value Δ. In the limit of infinite Δ, all values of enegy transfers are allowed so that L_Δ becomes equivalent to the stopping power i.e., $\lim_{\Delta \to \infty} L_\Delta = L_\infty = -dE/ds$.

The cutoff energy Δ is used to correct for the fact that the range of some of the secondary electrons (those electrons produced in the ionization process) is large compared to the device SEU-sensitive volume. Hence, they do not deposit all their energy therein, because their range easily spans the sensitive volume during their traversal through the volume. If L_∞ is used in the energy deposition computation, then all of these electrons are assumed to deposit their energy in the sensitive volume. It is apparent that this difficulty is enhanced for very small sensitive volumes, especially for high-energy incident ions which produce fast electrons with correspondingly large ranges that easily traverse these volumes, leaving only some of their energy behind.

The cutoff energy used in the following is given by the electron energy whose range R is equal to the mean chord length $\bar{s}$ in the microcircuit SEU-sensitive volume. The justification for this choice for Δ follows from the classical formulation of the Bragg–Gray cavity theory, where the cutoff energy is characterized by the dimensions of a sensitive volume embedded in a surrounding material [14].

Now, a function f_{ion} can be defined, in the mean, as the fraction of energy deposited by the incident ion within the sensitive volume [12]. The remaining particle energy escapes the sensitive volume by the fast secondary electrons that have sufficiently large range, as discussed. Because the stopping power refers to all energy deposited as the unrestricted LET (viz. L_∞), the desired LET $L_\Delta = f_{ion} L_\infty$, where $0 \leq f_{ion} \leq 1$; that is, $f_{ion} = L_\Delta / L_\infty$ defines f_{ion}. To obtain f_{ion}, and thus L_Δ, one begins with the well-known version of the stopping power expression, which holds for both heavy and light incident ions. Here, it is applied to heavy ions transferring their energy by ionization to material electrons. It is in its nonrelativistic form, which can be shown to provide sufficient accuracy in the following computations [13].

$$-\frac{dE}{ds} = \left(\frac{2\pi (Z_{ion} e^2)^2}{m_{0e} v^2}\right) NZ \ln\left(\frac{2m_{0e} v^2 H}{I^2}\right) \quad (3.24)$$

where

H = unrestricted energy transfer
$T_{ion} = \frac{1}{2} M_{ion} v^2$, ion kinetic energy
M_{ion} = mass of the incident ion
Z_{ion} = atomic number of the incident ion
v = velocity of the incident ion
Z = atomic number of the absorber material
N = number of material atoms per cubic centimeter
m_{0e} = mass of the electron
I = ionization potential of the material electrons

From the conservation laws of momentum and energy, it is well known that the maximum energy transfer T_{max} to a (secondary) electron from an incident ion, as the former is being ionized from the atom, is given by

$$T_{max} = 4\left(\frac{m_{0e}}{M_{ion}}\right) T_{ion} \quad (3.25)$$

Also, H in (3.24) can be construed as the cutoff energy Δ, where $0 \leq \Delta \leq T_{max}$. Then inserting $H = \Delta$ and (3.25) into (3.24), with $T_{ion} = \frac{1}{2} M_{ion} v^2$ gives L_Δ as

$$-\frac{dE}{ds}\bigg|_\Delta = L_\Delta = \left(\frac{\pi (Z_{ion} e^2)^2 NZ}{(m_{0e}/M_{ion}) T_{ion}}\right) \ln\left(\frac{T_{max}\Delta}{I^2}\right) \quad (3.26a)$$

For the case in which the cutoff energy Δ is a maximum (i.e., $\Delta = T_{max}$), (3.26a) can be written to give L_∞ as

$$-\frac{dE}{ds}\bigg|_\infty = L_\infty = \left(\frac{\pi (Z_{ion} e^2)^2 NZ}{(m_{0e}/M_{ion}) T_{ion}}\right)\left[2\ln\left(\frac{T_{max}}{I}\right)\right] \quad (3.26b)$$

Then f_{ion} is obtained by dividing (3.26a) by (3.26b) to yield

$$f_{ion} = \frac{L_\Delta}{L_\infty} = \frac{\ln(T_{max}\Delta/I^2)}{2\ln(T_{max}/I)} \quad (3.27)$$

For a needed digression, it is desired to rewrite T_{max} from (3.25) in terms of atomic mass units (AMU) for computational convenience. This will be obtained with some pedagogical intent. One AMU by definition (today) is one-twelfth the mass of the ^{12}C atom. The mass of the carbon atom is given by its mass number $A = 12$ (atomic weight) divided by Avogadro's number, N_0 [i.e., A_{12} (g/mole)$/N_0$ (atoms/mole) $= 12/(6 \times 10^{23}) = 2 \times 10^{-23}$ g. Thus, 1 AMU $= 2 \times 10^{-23}/12 = 1.66 \times 10^{-24}$ g. Its Mc^2 energy equivalent is given by (1 AMU) c^2 (ergs) $= (1.66 \times 10^{-24})(9 \times 10^{20})/(1.6 \times 10^{-6}$ ergs/MeV$) = 938$ MeV; that is, the mass number as an integer A gives $A_{12}/12 = A_1 = 1$, so that the rest mass energy of 1 AMU $= 938$ MeV, that of a proton.

It is easily seen that the energy equivalent of m_{0e} (9.1×10^{-28} g) $= 0.511$ MeV. Because the energy equivalent of an ion of mass M_{ion} is $M_{ion} = 938 A_{ion}$ (MeV), $M_{ion}/A_{ion} = 938$ (MeV per nucleon). Now, define $t_{ion} = T_{ion}/$AMU as the ion kinetic energy in AMU. Then from (3.25), dividing it inside by 1 AMU (938 MeV), T_{max} in units of AMU is given by the sought-after expression

$$T_{max} = 4 \left(\frac{m_{0e}}{M_{ion}/AMU} \right) \left(\frac{T_{ion}}{AMU} \right) = 4 \left(\frac{0.511}{938} \right) t_{ion} = \frac{t_{ion}}{459} \quad (3.28)$$

Countinuing, equation (3.27) is made more accurate by adding correction factors Δ_1 and Δ_2 to yield

$$f_{ion} = \frac{\ln(T_{max}(\Delta + \Delta_1 + \Delta_2)/I^2)}{2 \ln(T_{max}/I)} \quad (3.29)$$

where the ionization potential for silicon [2] is $I = 0.173$ keV. $\Delta_1 = \Delta \cdot (1 - \Delta/T_{max})$ is a correction to provide for the energy transfer in the sensitive volume by the fast electrons produced there but escape. Not all the energy remains, because the range exceeds the mean chord length $\bar{s}$, as discussed. $\Delta_2 = I(1 - \Delta/T_{max})$, a small correction factor, is added to account for excitation and binding energies when the fast electrons exit the sensitive volume. The derivational details for Δ_1 and Δ_2 are given in Ref. 12. The important remaining parameter to evaluate is the cutoff energy Δ. It is found from a range–energy relation for electrons in silicon depicted in Fig. 3.12. From the figure, the corresponding range–energy empirical fit is given by

$$R = 0.02493 (T + 0.2674)^{1.739} \quad (\mu m) \quad (3.30)$$

where T is in keV. Because the cutoff energy Δ is defined as the range $R = \bar{s}$ when $T = \Delta$, inserting Δ for T and $\bar{s}$ for R into the above yields

$$\Delta = 8.354 (\bar{s})^{0.575} - 0.2674 \quad (keV) \quad (3.31)$$

Equation (3.31) is valid for $0.25 \leq \Delta \leq 50$ keV [12].

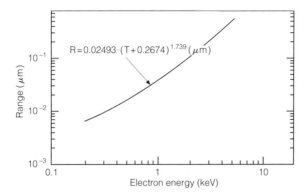

Figure 3.12. Electron range R as a function of energy T(keV) in silicon. From Ref. 12.

An example computation is now provided to appreciate the relative relationships among the parameters involved. Assume a 500-MeV/AMU ion incident on a rectangular parallelepiped sensitive volume of silicon of dimensions $5 \times 3 \times 0.5$ μm. Equation (3.28) yields a T_{max} of $500/459 = 1.089$ MeV/AMU of kinetic energy for this ion. To obtain Δ using (3.31), $\bar{s}$ must be known. From Section 1.4, for a convex body, $\bar{s} = 4V/S$, where V is its volume and S is its surface area. Thus, $\bar{s} = 4(5 \times 3 \times 0.5)/2[(5 \times 3) + (5 \times 0.5) + (3 \times 0.5)] = 0.739$ μm. Then, from (3.31), $\Delta = 7.02$ keV, where it is seen that it is well within the upper limit (50 keV) of validity. Computing the corrections Δ_1 and Δ_2 yields

$$\Delta_1 = 7.02\left(1 - \frac{7.02}{1089}\right) = 6.975 \text{ keV}, \quad \Delta_2 = 0.173\left(1 - \frac{7.02}{1089}\right) = 0.172 \text{ keV}$$

(3.32)

Inserting the preceding into (3.29) gives

$$f_{ion} = \frac{\ln[(1089)(7.02 + 6.975 + 0.172)/(0.173)^2]}{2\ln(1089/0.173)} = 0.752 \quad (3.33)$$

This implies that, in the mean, 75.2% of the incident ion energy is transferred to the microcircuit sensitive volume, whereas 24.8% escapes. It is seen by comparing the numbers in (3.33) that the energy corrections are in the ratios $\Delta : \Delta_1 : \Delta_2 = 7.02 : 6.975 : 0.172$. It is obvious that the first correction Δ_1 is comparable to the original cutoff energy Δ, whereas Δ_2 is very much less. Figure 3.13 depicts the fraction f_{ion} of ion energy remaining in the sensitive volumes versus their mean chord lengths, for various incident ion energies in MeV/

82 / *Single Event Phenomena*

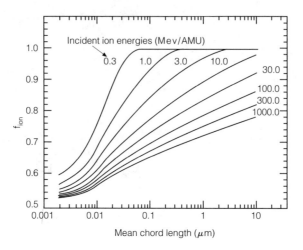

Figure 3.13. Fraction of ion energy loss within the sensitive volume as a function of its mean chord length for incident ion energy in MeV AMU^{-1}. From Ref. 12.

AMU. f_{ion}, LET, and the statistical aspects of energy deposition fluctuations (energy loss straggling) specifically affect the SEU error rate computations [12]. Briefly, straggling is the statistical dispersion in the energy lost by successive particles in their penetration in a material medium.

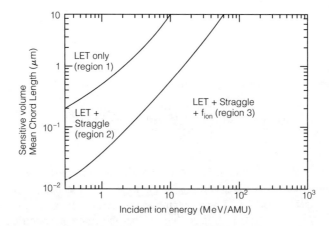

Figure 3.14. Regions of sensitive volume dimensions and ion energies per AMU where the energy deposition is significantly influenced by the parameters indicated. From Ref. 12.

At this point, some general results are depicted in Fig. 3.14. This figure [12] is separated into three regions labeled LET ONLY, LET + STRAGGLE, and LET + STRAGGLE +f_{ion}. The lines of demarcation are calculated using the "10%" criterion; that is, they are determined by when the individual effect of the three parameters causes a maximum of 10% variation in the energy deposited. For example, if the energy loss straggle is within 10% of the mean energy deposited in the sensitive volume as the ion penetrates it, then energy loss straggling can be disregarded in the energy deposition calculations. In actuality, these computations were made using the properties of He ions. However, the abscissa in the figure is energy per AMU, thus holding for essentially any ion. If refinements in these computations are desired to be made for other ion species, then ion scaling relations are available [12].

In Fig. 3.14, it is seen that region 3 requires not only LET for an adequate description of energy deposition but also incident particle straggling, and that not all of the energy deposited in the sensitive region remains there (i.e., $f_{ion} < 1$ as well). In region 2, in addition to LET, incident particle straggling is needed, whereas in region 1, LET is the sole adequate descriptor of the energy deposition process. This implies that regions where other factors besides LET influence incident particle energy deposition, occupy more than three-fourths of the area in the chord length–ion energy phase space, as in Fig. 3.14. As previously mentioned, an example is given whereby ions with the same LET deposit different amounts of energy [13]. In this case, the higher-energy ions deposit more energy, and more charge is collected than for the lower-energy ions. The precise reason for this discrepancy is not known, but further investigations along the lines of particle straggling and the $f_{ion} < 1$ condition, as they reflect ion track structure, should prove fruitful.

As mentioned earlier, straggling is the variation in the amount of energy lost by particles in passing through a material. The stopping power expression given in (3.24) is actually an average quantity; that is, it describes the mean energy loss by a particle penetrating matter. A more complete description involves the statistical dispersion of particle energy or corresponding particle range (track length). This is important in both the penetration of electrons and heavy particles, including heavy ions. It is then appreciated that two ions of different energy but with the same LET can result in the production of differing amounts of charge collected in a device SEU-sensitive region, due to their dispersion in track lengths and charge produced. This has been found in accelerator tests as discussed earlier. Figure 3.15 depicts energy straggling for 1414-keV electrons penetrating thin foils of aluminum [16]. Figure 3.16 shows range straggling for electrons versus dimensionless path length [16].

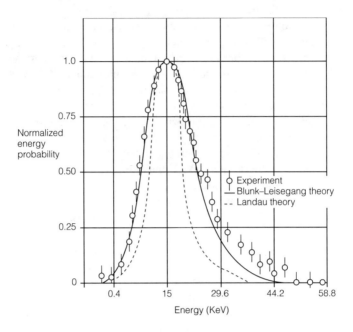

Figure 3.15. Energy straggling distribution for 1414-keV electrons after passing through 44 μm of aluminum. The ordinate is a normalized energy probability. From Ref. 16.

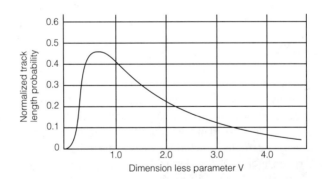

Figure 3.16. Range straggling distribution for 1414-keV electrons after passing through 44 μm of aluminum. The ordinate is a normalized track length probability. The abscissa is dimensionless parameter $v = 2\,[(4E/E_s)(x/\lambda_R)^{-1}]^2\,\Delta(x/\lambda_R)$; $E_s = 21$ MeV. λ_R is essentially the radiation length discussed in Section 3.5. $\lambda_R \cong 10$ cm for 1.414-MeV electrons in silicon, and $v = 32\,(E/E_s)^2\,(x/\lambda_R)^{-2}\,\Delta(x/\lambda_R) = 1.45\,\Delta x/x^2$. From Ref. 16.

Problems

3.1 The stopping power (energy loss) of an incident particle in a particular material, often equated to the particle LET in this material, is given by $-dE/dx$. When the energy loss from the incident particles is due to mainly to ionization, it is appreciated that the loss should be proportional to the electron concentration in the material because the interactions are between the atomic electrons being freed by the incident particles and so consume energy supplied by them; that is $|-dE/dx| \sim NZ$, where NZ is the electron concentration.

  (a) Show that, indeed, NZ does give the concentration, where $N = \rho N_0/A$. ρ is the material density, A is the material mass number (atomic weight), and N_0 is Avogadro's number.

  (b) Show that the stopping power is approximately constant to "first order" of approximation, irrespective of the material in which energy is being lost (deposited), if the stopping power is rewritten as $-dE/d\xi$, where the areal density $\xi = \rho x$.

3.2 Certain shielding situations, such as shielding against electrons, are spoken of as being bremsstrahlung limited. Explain.

3.3 What is the distinction between stopping power and LET in terms of the Bragg–Gray cavity theory?

3.4 Assume an incident heavy ion of energy E deposits all its energy in a device material by ionization, thus generating a charge Q in the material. Derive a relation between Q and E for both silicon and gallium arsenide, where their respective energy consumed to ionize one electron is 3.6 and 4.8 eV respectively. Show that $Q_{si}\,(\text{pC}) = E_{si}\,(\text{MeV})/22.5$ and $Q_{ga}\,(\text{pC}) = E_{ga}\,(\text{MeV})/30$.

3.5 $-dE/ds = \text{LET}$ (energy deposited per unit track length) can be expressed using a number of different kinds of units. If LET is given in units of pC/μm, what is its corresponding conversion in units of MeV cm^2/mg for Si and GaAs?. Use information from Problem 3.4.

3.6 That straggling should be taken into account is necessary for electrons as compared to heavy ions, because of the possibility that an incident electron may lose a large fraction of its energy in a single collision with an atomic electron. This does not exist for heavy particles because of the large mass difference between them and the colliding electron. Explain this statement in terms of the masses of the incident heavy ion and the atomic electron.

3.7 The range expression $R(E_0) = \int_0^{E_0} dE \, (-dE/ds)^{-1}$ accuracy increases as the incident particle energy increases because for a high-energy particle, the range is not very different than its random-walk distance. Explain this statement.

3.8 What is the major difference between the usual particle energy spectrum which depicts particle abundance or variations thereof versus energy and its LET spectrum?

3.9 The general definition of LET is normally very nearly given by $-dE/dx$, the rate of particle energy loss per unit penetration distance in the material of interest. However, it is well known that $-dE/dx$ can vary radically with penetration distance (e.g., as seen in Fig. 3.1). Then how can LET as a parameter be a valid descriptor of an incident particle energy loss in a particular device?

3.10 What is particle straggling, and what is its role?

References

1. H.A. Bethe, and J. Ashkin, in *Experimental Physics* (E. Segre, ed.) John Wiley and Sons, New York, 1943, Vol. 1.
2. R.D. Evans, *The Atomic Nucleus*, McGraw-Hill Book Co., New York, 1955, Chap. 22.
3. L.C. Northcliff, and R.F. Schilling, *Nuclear Data, Vol. A7*, Academic Press, New York, 1970.
4. R.E. Lapp, and H.L. Andrews, *Nuclear Radiation Physics*, Prentice-Hall, Englewood Cliffs, NJ, 1972, Chap. 10.
5. E. Segre, *Nuclei and Particles*, W.A. Benjamin, New York, 1964, Chap. 7.
6. J. Orear, A.H. Rosenfeld, and R.A. Schleuter, *Nuclear Physics*, University of Chicago Press, Chicago, 1949, Chap. 2.
7. M.J. Berger, and S.M. Seltzer, "Results of Some Recent Transport Calculations for Electrons and Bremsstrahlung," Second Symp. on Protection Against Radiation in Space, NASA SP-71, 1964.
8. B. Rossi, *High Energy Particles*, Prentice-Hall, Englewood Cliffs, NJ, 1952.
9. E.L. Petersen, J.C. Pickel, J.H. Adams, Jr., and E.C. Smith, "Rate Prediction for Single Events: A Critique," *IEEE Trans. Nucl. Sci.*, **NS-39** (6), 1577–1599 (1992).
10. W. Heinrich, "Calculation of LET Spectra of Heavy Cosmic Ray Nuclei at Various Absorber Depths," *Radiat. Effects*, **34**, 143–148 (1977).
11. W.J. Stapor, P.T. McDonald, A.R. Knudson, A.B. Campbell, and B.G. Glagola, "Charge Collection in Silicon for Ions of Different Energy But the Same Linear Energy Transfer (LET), *IEEE Trans. Nucl. Sci.*, **NS-35** (6), 1585–1590 (1988).

12. M.A. Xapsos, "A Spatially Restricted Linear Energy Transfer Equation," Radiat. Res., **132**, 282–287 (1992).
13. W.J. Stapor, P.T. McDonald, A.R. Knudson, A.B. Campbell, and B.G. Glagola, "Charge Collection in Silicon for Ions of Different Energy But the Same Linear Energy Transfer (LET), *IEEE Trans. Nucl. Sci.*, **NS-35** (6), 1585–1590 (1988).
14. E.L. Petersen, P. Shapiro, J.H. Adams, Jr., and, E.A. Burke, Calculation of Cosmic Ray Induced Soft Upsets and Scaling in VLSI Devices," *IEEE Trans. Nucl. Sci.*, **NS-29** (6), 2055–2063 (1982).
15. J.H. Adams, Jr., "The Variability of Single Event Upset Rates in the Natural Environment," *IEEE Trans. Nucl. Sci.*, **NS-30** (6), 4475–4480 (1983).
16. S. Flugge, *Encyclopedia of Physics*, Springer-Verlag, Berlin 1958. Vol. 34, pp. 87ff.

4

Single Event Upset: Experimental

4.1. Introduction

The discussion in this chapter centers around the major single event upset (SEU) simulation sources. Their importance lies in the fact that simulation methods are one of the few means by which microcircuit susceptibility to SEU can be measured. These sources and source types are few in number, principally because of the somewhat unusual properties of the corresponding particles required for substantive simulation. They must be heavy, as in the medium and high mass number end of the periodic table, ionized, and highly energetic to reasonably simulate cosmic rays incident on spacecraft avionics. For these reasons, the simulation sources are primarily particle accelerators such as synchrotrons.

The sources discussed include, besides the heavy-ion synchrocyclotrons, the Van de Graaf accelerator, californium-252 and the laser. As alluded to, the first-line sources in use are the particle accelerators. ^{252}Cf and the laser are important but are second-tier simulation means, as they do not consistently and very directly provide the salient SEU part parameters sought.

Also discussed is the manner by which the raw data are retrieved and transformed to useful information in order to evaluate piece parts for SEU vulnerability. It turns out that none of these simulators are in existence primarily for obtaining SEU parameter data, but were built for physics exploration needs. For example, they all antedate the emergence and recognition of single event phenomena.

Other details in this chapter include further discussion of critical charge, SEU cross section, and linear energy transfer (LET) in terms of how, and to what degree, the above sources can provide quantitative data of these and other important SEU parameters; that is, how well the sources are matched to determine some or all of them. All of these sources are presently being employed to obtain the sought-after SEU physical parameters and those that correspond to the manufacturing processes that bear on the approach to SEU hard devices.

4.2. Heavy-Ion Accelerators

The major SEU-inducing particles in microcircuit devices are cosmic ray heavy ions. It is fortuitous that there are some particle sources available that can be used to simulate them. Upon first blush, seemingly, one would be hard put to suggest a cosmic ray ion simulator. Recall that cosmic ray ions are typically among the most energetic species in the realm of nuclear particle physics. Their energies are measured in BeV (hundreds of MeV per nucleon), whereas their charge consist of high multiples of the electron charge e.g., $^{56}_{26}Fe^{6+}$.

It is also interesting that through ionization interactions with the device material, the incident cosmic ray ions usually cause only soft errors; that is, those that merely upset the informational charge state of the sensitive node of a memory cell, leaving it otherwise undamaged. Again, fortuitously, the cosmic ray flux levels are relatively small, such as the differential cosmic ray flux magnitude is of the order of $6000/L^3$ cosmic ray ions per square centimeter per second per LET [1]. This can be compared to the ionizing dose gamma ray flux in a ^{60}Co source cell, which is several orders of magnitude greater. This relegates the main cosmic ray ion impact to SEU as opposed to hard damage. This is not to say that they cause no damage, as they can cause device burnout, latchup, and other device crippling phenomena. It is the amount of ionization charge produced in the cosmic ray ion track within the device, with subsequent charge migration to, for example, memory cell nodes, that results in SEU being its principal radiation effect in microcircuit devices (integrated circuits).

As is well known, the major source that simulate cosmic ray ions that produce SEU, including protons and mesons (for simulating ground-level SEU), is high-energy accelerators. These are mainly cyclotrons and Van de Graaf accelerators. Some of the former are located at sites at Berkeley, Harvard, CERN (Switzerland), France, Germany and Russia. Some of the latter are sited at Brookhaven, Los Alamos, and Chalk River (Canada). Other source types include fission fragment ions from spontaneously fissionable transuranium nuclei, such as ^{252}Cf, as well as laser beams in the red and near-IR (infrared) wavelength spectrum.

From an electronics viewpoint, the cyclotron is essentially the output resonant circuit of a high-power radio-frequency oscillator operating in a frequency range of 10–30 MHz. Its output resonant "LC" circuit consists of two

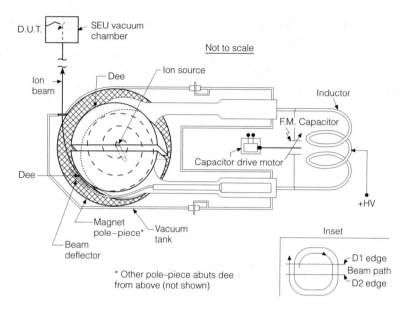

Figure 4.1. Basic cyclotron, when frequency modulated is labeled a synchrotron. From Ref. 2.

halves of a large, circular, hollow pillbox called "Dees," which make up the corresponding capacitor. This circuit also includes an inductor as in Fig. 4.1. The Dees are sandwiched between the poles of a very large electromagnet, weighing about 300 tons in the Berkeley 88-in. (Dee diameter) machine at 22 kilo-gauss. All except the magnet is enclosed in a vacuum tank. At the center between the Dees is the ion source, which can range from protons to ions whose masses are beyond uranium. When the ions begin their trajectories and are within the Dees, which act as a faraday cage, they are subject only to the magnetic field, which forces them into curvilinear paths. When the ions pass from one Dee to the other, they are also accelerated straight across by the radio-frequency electric field, thereby increasing their radii and speed. Their path is actually a barred spiral as in the inset. Due to the relativistic increase in the ion mass with increasing speed, they would soon fall out of synchrony with the radio-frequency **E** field across the Dees. To compensate, the oscillator is frequency modulated ($f_{mod} \sim 100$ Hz) by a motor-driven capacitor to form packets of ions, each packet in phase with the **E** field. The result is a current stream of ion pulses accelerated to very high energies measured in hundreds of MeV per nucleon. Fixed-frequency cyclotrons still can operate, but they yield much lower-energy ions but with constant higher beam currents.

The Van de Graaf accelerator (Fig. 4.2) consists essentially of a large spherical metal ball (terminal) supported on a large-diameter, hollow, insulated

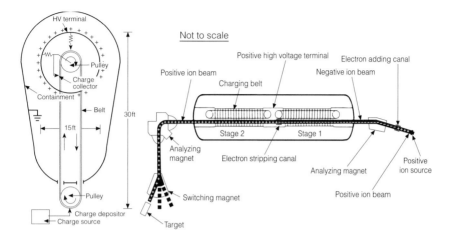

Figure 4.2. Van de Graaf accelerator (left) (From Ref. 3); two stage tandem accelerator (right) (From Ref. 4).

column (not shown). Charge is physically deposited onto and transported from ground up to the terminal on an endless insulated belt–pulley arrangement housed inside the column and ending inside the hollow terminal. The charge is retrieved onto the terminal, placing it at a very high potential V with respect to ground. $V = Q/C$, where C is the terminal capacitance with respect to ground and Q is the charge retrieved. V can easily be at a potential of millions of volts. Collinear with the belt–pulley system is the accelerating column (not shown in Fig. 4.2). It is made up of annular insulators interspersed with annular metal discs. Resistors are in series between the discs on the accelerating column, forming a voltage divider balancing network down to the ion beam exit. The ions are accelerated internally in the column, guided by the configured discs. In the two-stage tandem accelerator, higher ion energies can be reached by alternately switching their charge sign, beginning with a positive ion current beam, adding electrons to make it negative, then stripping the electrons off to make it positive again. This effectively quadruples the ion energy. The energy of the target for a proton is $2qV$. A two-stage tandem accelerator could produce 400-MeV uranium ions, with a 15-MV positive terminal. Three-stage accelerators are in existence at the University of Pittsburgh, the University of Washington (Seattle), and Brookhaven National Laboratories. Current activity in determining SEU parameters has been ongoing at the latter site.

4.3. Critical LET, Critical Charge

The heavy ion simulators principally measure L_c, the critical LET and/or SEU cross section σ_L as a function of L_{eff}, the effective LET. The experimental setup

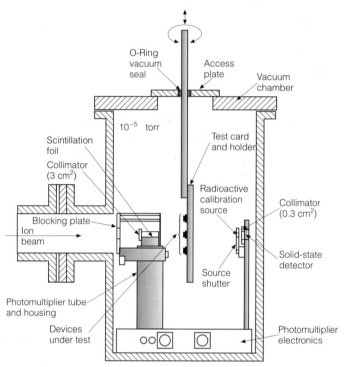

Figure 4.3. Vacuum chamber houses devices under test while it is exposed to an ion beam. From Ref. 5.

is shown in Fig. 4.3 [5]. As shown in the figure, the SEU vacuum chamber is connected to the ion accelerator beam tube through the port that admits the ion beam. A vacuum pump (not shown) maintains the chamber pressure at a typical value of about 10^{-5} torr. The access plate allows easy removal and insertion of the cards holding the devices under test and maintains a vacuum seal. The cards are electrically connected to the device tester and exerciser, whereas the card fixture is rotated to determine the critical angle, θ_c.

As alluded to earlier, to cause an SEU error in, say, a microcircuit memory, the incident SEU-inducing particle must deposit enough energy through ionization of the device material atoms to produce a critical charge Q_c at the memory storage cell. Q_c is that minimum charge incident at the memory storage node that causes a change of charge state—a bit flip (i.e., a SEU). In the SEU test using the accelerator ion beam as in Fig. 4.4, it is assumed that the device has not experienced an SEU at normal incidence ($\theta_c = 0$); that is, the available flux cannot deposit sufficient energy along the normal track length c to produce enough ionization to cause the deposition of critical charge Q_c at the device-sensitive node. However, when the device is rotated through

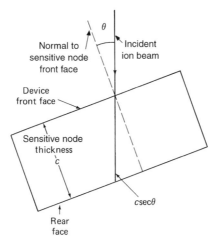

Figure 4.4. Edge view of parallelepiped model of sensitive node.

an angle θ_c, the increased track $c \sec \theta_c$ achieves a length such that SEU occurs at the node, as indicated by the associated instrumentation. This means that a corresponding critical energy ΔE_c has been deposited in the device from the beam ion-track length $c \sec \theta_c$.

In certain cases, the normal track length c can be of sufficient length to yield an SEU at $\theta_c = 0$. In this situation, experimental adjustments are made such as using absorbers to attenuate the beam flux so that θ_c becomes greater than zero. This assures a more accurate measurement whereby θ_c can be approached from above and below.

The linear energy transfer of a particular ion, L_{ion}, is a physical quantity of the ion parameters and the material in which it deposits its energy, as opposed to macroscopic parameters such as device geometry and/or electrical circuit properties. Then L_{ion} can be considered invariant in the sense that

$$L_{ion} \equiv L \cong \frac{\Delta E}{c \sec \theta} = \frac{\Delta E_c}{c \sec \theta_c} \tag{4.1}$$

where E is the ion energy and ΔE is the energy deposited. An effective LET is defined as $L_{eff} = \Delta E/c$, and the critical LET definition is $L_c = \Delta E_c/c$. It is seen that L_c is the L_{eff} normal to the device at the critical angle θ_c (i.e., at $\theta = \theta_c$, $L_{eff} = \Delta E_c/c = L_c$). Alternatively, within suitable conversion factors, $L_c = Q_c/c$. Then, from (4.1),

$$L = L_{eff} \cos \theta = L_c \cos \theta_c \tag{4.2}$$

94 / Single Event Phenomena

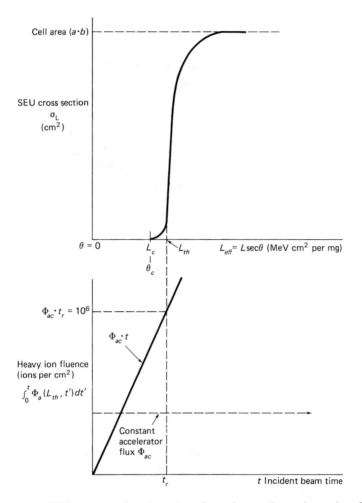

Figure 4.5a. SEU cross section (upper) and accelerator beam heavy-ion fluence (lower). From Ref. 6.

so that the critical LET $L_c = L \sec \theta_c$ is immediately derived from the accelerator experiment, because θ_c is measured and L is known.

From (4.1), $L_{\text{eff}} = L/\cos \theta$ implies that for any energy transfer per unit track length (LET) at any angle $\theta > 0$ (viz. $L/\cos \theta$), there is a corresponding effective energy transfer per unit *normal* track length to the sensitive node face (viz. L_{eff}). This is the principal reason that it is called the effective LET.

The second main parameter that is measured in SEU testing programs is the SEU cross section σ_L usually plotted as a function of $L_{\text{eff}} = L \sec \theta$. Ideally, σ_L would be depicted as in Fig. 4.5a (upper).

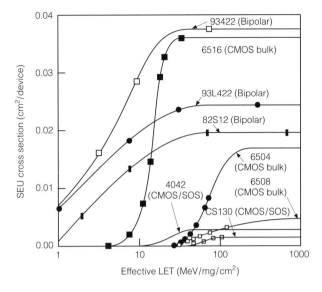

Figure 4.5b. Single event cross section versus LET for a variety of IC (integrated circuit) devices. From Ref. 7.

Unfortunately, SEU onset threshold and the asymptotic or saturation cross-section region are frequently not clearly delineated, as seen in Fig. 4.5b. Generally, the SEU cross section, like most cross sections, is a semiempirical parameter normally obtained by experiment. SEU cross-section data can be fitted by various analytical functions that represent probability density distributions, as discussed in Section 1.3. Experimentally, using accelerator beam ion sources, the SEU cross section is actually measured by computing

$$\sigma_L(\theta) = \lim_{t \to t_0} \frac{E_r(t)}{\Phi_a(t, L_{\text{th}}) \cos \theta} \qquad (4.3)$$

$E_r(t)$ is the SEU error rate measured by the instrumentation; L_{th} is the threshold LET as shown in Fig. 4.5a; t is the ion beam time measured from a reasonable starting fiducial; t_0 is the time required for the measurement to achieve a reasonably stable level for σ_L at a given θ. $\Phi_a(t, L_{\text{th}}) \cos \theta$ is the accelerator beam flux incident on the device SEU-sensitive node area. As the devices are rotated out of the normal to the beam line, their sensitive areas are foreshortened with respect to the incident beam flux. Hence, $\cos \theta$ is needed to represent this non-normal area facing the ion beam. Thus,

$$\Phi_a(t, L_{\text{th}}) \cos \theta \quad \text{(particles s}^{-1} \text{ per node area}/\cos \theta) \qquad (4.4)$$

is in the denominator of (4.3) instead of merely $\Phi_a(t, L_{\text{th}})$.

A third parameter, important for SEU accelerator and other heavy-ion source testing, is the threshold LET, $L_{th} \geq L_c$, and is usually given in SEU LET tables [8]. L_{th} is defined as [8] the minimum LET for which SEU occurs after the incident ion beam fluence $\int_0^{t_r} \Phi_a(t', L_{th}) dt'$ has reached a level of 10^6 ions cm^{-2}. t_r is the corresponding device exposure time from beam flux onset. This implies an approximate cross-section behavoir [6] as shown in Fig. 4.5a (lower); namely, at normal incidence ($\theta = 0$),

$$\sigma_L = \begin{cases} \lim_{t \to t_0} \dfrac{E_r(t)}{\Phi_a(t, L_{th})}, & t_r \leq t \leq t_0; \quad \int_0^{t_r} \Phi_a(t', L_{th}) dt' = 10^6 \text{ ions cm}^{-2} \\ 0, & t < t_r; \quad \int_0^{t_r} \Phi_a(t', L_{th}) dt' < 10^6 \text{ ions cm}^{-2} \end{cases} \quad \textbf{(4.4a)}$$

which is depicted in Fig. 4.5a (lower). For the case of pulsed ion beams [9], insert

$$\lim_{t \to t_0} \left(\dfrac{\int_{t_r}^{t} E_r(t') dt'}{\int_{t_r}^{t} \Phi_a(t', L_{th}) dt'} \right), \quad t_r \leq t \leq t_0$$

for the upper left side of (4.4a). For constant ion beam flux Φ_{ac}, $\Phi_{ac} t_r = 10^6$ ions cm^{-2}. L_{th} also sometimes corresponds to the intercept on the L_{eff} axis of the projection of the inflection point of σ_L, as shown in Fig. 4.5a.

The utility of the threshold LET is that it implies a statistical smoothing of the fluctuations of the ion beam flux level and also that of any associated instrumentation. An alternate definition of L_{th} is the L_{eff} corresponding to a measured σ_L that is, for example, equal to 1% of its asymptotic or saturation value [8]. These two definitions sometime conflict, resulting in differing values of L_{th}.

As microcircuits shrink in size toward submicron feature regions, accelerator ion measurements of SEU cross sections versus L_{eff} reveal discontinuities in σ_L for various L_{eff} as seen in Fig. 4.6 [2, 10]. This is especially the case for the newer CMOS-bulk and CMOS-epi integrated circuits. To aid in the explanation of the causes of these discontinuities, certain details of the manner by which accelerator beam ions are used in cross-section measurements are germane. Consider the process of how the σ_L versus L_{eff} curve in Fig. 4.6 is constructed. Starting with a low-LET ion such as ^{15}N (Table 4.1) whose $L_{ion} = L = 3$ MeV cm^2 mg^{-1}, the device is initially at normal incidence to this ion beam ($\theta = 0$). The device angle θ with respect to the beam line is then increased, taking cross section measurements along the arc, to a de facto safe maximum of about 60°. Because $L_{eff} = L/\cos \theta$, it is seen that initially $\theta = 0$, so

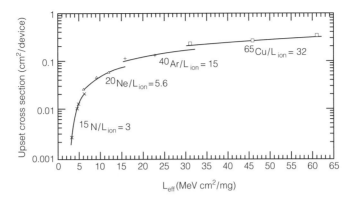

Figure 4.6. σ_L measurements for a 64K 1.3-μm SRAM using a typical set of accelerator ions. From Ref. 10.

the corresponding $L_{eff} = L$. Then a measurement of σ_L at 60° corresponds to a $L_{eff} = L/\cos 60° = 2L$, or $L_{eff} = 6$ in this case. Beyond 60°, it is felt that the results usually yield poor data, because the device packaging will begin to shadow the device node area from the ion beam flux. Also, the effective flux, $\Phi_a \cos \theta$, will drop to values less than half its $\theta = 0$ value, as seen from (4.3).

At this juncture, the ^{15}N ion beam is replaced with a higher-LET ion beam whose LET is as close to $L_{ion} = 6$ as practicable to maintain a continuous L_{eff}. Then the device rotation starts anew at $\theta = 0$ again. For Fig. 4.6, ^{20}Ne was used with LET = 5.6 (Table 4.1) so that now, initially, $L_{eff} = 5.6/\cos 0° = 5.6$. Taking $\sigma_L(\theta)$ data as the device is rotated to 70° in this segment yields $L_{eff} = 5.6/\cos 70° = 16$. Continuing in this vein by choosing ions of increasingly greater L allows the σ_L curve in Fig. 4.6 to be pieced together, including its asymptotic or saturation region. The σ_L in this figure used four different ion species in its construction.

Each time the beam is changed to the next ion type, the device is rotated back from its 60° slant position to its $\theta = 0$ restart position. In the 60° position, it can be appreciated that the sensitive node parallelepiped-model edge areas will present substantial projections normal to the incident beam direction. This will affect the amount of beam ion flux incident on the device, by altering the node area as seen by the beam. When the device is repositioned to $\theta = 0°$ for the next ion, the sensitive node edge projections disappear in a "discontinuous" manner. It is asserted that this is the cause of the σ_L discontinuities, as shown in Fig. 4.6.

One geometric explanation of the σ_L discontinuities is that the device-sensitive region parallelepiped projected edges alter the node area presented to the ion beam flux, especially for large θ. In an initial explanation [7], it is asserted that σ_L implies that all incident ions of a given LET that cause SEU

Table 4.1. Accelerator Ion Max Energy and LET Availability at Facilities in North America

Ser	Ion ^AX	BERK[a] MeV	LET[f]	BROOK[b] MeV	LET[f]	LASL[c] MeV	LET[f]	McMAS[d] MeV	LET[f]	CHALK[e] MeV	LET[f]
1	^{1}H	50	0.01	11	g	22	0.02	20	0.02	30	0.015
2	^{2}H	65	g	21	g						
3	^{4}He	12	0.35	41	g	33	0.16	30	0.18	45	0.13
4	^{7}Li					44	0.45	30	0.60		
5	^{9}Be					50	0.87				
6	^{12}C			105	1.4	65	1.9	70	1.85	600	0.36
7	^{15}N	67	3.							700	0.49
8	^{16}O	428	1.			75	3.6	70	3.78	850	0.67
9	^{19}F			150	3.0					850	0.89
10	^{20}Ne	90	5.6								
11	^{24}Mg					85	8.3				
12	^{28}Si			195	7.7	105	10.4			1260	2.14
13	^{35}Cl	100	15.2	210	11.5			125	14.2	1550	3.15
14	^{40}Ar	180	15.								
15	^{48}Ti					110	22.7			1800	5.9
16	^{56}Fe					110	28.3				
17	^{58}Ni			255	27.					2000	9.9
18	^{63}Cu									2050	11.
19	^{64}Cu							125	31.9		
20	^{65}Cu	290	32.								
21	^{72}Ge									2160	14.
22	^{79}Br			285	37.3			110	38.6	2100	17.6
23	^{86}Kr	380	41.								
24	^{107}Ag			300	53.1					2550	30.
25	^{127}I			320	59.7	150	54.5	80	44.9	2650	38.7
26	^{131}Xe	603	63.								
27	^{197}Au			345	82.3	150	63.8	80	47.	2900	77.4
28	^{238}U									2850	98.1

[a] Univ. of California, Berkeley, 88-in. cyclotron.
[b] Brookhaven Nat. Labs., twin tandem Van de Graaf.
[c] Los Alamos Sci. Labs., tandem Van de Graaf.
[d] McMaster Univ. Ont., Can. tandem Van de Graaf.
[e] Chalk River Nat. Labs., Ont. Can., tandem Van de Graaf/cyclotron.
[f] Units of MeV cm^2/mg in silicon.
[g] LET not given.

Source: Refs. 11 and 12.

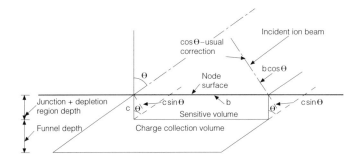

Figure 4.7. Parallelepiped-model edge effects influencing SEU cross sections. From Refs. 7 and 13.

must, at least, penetrate through the full junction depth c in Fig. 4.7 to produce the required critical charge Q_c from their ionization tracks. The ions of this LET through the node corners of the leftmost $c \sin \theta$ projection produce track lengths too short to provide Q_c. Hence, their corresponding projected parallelepiped area to the incident ion flux should be subtracted from the main node area to account for this edge effect; that is, the flux Φ_a should now be written as, using (4.4),

$$[\Phi_a] = \text{particles s}^{-1} \text{ per} \left(\frac{A_n}{b \cos \theta - c \sin \theta} \right) = \frac{\text{particles s}^{-1}}{(b/A_n)[\cos \theta - (c/b) \sin \theta]} \quad (4.5)$$

where the usual dimensions of the node area are $A_n \cong ab$ for $c \ll a, b$. Now $A_n/b \cong a$, and let $a = 1$ to normalize to a per unit area basis. Then, inserting (4.5) in (4.3) yields

$$\frac{\sigma_L(\theta) \cos \theta}{\cos \theta - (c/b) \sin \theta} = \lim_{t \to t_0} \frac{E_r(t)}{\Phi_a(t, L_{\text{th}})} \quad (4.6)$$

From (4.6), it is seen that a σ_L (corrected; θ) can be defined as

$$\sigma_L(\text{corrected}; \theta) = \frac{\sigma_L(\theta) \cos \theta}{\cos \theta - (c/b) \sin \theta} = \lim_{t \to t_0} \frac{E_r(t)}{\Phi_a(t, L_{\text{th}})} \quad (4.7)$$

Equation (4.7) yields the σ_L expression corresponding to the preceding initial interpretation of the edge effect.

A second explanation [13] for a σ_L(corrected; θ) is that of including the charge collection volume below the sensitive node in Fig. 4.7, in which funneling charge is the main contributor therein. Figure 4.7 is a composite drawing used for both explanations. Adding the funnel charge amounts to adding the $c \sin \theta$ projection to the main node projected area, instead of subtracting as in the preceding. The cross-section correction is found to be the same as that given in (4.7) except for a sign change to positive in its denominator. So that here [13], for the second explanation, is the σ_L(corrected; θ), namely

$$\sigma_L(\text{corrected}; \theta) = \frac{\sigma_L(\theta) \cos \theta}{\cos \theta + (c/b) \sin \theta} \qquad (4.8)$$

It is readily apparent that for, both interpretations,

$$\sigma_L(\text{corrected}; \theta) = \frac{\sigma_L(\theta)}{1 \pm (c/b) \tan \theta} = \sigma_L(0) \qquad (4.9)$$

where the sign in the denominator corresponds to the interpretation chosen, and that (4.9) holds for all $\theta \geq 0$ to account for its far right-hand side. Thus solving for $|c/b|$ gives

$$\left|\frac{c}{b}\right| = \frac{1 - \sigma_L(\theta)/\sigma_L(0)}{\tan \theta} \qquad (4.10)$$

There was sufficient data to determine $|c/b|$ for 13 1.3-μm 64K SRAMs [10] with the result that $0.1 \leq |c/b| \leq 0.2$ and $|c/b|_{av} = 0.16 \pm 0.04$. From the known asymptotic SEU device cross section $\sigma_{asy} = 0.5 \text{ cm}^2$ per device for these SRAMs [10], and each device contains 65,536 bits and 65,536 $\times$ (4 transistors per bit), the corresponding $\sigma_{xstr} = 0.5/(65,536 \times 4) = 190 \ \mu\text{m}^2$. Then $a \approx b = \sqrt{190} = 13.8 \ \mu\text{m}$. $c \cong 1.3 \ \mu\text{m}$ for 1.3 μm technology devices so that $|c/b| = 1.3/13.8 = 0.094$, which compares favorably with the experimental value of $|c/b| = 0.16 \pm 0.04$. This comparison can be construed as favoring the interpretation that the cross-section discontinuities are due to the geometrical edge effect of the device-sensitive node volume. Arguments involving the applicability of LET as the salient single parameter in lieu of the geometric interpretation are discussed in Section 3.9.

4.4. Californium-252 Source

As a seemingly reasonable alternative to the use of particle accelerator systems for SEU, single event latchup (SEL) and other single event phenomena testing, ^{252}Cf sources promise another quite different means for simulating SEU-inducing

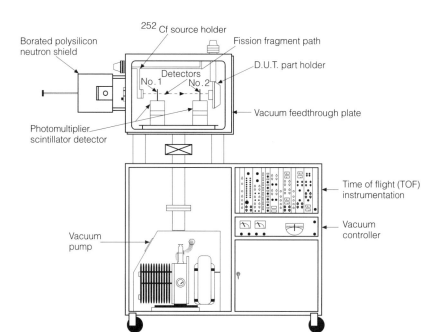

Figure 4.8a. Californium-252 SEU test system. From Ref. 15.

particles. It has definite advantages and disadvantages for determining SEU characteristics of semiconductor devices of interest. It should be realized that to mount an SEU testing program at a particular accelerator site, much time and wherewithal must be available; that is, not only for the initial equipment setup and experimental "shakedown" time and expense but also for a maintenance staff commuting to and from the site to provide the flow of devices to be exposed to the accelerator beam and, subsequently, posttest SEU characterized. Certain nonprofit enterprises on contract with the U.S. Air Force are presently conducting such ongoing programs in the United States.

Californium-252 SEU testing is, on the contrary, essentially a college laboratory table-top setup. It consists of an evacuated bell jar containing, in addition to the ^{252}Cf source, devices under test (DUTs), radiation detectors with their instrumented fixtures, umbilicals to vacuum pumps, and electronics, to exercise the DUTs while exposing them to the source radiation, as well as to process the SEU data in real time. Presently, there are compact commercial "turnkey" ^{252}Cf systems that will fit into a steamer trunk. Their costs are hardly 5% of the annual operating budget of a respectable cyclotron accelerator SEU testing program. A ^{252}Cf system is shown in Figs. 4.8a and 4.8b.

Californium-252 is one of a number of trans-uranium nuclei that, besides radioactively decaying by alpha particle loss, fission spontaneously. The fission

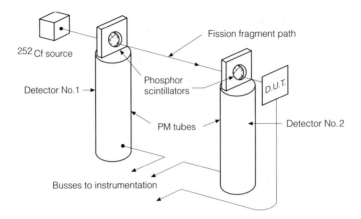

Figure 4.8b. Source–detector–DUT detail. From Ref. 16.

fragment nuclei divide into the characteristic two groups of fission products with respect to their mass number and energy, as shown in Figs. 4.9a and 4.9b. This is quite similar to neutron-induced fission of uranium nuclei in nuclear reactors, nuclear weapons, and spallation reactions in nuclei of far lesser mass number, such as silicon. Fission also releases gamma rays, electrons, neutrinos, and other light particles whose masses and energies can be neglected in this context. ^{252}Cf fission fragments are the particles used to simulate the incident individual cosmic ray ions and other heavy particles that produce SEU. Salient

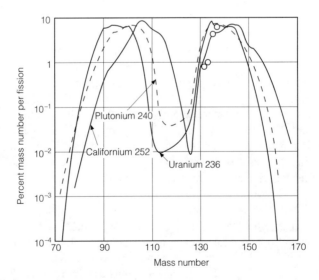

Figure 4.9a. ^{252}Cf, ^{240}Pu, and ^{236}U fission fragment mass number yield. From Ref. 33.

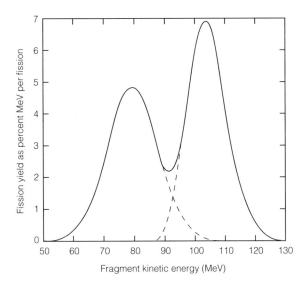

Figure 4.9b. ^{252}Cf energy distribution of fission fragments. From Ref. 14.

characteristics of ^{252}Cf are given in Table 4.2. As mentioned, the ^{252}Cf fission product nuclei comprise the well-known double-humped mass number and energy distributions falling into two distinct populations. However, their individual nuclei, on a per fission basis, as emerging from ^{252}Cf, cannot normally be identified, much less predicted. This is typical of radioactive decay which is a random process, because spontaneous fission is a form of such decay. In ^{252}Cf testing, it is desired to match the identity of the individual SEU-inducing fission fragment nucleus with that of the actual SEU event as closely as possible in real time while the experiment is ongoing. Fig. 4.10 shows the LET spectrum of ^{252}Cf fission products [35].

Incorporating time-of-flight techniques, whereby the individual fission fragment velocity is measured with high accuracy, into latter day ^{252}Cf SEU test instrumentation aids in accomplishing this matching task. With the fragment identity match, the SEU parameters such as L_c and the cross section can be extracted, analogous to heavy-ion accelerator SEU testing where the SEU-inducing beam ion identity is known a priori.

As seen in Fig. 4.8b, trajectories of the fission fragments that are incident on Detector No. 1, and that pass through Detector No. 2 on their way to the device under test are timed between the two detectors to measure their velocity. Velocities are on the order of 1 cm ns^{-1}, as compared to the velocity of light at 30 cm ns^{-1}. Hence, $\sqrt{1 - (v/c)^2}$ is about $\sqrt{1 - 10^{-3}} \cong 1$, so the kinematics are nonrelativistic. Velocity data on individual fission product fragments that trigger a corresponding SEU event are retained in the instrumentation memory

Table 4.2. ^{252}Cf Parameters

Decay mode	Half-life (years)	Branching ratio (%)	Decay products		Frequency (%)
Alpha particles	2.64	96.9	5.90 MeV	alphas	0.2
			6.08 MeV	alphas	15.2
			6.12 MeV	alphas	81.5
					96.9
Spontaneous fission	85	3.09			
		99.99			

Fission fragments	Mean mass number	Mean LET (MeV cm² mg⁻¹)	Mean range (μm)	Representative group nucleus
Heavy group	142.2	42	14	^{142}Ba (79 MeV)
Light group	106.2	42	14	^{106}Mo (104 MeV)

1-μCi ^{252}Cf source yields: 3.6×10^4 alpha particles s^{-1}
1.1×10^3 fission fragments s^{-1}
4.0×10^3 fast neutrons s^{-1}
The mean number of neutrons per fission, $\bar{\nu} = 3.85$ ($\bar{E}_n = 2.33$ MeV).

Source: From Ref. 17.

and are subsequently processed, whereas such data that do not cause SEU are counted and cleared. This is done through the use of multichannel and pulse height analyzer electronics. This system contains thousands of storage registers or channels, each representing a particular time of flight, velocity, or energy. The detector output pulse heights are proportional to the velocity (energy) of the fission fragments transiting between them. The range of pulse amplitudes are fed into a peak detector-A/D converter transforming the pulse heights (scope ordinate) to register addresses (scope abcissa). This yields a time-of-flight histogram of pulse heights—an energy spectrum as in Fig. 4.11. The registers (channels) are electronically swept, thus smoothing their contents and providing a continually increasing signal-to-noise ratio, making for enhanced accuracy of the resulting spectra.

To investigate the kinematics involved in the analytical determination of ^{252}Cf-induced SEU, an approach is provided by way of the conservation laws, namely the conservation of mass, momentum, and energy in the spontaneous

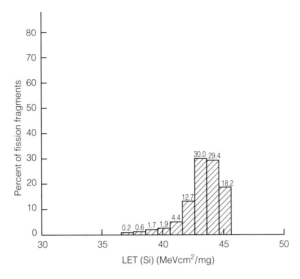

Figure 4.10. LET spectrum of ^{252}Cf fission products from fission product data. From Refs. 35 and 36.

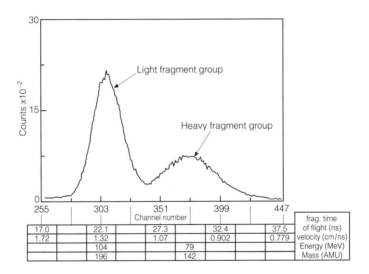

Figure 4.11. Californium-252 time-of-flight spectrum. From Ref. 16.

fission process. For an individual fission, the conservation of mass consists of summing the mass M_L of a representative fission fragment of the light mass group, plus M_H of the corresponding heavy mass group of that very fission, and the mass of the neutrons released in the fission. Fission products also include very low-mass particles such as electrons, neutrinos, and so forth, which have relatively small mass and thus are neglected, in contrast to their energies which are not small.

For a mean number of 3.8 neutrons released per ^{252}Cf fission, the conservation of mass yields

$$M_H + M_L + 3.8 M_N = M(^{252}Cf) \qquad (4.11)$$

Rewriting (4.11) in atomic mass units (AMU), discussed in Sections 2.2 and 3.9, where A is the corresponding mass number, results in,

$$A_H + A_L + 3.8 A_N = 252 \qquad (4.12)$$

Actually, the neutron $A_N = 1.008665$, where the least significant figures are dropped. However, they are used in nuclear binding energy computations, where small differences are important.

The conservation of energy, here the kinetic energy $T = \frac{1}{2}MV^2$ where M is the rest mass, can be written as

$$T_H + T_L + 3.8(2.33 \pm 1.08) = 183.23 \pm 11.25 \quad (\text{MeV}) \qquad (4.13)$$

The average energy of the fission neutrons is 2.33 MeV, with a standard deviation of 1.08. Similarly, the total kinetic energy released in the spontaneous fission process is 183.23 MeV $\pm$ 11.25 MeV. The conservation of momentum ($P = MV$) is written as

$$P_H - P_L = 0 \quad \text{or} \quad A_H T_H - A_L T_L = 0 \qquad (4.14)$$

because the ^{252}Cf nucleus splits into two fragments traveling in opposite directions. Also, $T_H = \frac{1}{2}A_H V_H^2 = P_H^2/2A_H$, or $P_H = (2A_H T_H)^{1/2}$. So $P_H - P_L = 0$ yields the right-hand side of (4.14) after squaring and canceling extraneous factors.

Rewriting the three conservation equations plus the time-of-flight measurement equation as follows yields,

Conservation of:

1. Fragment mass:

$$A_H + A_L = 248.2 \qquad (4.15)$$

2. Fragment momentum:

$$A_H V_H - A_L V_L = 0 \qquad (4.16)$$

3. Fragment energy:

$$\frac{1}{2}\left[938 A_H\left(\frac{V_H}{c}\right)^2\right] + \frac{1}{2}\left[938 A_L\left(\frac{V_L}{c}\right)^2\right] = 183.23 \pm 11.25 - 3.8\,(2.33 \pm 1.08)\ (\text{MeV}) \quad (4.17)$$

4. Fragment time-of-flight measurement:

$$V_L - rV_H = 0, \quad r = 1.303\ (\text{measured}) \quad (4.17')$$

where, in (4.17), the kinetic energy $T = \frac{1}{2}MV^2 = \frac{1}{2}Mc^2(V/c)^2 = \frac{1}{2}[938A\,(V/c)^2]$, as 1 AMU = 938 MeV.

For an ideal deterministic situation where there are no dispersions (i.e., if all standard deviations vanished), (4.15)–(4.17′) would be an exactly determined system, as there are four equations in the four unknowns: A_H, A_L, and the time-of-flight velocities V_H and V_L. If all quantities were known with certainty and they could be identified and tagged with each ^{252}Cf fission-fragment-induced SEU, then the time-of-flight experiment would be essentially the equivalent of the particle accelerator SEU experiment, because the "beam types" would be known exactly. Then, spectra and SEU cross sections of individual A_H and A_L fragments, whose identity is certain, could be measured.

A typical experimental value [18] for the time-of-flight resolution is about 0.5 ns, which results in a resolution of about 5 AMU for the fragments (i.e., $A_L \pm 2.5$ and $A_H \pm 2.5$). Corresponding energy resolutions are about $T_L \pm 2.5$ MeV and $T_H \pm 1.5$ MeV. Therefore, (4.15)–(4.17′) do not constitute a solvable system because their parameters are not known with certainty. However, the set of possible values of A_H and A_L are known quite well. This is certainly the case for their corresponding LETs [18]. Thus, maximum likelihood methods can be used to determine A_H and A_L.

It is instructive to let all the dispersions vanish and compute A_H and A_L from the resulting equations. Consider the conservation of energy equation (4.17) as a side condition, and solve (4.15) and (4.16) for A_H and A_L yielding

$$A_H = \frac{248.2}{1 + V_H/V_L}, \quad A_L = \frac{248.2}{1 + V_L/V_H} \quad (4.18)$$

Inserting the above into the conservation of energy equation (4.17) results in

$$V_H V_L / c^2 = 0.0017 \quad (4.19)$$

Experimental time-of-flight results [16] yield $V_L = 1.29$ cm ns^{-1} and $V_H = 0.99$ cm ns^{-1}, so that the corresponding

$$V_H V_L / c^2 = 0.0015 \tag{4.20}$$

which is about 14% less than the above computed values.

An ingenious time-of-flight experiment [19] was conducted to measure the fission fragment range–energy, $E(\xi)$, behavior in air, and its LET, $E'(\xi)$, where $\xi = \rho x$ is the areal density in air and x is the air penetration. This was done by metering the air entering the test chamber and varying the internal air pressure by varying its air density. The results yield [19] the following for units of ξ in milligrams per square centimeter of air ($0 \leq \xi \leq 1.235$ mg cm^{-2}) in terms of a quadratic polynomial in ξ:

For the light group of fission fragments,

$$E_L = (104.4 \pm 0.26) - (63.69 \pm 1.08)\xi + (7.04 \pm 0.65)\xi^2 \text{ MeV},$$
$$0 \leq \xi \leq 1.235 \text{ mg cm}^{-2} \tag{4.21}$$

For the heavy group of fission fragments,

$$E_H = (79.76 \pm 0.15) - (72.89 \pm 0.65)\xi + (20.35 \pm 0.50)\xi^2 \text{ MeV},$$
$$0 \leq \xi \leq 1.235 \text{ mg cm}^{-2} \tag{4.22}$$

and the corresponding LETs in air for the heavy and light groups are given by their derivatives [viz., LET$_L = E'_L(\xi)$ and LET$_H = E'_H(\xi)$]. These results compare well with experimental tabulations [19].

Single event upset cross sections as a function of $L_{\text{eff}} = L_{\text{ion}} \sec \theta$, like those obtained from the cyclotron accelerators, can be obtained from ^{252}Cf time-of-flight data, with the proviso that the LET values are those available from fission fragments. Generally, fission fragment LETs are relatively high, up to 43–44 MeV cm^2 mg^{-1}. This level corresponds to nearly the cosmic ray LET spectra cutoff. For cross section work, the fission fragment flux mean LET can be lowered to about 20 MeV cm^2 mg^{-1} by using absorbers between the incident flux and the device of interest. However, it is often desired to go lower as with particle-accelerator-obtained LETs. This is difficult due to fission alpha particle "noise," limited fragment range in actual devices, and other difficulties [20].

By varying the angle of incidence of the fragment flux by rotating the device under test, or using air density as an absorber to lessen the flux energy, corresponding "LET spectrum" curves are displaced vertically, as shown in Figs. 4.12a and 4.12b, respectively. These vertical shifts, which are, importantly, implicit functions of the SEU cross section, can be controlled as above,

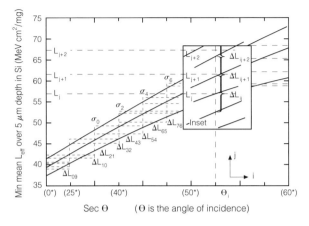

Figure 4.12a. ^{252}Cf measured LET spectra versus angle of incidence of the device under test. The figure shows only even-index σs. ΔL_{ij} is a small segment of fission fragment spectrum used in the weighting computation as in (4.24). From Ref. 21.

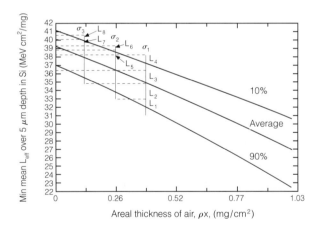

Figure 4.12b. ^{252}Cf fission fragment effective LET versus test chamber air density. Average 90%, and 10% cumulative LET fractions are shown. From Ref. 21.

so that a family of such curves can be generated as seen in the figures. For example, in Fig. 4.12b, an average L_{eff} over a 5-μm depth in silicon is plotted versus the areal thickness of the test chamber air to obtain this family [21]. These spectral shifts will be used in the computational method described below to obtain the SEU cross sections. In this case, the data of Figs. 4.12 will be employed. The following method supplies a computational algorithm for translating SEU cross-section data as a function of incidence angle, or chamber air density, to a cross section versus LET (i.e., a LET spectrum) curve. The computations are straightforward. However, cross-section data for the device of interest for a family of LET spectra (viz., $\sigma_m(\theta_i)$ as a function of incidence angle θ_i here), must be available.

With the computations used to draw Figure 4.12a, σ_L versus L_{eff} can be obtained from the set of measured SEU cross sections $\{\sigma_m(\theta_i)\}$, at each θ_i, giving n equations,

$$\sigma_m(\theta_i) \equiv \sigma_{mi} = \sum_{j=1}^{n} \left(\frac{\Delta L_{ij}}{L_j}\right) \sigma(L_j) = \sum_{j=1}^{n} w_{ij}\sigma(L_j), \quad i = 1, 2, \ldots, n \quad (4.23)$$

Equation (4.23) provides a transformation from cross sections measured at given angles of incidence, $\sigma_{mi}(\theta_i)$, to those at corresponding LETs, $\sigma(L_j)$. Thus, the measured cross section σ_{mi} at a given angle θ_i is obtained by the weighted sum of the sought-after cross sections $\sigma(L_j)$, each of which corresponds to a given LET L_j from the data used to plot the figures. The transformation coefficients (i.e., the weights w_{ij}) are given by

$$w_{ij} = \Delta L_{ij}/L_j \quad (4.24)$$

the ratio of that increment of the fission fragment spectrum ΔL_{ij} corresponding to LET L_j shown pictorially in Fig. 4.12a (inset). n is the desired number of terms in the sum, depending on the fineness of the resolution of the data. Equation (4.23) can be written in vector matrix form as

$$[\sigma_m] = [w_{ij}] \cdot [\sigma(L_j)] \quad (4.25)$$

where $[\sigma_m]$ and $[\sigma(L_j)]$ are the respective vectors and $[w_{ij}]$ is the transformation matrix. Because $[\sigma(L_j)]$ is the desired vector, inverting (4.25) by means of $[w_{ij}]^{-1}$ gives the $[\sigma(L_j)]$ vector (i.e., the $\{\sigma_L\}$ cross section set of values) as

$$[\sigma(L_j)] = [w_{ij}]^{-1} \cdot [\sigma_m] \quad (4.26)$$

With the data used to plot the LETs in Fig. 4.12a, $[\sigma(L_j)]$ is computed using (4.26). This $\{\sigma_L\}$ is plotted in Fig. 4.13 and compared with that obtained from a Van de Graaf 140-MeV bromine ion beam, as well as that from the standard ^{252}Cf cross-section measurements discussed earlier [21].

As can be seen from Fig. 4.13, the agreement with the Van de Graaf 140-MeV accelerator ions and the transformation method of cross-section calculations is quite good, in contrast to the standard method (i.e., $\sigma \cong$ SEU error rate/fragment flux).

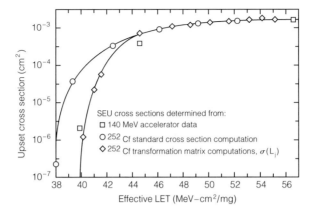

Figure 4.13. SEU cross section using ^{252}Cf for 16K SRAM compared with accelerator ions, standard cross-section computations, and transformation matrix computations. From Ref. 21.

As mentioned earlier, heavy ions can cause latchup in semiconductor devices. Latchup is an anomolous state of a device, triggered by an electrical or radiation transient, whereby the device acts as an "on" silicon-controlled rectifier (SCR) because of built-in parasitic electrical paths forming an SCR due to construction [22]. Many otherwise serviceable devices are subject to latchup. Latchup is usually detected by a sudden drop in V_{cc} in bipolar transistors and/or drain current surges in FETs. If the device is immediately powered down at the onset of latchup, it merely reverts to its state prior to the latchup. If it is not powered down quickly, irreversible damage (burnout) can result.

Using the ^{252}Cf source with appropriate latchup detection circuitry, and without the time-of-flight feature, the SEU latchup cross section was obtained for a number of devices [23]. The measured cross section as a function of LET for an HM6504 SRAM is shown in Fig. 4.14. The LET was varied during the test by changing the air pressure in the bell jar which varies its density as discussed earlier.

Californium-252 latchup capability is comparable with that measured by accelerator ions for LET values compatible with ^{252}Cf fission fragment LETs (i.e., $20 \leq$ LET ≤ 45 MeV cm^2 mg^{-1}). If the L_{th} of the device of interest is less than 20 MeV cm^2 mg^{-1}, then ^{252}Cf testing may not be applicable, because to reduce the ^{252}Cf LET to less than 20 MeV cm^2 mg^{-1} by using sufficient absorbers implies a fission fragment residual energy too small to penetrate the sensitive regions of many devices. That SEU latchup has been detected in spacecraft avionics is well known [24, 25].

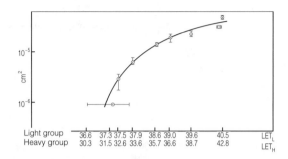

Figure 4.14. Microcircuit HM-6504 latchup cross section. From Ref. 23.

4.5. Lasers

Another way of simulating SEU uses laser beams. Prior to discussing laser photon interactions with semiconductor devices to simulate SEU, it is a good idea to briefly examine the features of laser light, without going into the details of laser operation.

There are four basic properties of laser light that make it unique as a source. They are (a) light beam intensity (power density), (b) beam directionality, (c) beam monochromaticity, and, last but by no means least, (d) beam coherence. It is the characteristic of laser energy that makes it unique compared to all other SEU simulators, as well as essentially all other light sources.

To exhibit the properties of laser light, a comparison with thermal (blackbody) light from a very hot (1000°K) source will be made. These properties will be described in the same order [(a)–(d)] as above.

(a) Power density: Laser light beam high intensity can be appreciated by its power output in the visible ($\lambda \sim 0.6\,\mu\mathrm{m}$) which can range from 10^{-3} to 10^9 W. At $\lambda = 0.6\,\mu\mathrm{m}$, each photon energy $E_p = h\nu = hc/\lambda = ((6.6 \times 10^{-27})(3 \times 10^{10})/(6 \times 10^{-5}))(1.6 \times 10^{-12}\,\mathrm{ergs/eV}) = 2\,\mathrm{eV} = 3 \times 10^{-19}\,\mathrm{J}$. The corresponding number of photons emitted per second for milliwatt to gigawatt lasers are respectively $10^{-3}/3 \times 10^{-19}$ to $10^9/3 \times 10^{-19}$ or 10^{16}–10^{28}. Compare this to a thermal planar light source of $\Delta A \cong 1\,\mathrm{cm}^2$ area, emitting in a band $\Delta\lambda = 0.4\,\mu\mathrm{m}$ about $\lambda = 0.6\,\mu\mathrm{m}$, which gives

$$\left(\frac{2\pi}{\lambda^2}\right)\left[\exp\left(\frac{h\nu}{kT}\right) - 1\right]^{-1} \Delta\nu\Delta A \cong 2 \times 10^{13}\ \text{photons s}^{-1} \qquad (4.27)$$

This is only $2 \times 10^{13}/10^{16} = 2 \times 10^{-3}$ as much light as the milliwatt laser.

(b) Beam directionality: The relation between laser beam size w_0 (beam diameter) and its divergence angle $\theta \cong \lambda/\pi w_0 = 0.5 \times 10^{-4}/0.1\pi = 1.6 \times 10^{-4}$ radians for a 1-mm beam spot size. This corresponds to a solid angle $\Omega \simeq \theta^2 = 2.5 \times 10^{-8}$ sr, which is a very highly collimated beam. With associated optics, this narrow beam can be used to probe the sensitive regions in a microcircuit. It is well known that the laser beam can be focused into an area $\Delta A \sim \lambda^2$, where λ is the laser wavelength. Inserting this area (i.e., $\Delta A \sim \lambda^2$) into (4.27), the expression for the thermal light photon output gives, now with the same aperture area, $2\pi[\exp(h\nu/kT) - 1]^{-1}\Delta\nu = (2 \times 10^{13})(0.5 \times 10^{-4})^2 \sim 5 \times 10^4$ photons s^{-1} compared to 10^{16} photons s^{-1} from a 1-mW laser.

(c) Beam monochromaticity: The coherence property of laser light ensures that the laser output bandwidth is extremely narrow. A typical laser bandwidth expression is

$$\Delta\nu \cong (h\nu/P)(\Delta\nu_n)^2 \tag{4.28}$$

For $h\nu = 2$ eV $= 3 \times 10^{-19}$ J, $P = 10^{-3}$ W and the spontaneous emission "noise" bandwidth $\Delta\nu_n \sim 10^3$ MHz, then $\Delta\nu \cong (3 \times 10^{-19}/10^{-3})(10^9)^2 = 300$ Hz, or, correspondingly, $|\Delta\lambda| = |c\Delta\nu/\nu^2| = (3 \times 10^{10})(300)/(6 \times 10^{14})^2 = 10^{-9}$ Å. An actual laser oscillates in many modes simultaneously, so the gross bandwidth is larger, but within an order of magnitude or so of the above $\Delta\lambda$.

(d) As mentioned, the salient property that distinguishes laser light is coherence. It is at the heart of laser function and catalyzes essentially all the other characteristics of its light. Coherence ensues because the main laser light output is the result of laser medium-energy transitions that are *stimulated* by the excitation (pumping) energy. Other transitions are spontaneous, producing photons due to the natural radioactive decay of the excited states, and are considered extraneous "noise." The level of stimulated emission light output is proportional to the incident (pumping) radiation, but more importantly is in exact phase with this radiation. Quantum mechanically, the exciting radiation and the main output radiation have identical momentum vectors. This means that as part of the laser output light is fed back to keep exciting the laser oscillator medium. The resulting stimulated emission produced thereby is in exact phase with the excitation. The output light wave stays exactly in train (i.e., is coherent with itself). Stimulated emission is also highly polarized.

Unfortunately, even with all the above characteristics of laser light, it suffers from a lack of deep penetration ability through opaque materials as does ordinary light. This curtails its effectiveness as an SEU-inducing source. However, with suitable optics, it can provide nanosecond to near femtosecond pulses of very small-diameter beam spots of intense power. Their size is small enough $\sim \lambda^2$ to probe portions of microcircuits for SEU investigations [26], as long as the overlayers are transparent to laser radiation, or are absent. Of course, devices are de-lidded, as they are using particle accelerators, during testing.

Again, without elaborating the details, it is appropriate to discuss the phenomenology of the penetration interactions of light with semiconductor materials. This reveals how the laser is used in depositing its energy and thus simulating the production of SEU in these materials. The wavelength of the laser beam light is selected with the intent to maximize its penetration and simultaneously to obtain a uniformity of energy deposition with depth in the material. The resulting charge collection phenomena will then approach that of heavy ion accelerators, the accepted SEU simulation source. The charge collection depends on the device structure as well as the laser light–material interaction processes. For SEU in shallow junctions, such as in GaAs semiconductors, the collection depth need be but a few microns. This allows somewhat more latitude in laser wavelength selection, therefore other criteria can be used in this selection. However, for latchup studies, charge collection depths can be at least an order of magnitude deeper to aid in ferreting out the possible charge paths that result in latchup.

GaAs is a direct-band-gap material, which means that a valence-band electron can be easily excited by laser light directly to the conduction band, called band-to-band light absorption. However, silicon is an indirect-band-gap material which requires a third particle (phonon) to balance the momentum vector to aid in the analogous band-to-band transition. This is reflected in the laser light GaAs absorption coefficient $\alpha (\text{cm})^{-1}$ in that its value changes very rapidly with laser light wavelength, and so is localized to a very narrow wavelength spread [28], as shown in Fig. 4.15. This makes the problem of selecting an appropriate wavelength more difficult because of the relative paucity of laser wavelengths (i.e., lasers available). This is in contrast to silicon whose α is comparatively slowly varying with wavelength, as in Fig. 4.15. This allows greater latitude in laser selection with which to induce SEU in devices of interest.

For laser energy absorption, there are a number of forms applicable in which to express LET in semiconductor material. One such follows from the assumption that the optical photon processes are linear; that is, all the interactions imply that an optical effect is proportional to its cause. The proportionality factor can be complicated but dependent on the system parameters only, not on the cause or effect variables. This is analogous to the theory of

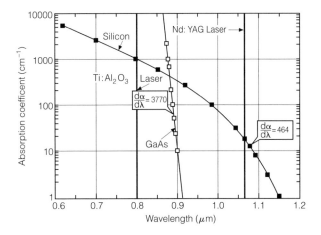

Figure 4.15. Absorption coefficient in lightly doped silicon and gallium arsenide. From Ref. 27.

communication signals, where linearity embodies the transfer function idea. However, nonlinear effects also play a role in laser light interactions and will be discussed.

For linear systems, Beer's law of attenuation holds, in that the laser pulse energy $E(x)$ satisfies $dE(x)/dx = -\alpha E(x)$, where x is the depth of penetration of the laser beam into the material. α is the absorption coefficient, assumed to be wavelength dependent only (i.e., not dependent on the penetration depth). Thus, the above equation implies that $E(x) = E_0 \exp(-\alpha x)$ at a particular wavelength, and laser output is assumed to occur at a single wavelength. Then, the LET can be written in dimensions of picocoulombs per micron within a constant as,

$$\text{LET} = \frac{e\,\overline{|dE/dx|}}{h\nu} = \frac{e\alpha E_0}{hc/\lambda} \quad (\text{pC}/\mu\text{m}) \qquad (4.29)$$

where the laser light frequency and wavelength satisfy $\lambda\nu = c$, the speed of light. $\overline{|dE(x)/dx|}$ is the mean vertical gradient of laser pulse energy deposition (i.e., the mean stopping power). It is seen that $\overline{|E'(x)|} = |(1/c) \int_0^c E'(x)\,dx| \cong \alpha E_0$, where c is the thickness of the SEU-sensitive region, assumed small compared to the mean depth of penetration (α^{-1}) (i.e., $\alpha c \ll 1$ in equation 4.29). Context will assert whether c is the speed of light or SEU-sensitive region thickness herein. Because $h\nu$ is the quantum of energy of the individual photon, $|E'|/h\nu$ is the mean number of photons deposited per unit depth. Linearity also implies that when a photon is annihilated in the band-to-

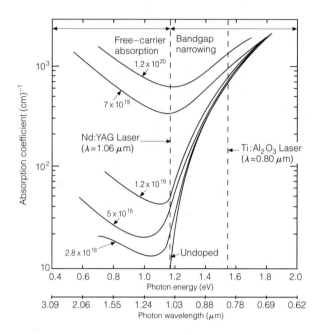

Figure 4.16. Bandwidth narrowing and free carrier absorption in silicon. From Refs. 27–29.

band absorption process, its energy produces one each electron–hole pair. Hence, $\alpha E_0/h\nu$ is also the number of charged carriers produced upon light penetrating into the semiconductor. Multiplying $\alpha E_0/h\nu$ by the electron charge e (1.6×10^{-7} pC) yields the amount of charge produced per micron depth, the LET in pC/μm^{-1}, where α units are cm^{-1}. In silicon [28], for a pulse of energy $E_0 = 7.3 \times 10^{-11}$ J, $\alpha \cong 15$ cm^{-1}, and using a neodymium ion, yttrium–aluminum–garnet (Nd^{3+}:YAG) laser whose $\lambda = 1.06$ μm yields, from (4.29), where 1 pC μm^{-1} = 96.6 MeV cm^2 mg^{-1} for silicon.

$$\text{LET} = 0.094 \frac{\text{pC}}{\mu\text{m}} = (0.094)\left[96.6 \frac{\text{MeV cm}^2}{\text{mg}} / \left(\frac{\text{pC}}{\mu\text{m}}\right)\right] = 9.21 \frac{\text{MeV cm}^2}{\text{mg}} \quad (4.30)$$

A laser wavelength of $\lambda = 1.06$ μm is a near-optimum choice for SEU in silicon in terms of high mean penetration distance $\bar{x} = \alpha^{-1} = 670 \mu$m, because it corresponds to nearly the minimum absorption coefficient in this wavelength regime, as in Fig. 4.16. Laser emission at wavelengths less than about 0.8μm (red) have small absorption coefficients corresponding to shallow penetration depths and can be used for GaAs SEU investigations as mentioned previously. However, latchup probing needs greater depths to get at the four-layer *pnpn* or *npnp* SCR paths that are latchup-prone.

Laser pulses used in modern submicron device SEU investigations have high power densities where nonlinear optical phenomena are in evidence. These translate into modifications of laser analytical computations such as for LET. At high power densities, band-to-band transitions where one photon results in one electron–hole pair no longer hold. These free–free transitions as opposed to bound-atom transitions, consist of at least two photons being absorbed by a free carrier. The energy deposition equation above is altered by the addition of a correction term, resulting in the nonlinear equation

$$E'(x) = -\alpha E(x) - \beta E^2(x), \qquad E(0) = E_0 \qquad (4.31)$$

with an integral

$$E(x) = \frac{(\alpha/\beta)}{\{1 + [(\alpha/\beta E_0)]\}\exp(\alpha x) - 1} \qquad (4.32)$$

This can be used in (4.29) to recalculate the LET. Also, $\bar{E}' = -\alpha \bar{E} - \beta \bar{E}^2 \cong -\alpha E_0 - \beta E_0^2$ for small αc.

The absorption coefficient α is a complicated function of the laser wavelength, beam power density, semiconductor dopant density, and other parameters. Figure 4.16 shows the variation of α as a function of laser photon energy and wavelength for various silicon dopant densities. It is seen in the figure that for increasing photon energies (decreasing laser wavelengths), band-gap narrowing influences α greatly. Band-gap narrowing depends in a complex manner on many parameters from semiconductor temperature to dopant density, the detailed discussion of which is beyond the scope of this treatise.

Suffice it to say that band-gap narrowing by definition is a decrease in the potential energy difference between the upper edge of the valence band and the lower end of the conduction band. This implies an increase in the band-to-band transition probability. Thus, carriers will be more readily excited by the laser photons to make such transitions. Hence, the number of exciting photons absorbed by them will increase, which will be reflected as an increase in the absorption coefficient, as seen in Fig. 4.16.

In some quarters, it is felt that, because of band-gap narrowing, the optimum laser wavelength is not necessarily confined to wavelengths near 1.0 μm. For example, an 0.8-μm titanium sapphire (Ti : Al$_2$O$_3$) laser is favorable being remote from the varying α region near 1.0μm, as in Fig. 4.16. In a device, by virtue of its manufacture as layers of various materials of various dopant levels, it is claimed that it is more difficult to predict the amount of SEU-inducing charge, and also the correspoding LET, where α varies rapidly with λ near 1.0 μm. By using lasers at lesser wavelengths (greater laser energies), $\alpha(\lambda)$ becomes less sensitive to dopant levels while maintaining reasonable penetration depths, so more accurate estimates of LET can be achieved [28].

118 / *Single Event Phenomena*

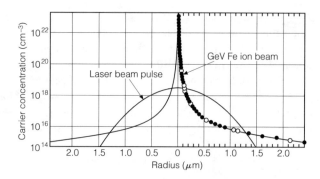

Figure 4.17. Identical LET radial charge density profiles of 1-GeV iron ion beam and Gaussian profile laser pulse at a depth of 0.125 μm. From Ref. 28.

Another aspect of the comparison between the laser and the accelerator ion in terms of charge deposited is their charge density beam profiles. As is well known from physical optics, the minimum diameter of the laser spot is of the order of the laser wavelength λ (Rayleigh criterion). It is about 100 times greater than the heavy-ion track diameter. Further, the laser beam peak charge density is about a few thousand times less than that of the ion beam maximum, as can be seen in Fig. 4.17. Among other items, this affects the beam-induced funneling charge, as it can be appreciated that the ion beam will produce much more charge at greater depths (where the bulk of the funnel lies) than the laser beam. The laser beam radius corresponds to that of the ion beam a long time after ion-track charge production. For lightly doped semiconductors, this difference in charge profiles is of less importance than it is in the case of heavily doped devices [27]. Also, the optics contribute to a degradation of the laser-charge production. To obtain a focused small spot size at a given region in the device, the corresponding laser optics must also cause the beam to diverge beyond the spot, as in Fig. 4.19c. The resulting diminution of the charge density generated by the laser deeper into the device reduces the charge that can be attributed to funneling, in the sense of simulating funneling. This is equivalent to an unwelcome reduction in LET in the deeper regions of the device. Also, with the advent of submicron devices, the laser spot size being limited as discussed poses a difficulty in probing such devices, as its spot can now almost cover a whole transistor. Hence, its effectiveness as a probe is greatly decreased.

To simulate charge funneling, which contributes additional charge due to the early collapse of the device regions following the ion strike, laser simulators must be able to cause charge production well below the semiconductor surface, which is the funnel region. This implies laser pulses of less than 100 ps duration [27] and that have sufficient energy to yield LET values of from 1 to 100 MeV cm^2 mg^{-1} with increasing dopant concentrations corresponding to

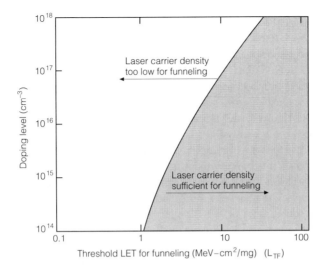

Figure 4.18. Doping levels versus laser funneling LET threshold showing allowable and nonallowable laser energy density regions to achieve funneling. From Ref. 27.

the higher LET values. Using the conservative criterion that funneling cannot occur until the excess carrier density exceeds the background concentration, computer program (PISCES) calculations are performed that yield a funneling LET (L_{TF}) threshold as a function of dopant density (N_d); that is, for a given dopant density, there is a minimum L_{TF} that will cause funneling, according to the computations [27]. This is depicted as a boundary curve in the $L_{TF} - N_d$ plane as shown in Fig. 4.18. For "vintage" devices whose feature sizes are greater than about 2 μm which are usually relatively lightly doped, $L_{TF} \gtrsim L_{th}$ in general. L_{th} is the threshold LET for the SEU of accelerator produced heavy ions. For these devices, the laser is probably capable of reasonably simulating SEU in their pertinent substrate regions. However, for modern devices with reduced feature size and concomitant increased dopant densities, the threshold criterion will probably be reversed (i.e., $L_{TF} < L_{th}$); that is, the laser pulse probably cannot produce the required pulse energy in the pulse time to correspond to a L_{th} of a heavy ion in the deeper substrate regions. For example, for dopant concentrations greater than 10^{17} cm^{-3}, L_{TF} must be greater than 10 MeV cm^2 mg^{-1} for funneling to be produced by a laser, as seen in Fig. 4.18. It is therefore felt that in certain situations where funneling contributes the preponderance of SEU-inducing charge, a larger laser L_{TF} will be needed. Whether this can be achieved in a practical fashion with available lasers is presently under study. Other aspects of the utilization of lasers for SEU simulation include the effects of device surface reflections and carrier lifetime dependencies, which are covered elsewhere [28].

The profile of the charge deposited with depth of penetration by the laser is important even for older devices with larger feature sizes and lower dopant concentrations than modern microcircuits. Figure 4.19 [27] compares the track profiles of a Nd:YAG laser with those of galactic cosmic ray particles and an accelerator heavy ion whose range is 40 μm. This latter range is specified in ASTM Test Guide F1192-90 [30] as a minimum for SEU investigation. Also depicted in the figure is an approximation to the mean prompt charge collection range due to the heavy-ion impact.

It is seen that the range of the heavy ion is marginal for the production of diffused charge. This implies that the laser pulse may produce more SEU-inducing charge than the heavy ion for the same LET for latchup and multiple-bit SEU processes that depend on diffused as well as prompt charge. Thus, lasers may produce more or less charge than an equivalent heavy ion depending on dopant levels, as also shown in Fig. 4.18.

The numerical value of the mean penetration distance $\bar{r}_{pr}$ to produce prompt charge can be written as a function of the heavy-ion range, consistent with the other parameters in the figure; that is, from (1.42), rewritten here but using the above terminology yields

$$\frac{1}{\bar{r}_{pr}} = \frac{1}{\bar{R}} + \frac{1}{\bar{r}} \tag{4.33}$$

where $\bar{R}$ is the heavy-ion range and $\bar{r}$ is the mean chord length. $\bar{r}_{pr}$ is the mean travel of the heavy-ion track segment, stopped in the medium after producing prompt charge along the way. As discussed in Section 1.4, (4.33) is derived under the assumption of isotropic fluence but can be shown to hold for a uniform bidirectional fluence as well.

Assume that each of Figs. 4.19a–4.19c is a plan view of the same parallelepiped volume with dimensions $a = b = 100$ μm, and depth c to be determined. This volume is similar to but larger than the usual SEU-sensitive region volume in most modern devices. From Fig. 4.19, it is seen for this "wafer" that a typical heavy ion range $\bar{R} = 40$ μm for a corresponding ion energy. Also, $\bar{r}_{pr} = 20$ μm from the figure. From (4.33), this results in a mean chord length $\bar{r} = (1/\bar{r}_{pr} - 1/\bar{R})^{-1} = 40$ μm. The PISCES code [27] as run for this lightly doped (10^{15}-cm^{-3}) wafer with the above dimensions yields $c = 33$ μm. From Section 1.4 and for $a = b$, this volume also gives an $\bar{r} = 4V/S = 2ac/(a + 2c) = 39.76$ μm, which agrees very well with $\bar{r}$ from (4.33). It is seen that c approaches the thickness criterion for the domino-thin parallelepiped dimensions ($a, b \gg 3c$) discussed in Section 5.2. The above agreements lend credibility and consistency to these calculations and the associated parameters. Even though the above dimensions are much larger than the usual SEU-sensitive volumes, it is simply a matter of linear scaling (Section 6.5) to the desired sensitive region dimensions, as the corresponding dimensional ratios would remain the same.

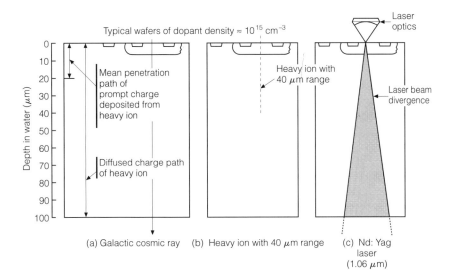

Figure 4.19. Comparison of charge deposition profiles of galactic cosmic rays, accelerator ions, and 1.06-μm focused laser pulse versus depth, in typical wafers. From Ref. 27.

Another aspect of the laser penetration, shown in Figure 4.19, and mentioned previously, is that the laser pulse spreads laterally as it penetrates. This lack of localizability limits the laser capability to probe submicron feature size devices.

Fundamental aspects of diffraction optics place a lower limit of ~ 1.5 μm on a focused laser spot incident on a device. The corresponding carrier density is less than that of a heavy ion by a factor of about 200 [27]. This lesser carrier density reduces funneling at high values of the former which impact calibration of laser effects. However, it also implies a contradiction to the claim that funneling effects are essentially the same for lasers and heavy ions [27]. Reduced laser funneling is significant for scaled devices whose $L_{th} < 10$ MeV cm^2 mg^{-1}, because of their high dopant densities. At high laser power, nonlinear effects become manifest, including nonlinear absorption. This also complicates the analysis of laser experiments.

Usually, lasers can only supply LET threshold cross sections and/or the saturation (asymptotic) cross section, because of the transparency difficulty in probing all parts of the sensitive regions because some are shadowed. Accelerators usually must be employed to obtain the cross-section curve in detail for most devices. Generally, lasers can be used in certain part screening tests for SEU. Laser wavelengths should be tunable so that the corresponding photon energy can be put above the device material band-gap energy. For example, photon energies should lie just above the band gap for adequate penetration [31]. This corresponds to wavelengths in the near-infrared region ($\sim 1\mu m$). Perhaps in the future, the coherence property of laser light will be used directly in some form of interference technique in SEU investigation.

A short list of laser source attributes for SEU work is given in the description of a laser facility [32], paraphrased as follows:

1. Identification of unanticipated SEU as a result of modeling
2. Comparison of SEU thresholds in cells and their peripheral circuitry
3. Bit maps for use in logical and physical multiple-bit SEU investigations
4. Investigation of SEU and transient propagation in devices through to output pins
5. Testing of multichip models with thick dielectric overlayers
6. Confirmation of error detection and correction (EDAC) methods and fault tolerant schemes
7. SEU hardness assurance measurements
8. Validation of SEU testing procedures
9. Dynamics of charge collection in test structures
10. Determination of SEU clocked-circuit timing vulnerabilities

Problems

4.1 A cyclotron has an oscillator frequency of 15 MHz, a Dee radius of 44 in., and a magnetic field $H = 22$ kilogauss. The beam ion used is $^{131}_{54}Xe$.

  (a) What is the ion charge state Z' that the electron stripped xenon ion beam has, consistent with the given parameter values? Are the electrons completely stripped from the $^{131}_{54}Xe$ ions?

  (b) What is the corresponding energy of the xenon ion beam?

  (c) Is the speed of the xenon ion beam in the relativistic region i.e., $(1 - v^2/c^2)^{1/2}$, near zero?

4.2 For the Van de Graaf accelerator shown in Fig. 4.2, its system yields beam energies of $2QV$ (J), Q (C) and V (Volts), by alternately stripping and adding charge to the beam ions as discussed. Show that this energy level yields a factor of 4 over particle energies E_p obtained with a "no charge alteration" accelerator.

4.3 Modern Van de Graaf accelerators can provide beams of 400-MeV uranium ions using an accelerating potential of only 15 MV. What must the charge state of the uranium ion be?

4.4 A parameter of major importance in the determination of "SEU-hard" parts is the critical LET L_c. Is an "SEU-hard" part characterized as one with a high or low L_c, and why?

4.5 Using the ^{252}Cf source for the determination of SEU parameters, three conservation equations plus the time-of-flight measurement equation, (4.15)–(4.17'), describe the corresponding fission fragment kinematics. They are ($C_1 = 248.2$ AMU, $C_3 = (3.35 \times 10^{20}$ AMU$)c^2$, and $r = 1.303$)

(1) $A_L + A_H = C_1$ (conservation of mass)
(2) $V_L A_L - V_H A_H = 0$ (conservation of momentum)
(3) $V_L^2 A_L + V_H^2 A_H = C_3$ (conservation of energy)
(4) $V_L - rV_H = 0$ (time of flight measurement)

Solve this system for the four unknowns A_L, A_H, V_L, and V_H and compare with Fig. 4.11.

4.6 ^{252}Cf is also used to obtain the range–energy relation for its fission fragments in air as discussed. The results for the light group are given in (4.21), whereas that for the heavy group is given in (4.22).

(a) Use the expressions in (4.21) and (4.22) to obtain the corresponding LETs.

(b) The preceding are valid between $0 \leq \xi \leq \xi_{max} = 1.235$ mg/cm^{-2} only. Convert ξ_{max} to x_{max} (i.e., $0 \leq x \leq x_{max}$). What is the value of x_{max} when the density of air $\rho_{air} = 0.247$ mg cm^{-3}?

(c) The galactic cosmic-ray-ion LET spectrum (Fig. 5.1) provides values of their integral flux versus their LET values. For example, an ion whose LET is 10 MeV cm^2 mg^{-1} has an integral flux of 1.5 ions cm^{-2} da^{-1}. For the fission-fragment-ion LET in air, what are the LET values for the light and heavy fission-fragment-ion groups respectively?

4.7 In Section 4.3, the relationship between the incident particle LET L, the device effective LET L_{eff}, and the device critical LET L_c can be rewritten as $L = L_{eff} \cos \theta = L_c \cos \theta_c$, where $L_{eff} = \Delta E/c$ and $L_c = \Delta E_c/c$. There is also an inequality relation between the three LETs. Is it:

(1) $L \leq L_{eff} \leq L_c$
(2) $L \geq L_{eff} \geq L_c$
(3) $L_{eff} > L \geq L_c$
(4) $L_c \leq L \leq L_{eff}$
(5) All of the above
(6) None of the above

4.8 (a) $\bar{r}_{pr} = 20\,\mu m$ as seen in Fig. 4.19 is an approximation. To yield exactly $\bar{r}_{pr} = 20\,\mu m$ for a square parallelepiped whose edge $l = 100\,\mu m$ as in the figure, what must its thickness w be?
(b) Show that for a parallelepiped of dimensions l, w, and h that $h^{-1} + w^{-1} + l^{-1} = 2(1/\bar{r}_{pr} - 1/\bar{R})$.

4.9 In Section 4.5, the linear energy transfer for a laser system is given as LET $= \overline{|(dE/dx)|}(e/h\nu)$ in units of pC μm^{-1}, within a constant of proportionality. Explain this dimensionally, where dE/dx is the laser energy loss per micron; $h\nu$ is a quantum of energy per photon in this case. Both have the same energy dimension. e is the electron charge in picocoulombs.

4.10 Using the cyclotron to determine the device critical LET, show that the fractional change in the effective LET, L_{eff}, where $L_{\text{eff}} = L/\cos\theta$, is given by $dL_{\text{eff}}/L_{\text{eff}} \cong (\Delta\theta)^2$ for small angles $\Delta\theta$. $\Delta\theta$ is the incremental angle that the device normal makes with the ion beam. Recall that the sought-after critical LET from this test for a device is $L_c = L_{\text{eff}}(\theta_c) = L/\cos\theta_c$, where θ_c is the critical angle and L is the (constant) LET of the incident ion.

4.11 The usual relationship between the ion LET, L, and the effective LET, L_{eff}, of a device is given in (4.2) as $L_{\text{eff}} = L/\cos\theta$. Let the relation $L = \Delta E/c \sec\theta \equiv L_{\text{eff}}/\sec\theta$ be modified as $L = \Delta E(\eta/c \sec\theta)$, where a charge collection efficiency η is given by the ratio of charge collected along the track to that produced along the track. If $\eta \sim \cos\theta$ as well it might [34], because η would fall off due to passage through various portions of the device with increasing track length $(c/\cos\theta)$, show that this results in $L_{\text{eff}} \cong L\sec^2\theta$.

References

1. E.L. Petersen, J.B. Langworthy, and S.E. Diehl, "Suggested SEU Figure of Merit," *IEEE Trans. Nucl. Sci.*, **NS-30** (6), 4533–4539 (1993).

2. R.M. Bescanson (Ed.), *The Encyclopedia of Physics*, Van Nostrand Reinhold, New York, 1974.

3. R.J. Van de Graaf, J.G. Trump, and W.A. Buechner, *Rep. Progr. Phys.*, **11**, 1 (1948).

4. R.J. Van de Graaf, *Nucl. Instrum. Methods*, **8**, 195–202 (1960).

5. NASA Tech Briefs, Winter 1985.

6. G.C. Messenger, and M.S. Ash, *The Effects of Radiation on Electronic Systems*, 2nd ed., Van Nostrand Reinhold, New York, 1992.

7. E.L. Petersen, J.C. Pickel, J.H. Adams, Jr., and E.C. Smith, "Rate Prediction for Single Events: A Critique," *IEEE Trans. Nucl. Sci.*, **NS-39** (6), 1577–1599 (1992).

8. D.K. Nichols, L.S. Smith, and W.E. Price, "Recent Trends in Parts SEU Susceptibility from Heavy Ions," *IEEE Trans. Nucl. Sci.*, **NS-34** (6), 1332–1337 (1987).
9. ASTM F1192-90, "Standard Guide for Measurement of Single Event Phenomena Induced by Heavy Ion Irradiation of Semiconductor Devices," p. 2., April 1990.
10. P.M. O'Neill, and G.D. Badhwar, "SEU for Space Shuttle Flights of New General Purpose Computer Memory Devices," *IEEE Trans. Nucl. Sci.*, **NS-41** (5), 1755–1764 (1994).
11. F.W. Sexton, "Measurement of SEU in Devices and ICs," *IEEE NSRE Conf. Short Course*, Chapter 3, New Orleans, July 13, 1992.
12. "Survey of Particle Accelerators," Radiation Effects Information Center, Battelle Mem. Inst. REIC No. 31, 1964.
13. F.W. Sexton, J.S. Fu, R.A. Kohler, and R. Koga, "SEU Characterization of a Hardened CMOS 64K and 256K SRAM," *IEEE Trans. Nucl. Sci.*, **NS-36** (6), 2311–2323 (1989).
14. J.T. Blandford, Jr., and J.C. Pickel, "Use of Cf-252 to Determine Parameters for SEU Rate Calculations," *IEEE Trans. Nucl. Sci.*, **NS-32** (6), 4282–4286 (1985).
15. Spectrum Sciences Model 5005-TF, Spectrum Sciences Inc., Santa Clara CA.
16. M. Reier, G. Swift, G. Soli, and B. Blaes, "Cf-252 Time-of Flight System for SEU Testing," Jet Propulsion Labs., Cal. Tech., Pasadena (unpublished).
17. J.H. Stephan, T.K. Sanderson, D. Mapper, J. Farren, R. Harboe-Sorensen, and L. Adams, "Cosmic Ray Simulation Experiments for the Study of SEU and Latchup in CMOS," *IEEE Trans. Nucl. Sci.*, **NS-30** (6), 4464–4469 (1983).
18. M. Reier, G. Swift, G. Soli, and B. Blaes, "Cf-252 Time-of Flight System for SEU Testing," Jet Propulsion Labs., Cal. Tech., Pasadena (unpublished).
19. M. Reier, "An Experimental Measurement of the Energy Loss of Cf-252 Fission Fragments in Air," *Nucl. Instrum. Methods Phys. Res. (Amsterdam)*, **B30**, 503–506 (1988).
20. D. Mapper, T.K. Sanderson, J.H. Stephan, and J. Farren, "An Experimental Study of the Effect of Absorbers on the LET of Cf-252 Fission Particles," *IEEE Trans. Nucl. Sci.*, **NS-32** (6), 4276–4281 (1985).
21. J.S. Browning, "Single Event Correlation Between Heavy Ions and CF-252 Fission Fragments." *Nucl. Instrum. Method. Phys. Res.*, **B43**, 714–717 (1990).
22. G.C. Messenger, and M.S. Ash, *The Effects of Radiation on Electronic Systems*, 2nd ed., Van Nostrand Reinhold, New York, 1992, Sec 7.7.
23. M. Reier, "The Use of Cf-252 to Measure Lathchup Cross Section as a Function of LET," *IEEE Trans. Nucl. Sci.*, **NS-33** (6), 1642–1645 (1986).
24. T. Goka, S. Kuboyama, T. Shimano, and T. Kawanishi, "The On-Orbit Measurements of Single Event Phenomena by ETS-V Spacecraft," *IEEE Trans. Nucl. Sci.*, **NS-38** (6), 1693–1699 (1991).
25. L. Adams, E.J. Daly, R. Harboe-Sorensen, R. Nickson, J. Haines, W. Schafer, M. Conrad, H. Griech, J. Merkel, T. Schwall, and R. Henneck, "A Verified Proton Induced Latchup in Space," *IEEE Trans. Nucl. Sci.*, **NS-39** (6), 1804–1808 (1992).

26. S. Buchner, J.B. Langworthy, W.J. Stapor, A.B. Campbell, and S. Rivet, "Implication of the Spacial Dependence of the SEU Threshold in SRAMs Measured With a Pulsed Laser," *IEEE Trans. Nucl. Sci.*, **NS-41** (6), 2195–2202 (1994).

27. A.H. Johnston, "Charge Generation and Collection in P-N Junctions Excited with a Pulsed Infra Red Laser," *IEEE Trans. Nucl. Sci.*, **NS-40** (6), 1694–1702 (1993).

28. J.S. Melinger, S. Buchner, D. McMurrow, W.J. Stapor, T.R. Weatherford, A.B. Campbell, and H. Eisen, "Critical Evaluation of the Pulsed Laser Method For SEU Effects Testing and Fundamental Studies," *IEEE Trans. Nucl. Sci.*, **NS-41** (6), 2574–2584 (1994).

29. S. Pantelides, A. Selloni, and R. Cor, "Energy Gap Reduction in Heavily Doped Silicon: Causes and Consequences," *Solid State Electron.*, **28**, 17 (1985).

30. ASTM F1192-90, "Standard Guide for Measurement of Single Event Phenomena Induced by Heavy Ion Irradiation of Semiconductor Devices," p. 2., April 1990.

31. S. Buchner, A. Knudson, K. Kang, and A.B. Campbell, "Charge Collection for Focused Picosecond Laser Pulses," *IEEE Trans. Nucl. Sci.*, **NS-35** (6), 1517–1522 (1988).

32. S. Buchner et al., "Pulsed Laser Facility for Single Event Effects Investigations," Naval Research Laboratories Advertisement.

33. Sci. Am., August 1983.

34. A.R. Knudsen, and A.B. Campbell, "Charge Collection in Bipolar Transistors," *IEEE Trans. Nucl. Sci.*, **NS-34** (6), 1246–1250 (1987).

35. J.H. Stephen, T.K. Sanderson, D. Mapper, J. Farren, R. Harboe-Sorensen, and L. Adams, "A Comparison of Heavy Ion Sources Used in Cosmic Ray Simulation Studies of VLSI Circuits," *IEEE Trans. Nucl. Sci.*, **NS-31** (6), 1069–1072 (1984).

36. J.F. Ziegler, *Handbook of Stopping Cross Sections for Energetic Ions in All Elements (in) the Stopping and Ranges of Ions in Matter.* Vol. 5 (J. Ziegler, Ed.), Pergamon Press New York, 1980.

5

Single Event Upset Error Rates

5.1. Introduction

This chapter together with Chapter 8 provides the major share of the discussion on the practical aspects of single event upset (SEU). This includes formulas for computing SEU in various particle environments. Section 5.2 discusses SEU calculations for heavy-ion cosmic rays at geosynchronous altitudes and Section 5.3 for Van Allen belt protons. Section 5.4 is on SEU-inducing neutrons at cruising altitudes for high-flying aircraft, and Section 5.5 is on alpha particles in microcircuit chip packages. Finally, Section 5.6 discusses ground-level SEU. The formulas are derived from the fundamentals given in the earlier chapters, and their limitations with regard to applicability are delineated. Also, their evolution as it pertains to an increase in understanding single event phenomena up to present-day concepts are discussed.

Detailing the preceding, Section 5.2 begins with the derivation of one of the early expressions for SEU error rate in a geosynchronous environment, mainly implying SEU induced by galactic cosmic rays. It is called the "figure of merit" estimation formula. It is given in terms of the critical charge of the device SEU-sensitive region as modeled by a rectangular parallelepiped volume of material. It uses the chord distribution functions discussed in Section 1.4.

Alternatively, the same formula is expressed in terms of its measured or known SEU cross section and its linear energy transfer (LET) for these cosmic rays. This formula can also be extended to include solar flare conditions by means of appropriate multipliers. Section 5.3 discusses the SEU error rate expressions for proton-induced SEU, such as that in the inner Van Allen belt.

Protons cannot transfer sufficient energy to cause SEU directly in today's microcircuits. However, SEU can be caused indirectly by secondary particles that are produced in proton-induced nuclear reactions in the device material. This essentially bars using an error rate expression analogous to that for galactic cosmic rays. Instead, the classical integral (over energy) of the product of the proton SEU cross section and the proton flux is used. This cross section is obtained from empirical data fits, involving a threshold energy for the production of SEU by these secondary particles.

Section 5.4 discusses semiempirical methods for obtaining neutron-induced SEU error rates, principally for aircraft cruising at 40,000 ft or higher. This involves the burst generation rate (BGR) method for computing the SEU error rate from the neutron-produced secondary recoil particles from the device material.

Alpha-particle-induced SEU, Section 5.5, is principally that caused by alpha particles penetrating into the chip proper from their origin in the chip package. The earth's crust contains all of the series of radioactive trace actinide elements from uranium to lead. Their decay products are alpha particles principally and are present in device materials from which the package is manufactured. In this case, remedial methods are available, as the range of alpha particles is very short. Two or three mils of paper thickness will stop most alpha particles in this context.

Ground-level SEU, Section 5.6, is mainly due to high-energy particles, such as cosmic rays and their secondaries that have penetrated through the atmosphere from extraterrestrial space down to the ground. They consist principally of neutrons, muons, pions, protons, and electrons. Their SEU error rates relative to those discussed previously are very much less but not negligible.

5.2. Geosynchronous SEU

As many spacecraft are in geosynchronous orbits, the appropriate flux Φ is due to galactic cosmic ray ions. Their altitudes are at the outer fringes of the Van Allen belts; therefore, their proton–SEU interactions are assumed to be negligible. For the geosynchronous environment, the SEU error rate expressions are obtained by using the chord distribution function for the cross section as discussed in Section 1.4. The chords correspond to the individual energy deposition tracks made by the individual incident cosmic ray ion flux particles threading the SEU-sensitive region volume in the device; that is, from Section 1.4, the mean number of chords $\bar{N}_{ch}$ threading the sensitive region, considered as a convex body, is given by

$$\bar{N}_{ch} = S\Phi/4 = \bar{A}_p\Phi \tag{5.1}$$

where S is the body area and $\bar{A}_p = S/4$ of the mean projected area of its volume. Also, $\bar{A}_p = \sigma_p$, where σ_p is an average SEU saturation cross section given as $\sigma_p = \int_{4\pi} \sigma_\infty (\theta, \phi) \, d\Omega/4\pi$ and $\sigma_\infty (\theta, \phi)$ is the asymptotic SEU cross section equivalent to $\lim_{L \to \infty} \sigma(L)$ where L is the LET of the incident ion, and θ and ϕ are the usual polar and azimuthal angles [1].

$f(s)$ is the chord distribution function i.e., $f(s) \, ds$ is the probability density, which yields the probability that a chord in the SEU-sensitive volume has an incremental length spread ds about chord length s. The definition of cosmic ray ion flux Φ is refined to be the isotropic homogeneous integral flux of incident cosmic ray particles, with LET $L(s)$, each of which is one-to-one with a chord threading the sensitive region. A differential cosmic ray flux $\varphi(L(s)) \equiv \Phi'(L(s))$ is called the flux per unit LET. Thus, the expected number of chords in the sensitive volume per unit time, which are those between s_{min} and s_{max} i.e., those that have tracks of sufficient length to deposit enough energy (ΔE_c) to produce SEU in the sensitive region, yield the SEU error rate E_r,

$$E_r = \int_{s_{min}}^{s_{max}} \bar{N}_{ch}(s) f(s) \, ds = \bar{A}_p \int_{s_{min}}^{s_{max}} \Phi(L(s)) f(s) \, ds \tag{5.2}$$

where $\bar{N}_{ch}$ is the mean number of chords (Section 1.4).

From (4.1), the critical energy ΔE_c required to be deposited by a chord track to cause SEU corresponds to a critical LET in the sensitive region, $L_c = \Delta E_c/c$. s_{min}, the integral lower limit, is the shortest chord that still can produce an SEU (i.e., $s_{min} = \Delta E_c/L_{max}$). L_{max} is the high LET spectral cutoff of the particular cosmic ray LET spectrum [i.e., $\Phi(L_{max}) \cong 0$ marks the spectrum cutoff bound]. s_{max} is the maximum chord in the sensitive region. For a parallelepiped sensitive region volume of dimensions a, b, and c, $s_{max} = (a^2 + b^2 + c^2)^{1/2}$, its main diagonal length. Integrating the rightmost Eq. (5.2) ($\int u \, dv$) by parts, where $u = \Phi(L(s))$ and $dv = f(s) \, ds$, gives

$$E_r = \bar{A}_p \left\{ \left[-\int_{s(L)}^{s_{max}} f(s') \, ds' \right] \Phi(L(s)) \bigg|_{s_{min}}^{s_{max}} + \int_{L_{min}}^{L_{max}} \varphi(L) \int_{s(L)}^{s_{max}} f(s') \, ds' \, dL \right\} \tag{5.3}$$

It is seen that the first term vanishes, because $\Phi(L(s_{min})) \equiv \Phi(L_{max}) = 0$. Equation (5.3) yields, with $\bar{A}_p = S/4$,

$$E_r = \left(\frac{S}{4}\right) \int_{L_{min}}^{L_{max}} \varphi(L) \, C(s(L)) \, dL \tag{5.4}$$

where $C(s(L)) = \int_{s(L)}^{s_{max}} f(s') \, ds'$ is the cumulative probability that a chord length lies between s and s_{max}. L_{min} is the minimum LET required to produce an SEU (i.e., $L_{min} = \Delta E_c/s_{max}$). Equation (1.37) is an approximation to $C(s)$ for the domino (thin)

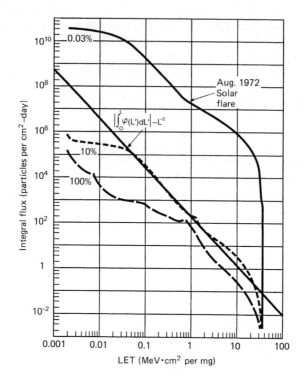

Figure 5.1. Heinrich curve in silicon of cosmic ray abundance versus LET for three different geosynchronous environments. From Ref. 3.

parallelepiped, (i.e., for $3c \ll a, b$, the de facto representation of the SEU-sensitive region volume), as discussed in Section 1.4. Because $L = \Delta E_c/s$ and $L_c = \Delta E_c/c$, then $c/s = L/L_c$, which is used to rewrite that expression for $C(s)$ to give

$$C(s) = \begin{cases} \dfrac{3}{4}\left(\dfrac{L}{L_c}\right)^{2.2}, & L \leq L_c \\ 1 - \dfrac{1}{4}\left(\dfrac{L_c}{L}\right), & L > L_c \end{cases} \quad (5.5)$$

An approximation to $\varphi(L)$ that corresponds to the "Adam's 10% Environment," where $\varphi(L)$ is only exceeded (worse) 10% of the time, as in the Heinrich curve of Fig. 5.1, is given by [2]

$$\varphi(L) = 5.8 \times 10^8/L^3 \text{ (particles cm}^{-2}\text{ da}^{-1}\text{LET}^{-1}) \quad (5.6)$$

Frequently, it is called the "Adam's 90% Environment", and LET is expressed as MeV cm^2/gm there as opposed to Fig. 5.1.

The Heinrich curve of Fig. 5.1 depicts the integral cosmic ray spectrum flux versus LET for three geosynchronous environments. The curve labeled "100%" refers to a solar maximum condition, and the cosmic ray environment is always worse. This is because the corresponding solar wind is at a very high level, thus maximally shielding the earth from galactic cosmic rays. The 10% curve combines solar minimum galactic cosmic rays, and solar particle (proton) activity, thus, the environment is worse only 10% of the time. This is the "Adams 10% environment". The 0.03% curve is that for an anomolously large flare, whose particle output temporarily dominates, by far, all the other near-earth cosmic ray sources. Van Allen belt protons and electrons are not included in these curves, as the geosynchronous region is considered at the outer fringes of the Van Allen belts.

Now, insert (5.5) and (5.6) in (5.4) and partition the limits of integration to accommodate the constraints on $C(s(L))$ in (5.5). Then, approximate the domino parallelepiped surface area as $S = 2(ab + bc + ac) \cong 2ab = 2\sigma_{sat}$. $\sigma_{sat} = ab$ is the asymptotic or saturation SEU cross section (maximum) which corresponds to the face area of the parallelepiped. From (5.4), this yields

$$E_r = 5.8 \times 10^8 \left(\frac{\sigma_{sat}}{2}\right)\left\{\left(\frac{3}{4}\right)\int_{L_{min}}^{L_c}\left(\frac{dL}{L^3}\right)\left(\frac{L}{L_c}\right)^{2.2} + \int_{L_c}^{L_{max}}\left(\frac{dL}{L^3}\right)\left[1 - \frac{1}{4}\left(\frac{L_c}{L}\right)\right]\right\} \quad (5.7)$$

After integrating, neglecting terms involving L_{max} because the corresponding flux contributes little to E_r, and using $L_c = Q_c/c$, dimensionally, $[L_c] = \text{pC } \mu\text{m}^{-1}$, and $[\sigma_{sat}] = \mu\text{m}^2$, finally gives the SEU estimation formula for geosynchronous SEU called the "figure of merit" formula [2]

$$E_r = 5 \times 10^{-10} \frac{\sigma_{sat}}{L_c^2} = 5 \times 10^{-10} \frac{abc^2}{Q_c^2} \quad (5.8)$$

with dimensions of SEU error rate per sensitive region per unit time, where a, b, and c are given in microns and Q_c in picocoulombs. Parenthetically, for silicon, a LET of 1 pC μm^{-1} is equivalent to a LET of 96.6 MeV cm^2 mg^{-1}, and for GaAs, 1 pC $\mu\text{m}^{-1} \cong 56.4$ MeV cm^2 mg^{-1}. The dimensions of E_r depend on the dimensions of the differential flux $\varphi(L)$. In (5.6), $\varphi(L)$ has dimensions of particles cm^{-2} da^{-1} per LET, and so E_r has the same time dimensions.

It is realized that (5.8) as derived from (5.2) yields an approximate expression for the SEU error rate, including the multipliers from Table 5.2 for extrapolation to other environments—mainly solar flares. As seen, the major approximations made were for the cumulative chord distribution function $C(s(L))$ for domino-type parallelepiped sensitive regions of small thickness compared to their faces, as well as an approximate differential LET expression $\varphi(L)$ for the 90% worst-case LET spectrum.

Table 5.1. CREME Program Environment Options

M	Flare conditions	Environment description
1	No flare	Basic condition, galactic cosmic rays only
2	No flare	Ml + fully ionized anomolous component
3	No flare	Ml + solar + interplanetary flux to reach 90% worst-case levels
4	No flare	Ml + singly ionized anomolous component
5	Flare	Peak ordinary flare flux, average heavy-ion composition
6	Flare	Peak ordinary flare flux, worst-case heavy-ion-rich composition
7	Flare	Peak 90% worst-case flare flux, average heavy-ion composition
8	Flare	Peak 90% worst-case flare flux, worst-case heavy-ion-rich composition
9	Flare	Peak Aug. 4, 1972 flare flux, average heavy-ion composition
10	Flare	Peak Aug. 4, 1972 flare flux, heavy-ion-rich composition
11	Flare	Peak composite worst-case flare flux, average heavy-ion composition
12	Flare	Peak composite worst-case flare flux, worst-case heavy-ion-rich composition

Source: Ref. 5.

As is well known, computer programs have been written for the SEU error rate to achieve greater accuracy than the above approximate formulas. A predominant program is CREME [5] (Cosmic Ray Effects on MicroElectronics). This program takes advantage of the exact multiterm expression for the chord distribution function $f(s)$ [4] by numerically integrating (5.2) without proceeding to (5.8), as discussed earlier. The program contains up to a dozen environment options as given in Table 5.1. Equation (5.2) is transformed into a more

Table 5.2. Multipliers (N_c) and Equivalent Days for Petersen SEU Error Rate Estimation, to Extrapolate to Various Environments (i.e., $E_r = 5 \times 10^{-10} N_c abc^2/Q_c^2$ Errors per Bit Day

SER (k)	Model	N_c	T_{eq} Equiv. days	Remarks
(1)	Petersen model	1.0	1.0	SEU figure of merit, Ref. 2
(2)	Galactic model	0.44	0.44	Solar minimum, Ref. 3
(3)	IMP-8 flare	11	9	Heavy-ion spectrum, Ref. 6
(4)	"Ordinary" flare	13	13	Ordinary solar flare, Ref. 3
(5)	90% Worst-case flare	33	61	Adams 10% Worst case Ref. 3
(6)	Anomolously large (A.L.) flare	5000	4000	Adams A.L. flare Ref. 3

Source: Ref. 6.

convenient expression for use in CREME by the following: (a) changing variables from the path length s to the LET variable through $L = Q_c/s$, so that $ds = -Q_c\, dL/L^2$. (b) The integral fluence environments $\Phi(L)$ have dimensions of particles cm^{-2} sr^{-1} s^{-1}. In (5.2), an integration is then made over all solid angles, eliminating the angular dependence. As $\Phi(L)$ is assumed isotropic, the integration yields a factor of 4π, which becomes a second multiplier in front of the integral. With $\bar{A}_p = S/4$ (Section 1.4), the corresponding coefficient is now $4\pi\bar{A}_p = \pi S = \pi\sigma_{\text{sat}}$, where σ_{sat} is the saturation cross section in the limit of the parallelepiped volume when its thickness c approaches zero in the usual approximation. (c) To convert from ionization energy to charge [i.e., 22.5 MeV pC^{-1} of charge produced is used to include the critical charge Q_c explicitly in (5.2)]. (d) Then the lower limit of the integal in (5.2), now $L_{\min}$ because of the above variable change, becomes $22.5Q_c/s_{\max}$, where $s_{\max}$ is in microns and Q_c is in picocoulombs. Correspondingly, $L_{\max}$ is chosen sufficiently high (~ 100 MeV/cm^2/mg^{-1}) because the $\Phi(L_{\max})$ spectrum cutoff occurs at a much lesser value of LET. The integral in (5.2) finally becomes

$$E_r = 22.5\,\pi\sigma_a Q_c \int_{22.5Q_c/s_{\max}}^{L_{\max}} \Phi(L) f(s(L))\, dL/L^2 \qquad (5.8')$$

Again, this very integral is used in the CREME code to compute the SEU error rates for the environments given in Table 5.2 [5].

Table 5.2 provides comparison factors between the Petersen [2] figure of merit error rate formula and other SEU environments. The N_c column gives multiplying factors for the Petersen formula for users to correspond to their environment of interest. For example, for an anomolously large solar flare environment—the worst possible in terms of the highest error rate—N_c is ~ 5000. The SEU error rate in this environment is given as $E_r \cong (5000)$ $(5 \times 10^{-10})\,abc^2/Q_c^2 = 2.5 \times 10^{-6}\,abc^2/Q_c^2$ errors per bit day.

The column "equivalent days" gives the number of days needed to integrate the Petersen model SEU flux (ϕ_p) to equal the actual fluence from any of the other models in lines (2)–(6); that is, the equivalent number of days T_{eq} is obtained from

$$\int_0^{T_{\text{flr}}} \phi_k(t)\, dt = \int_0^{T_{\text{eq}}} \phi_p(t)\, dt \qquad (5.9)$$

where

$\phi_k(t)$ is the flux corresponding to the models, $k = (2)$–(6), other than the Petersen model;

$\phi_p(t)$ is the Petersen model flux; Table 5.2, line (1);

T_{flr} is the number of actual days duration of the ϕ_k flux;

T_{eq} is the number of equivalent days for the Petersen model flux.

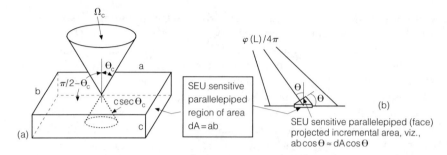

Figure 5.2. (a) Cosmic ray flux incident on sensitive region parallelepiped. Track LET L must satisfy $L \geq L_c$, where $L_c = c \sec \theta_c$, to produce SEU. (b) Cosmic ray flux per unit solid angle incident on parallelepiped sensitive region.

For example, for the anomalously large flare flux $\phi_{A.L.}$ and an assumed constant ϕ_p, using Table 5.2,

$$\int_0^{T_{flr}} \phi_{A.L.} dt = \int_0^{T_{eq}} \phi_p dt = 4000 \phi_p \qquad (5.10)$$

If $\phi_{A.L.}$ is also constant, integrating (5.10) gives $T_{flr} = 4000 \, (\phi_p/\phi_{A.L.}) = 4000/5000 = 0.8$ day.

For geosynchronous SEU error rate computations, there is a variant derivation and resulting E_r that does not use the chord distribution function $f(s)$ [7]. Instead, it uses the more usual flux–cross section product integral as discussed previously. It is based more directly on the incident cosmic ray flux particles that cause SEU by virtue of their direction and track lengths. Referring to Fig. 5.2a, it is evident that the differential cosmic ray flux that has the required energy to deposit to produce a critical charge Q_c and incident radiation to cause SEU lies outside of a cone of solid angle Ω_c whose half-apex angle θ satisfies $0 \leq \theta \leq \theta_c$. This is merely a restatement of the fact that to produce an SEU, the incident particle track must be equal to or greater than $L \sec \theta_c$ as seen in Fig. 5.2a.

Let $\varphi_e(L)$ be the differential flux that corresponds to incidence on the parallelepiped at angles $\theta \geq \theta_c$. Assuming that the total differential flux $\varphi_L(L)$ is isotropic, homogeneous, and incident on both parallelepiped faces for all solid angles, then,

$$\varphi_e(L) = 2 \int_{\theta_c}^{\pi/2} \left(\frac{\varphi_L(L)}{4\pi} \right) \cos \theta \, d\Omega \qquad (5.11)$$

where the factor of 2 accounts for both parallelepiped faces, and $\cos \theta$ allows for the projection of its faces normal to the incident flux direction, as in Fig. 5.2b.

Integrating (5.11) gives

$$\frac{\varphi_e(L)}{\varphi_L(L)} = \left(\frac{1}{2\pi}\right) \int_0^{2\pi} d\phi \int_{\theta_c}^{\pi/2} \cos\theta \sin\theta \, d\theta = \tfrac{1}{2}\cos^2\theta_c \qquad (5.12)$$

With $\cos\theta_c = L/L_c$ from (4.2),

$$\varphi_e(L) = \begin{cases} \tfrac{1}{2}\varphi_L\left(\dfrac{L}{L_c}\right)^2, & L < L_c \qquad (5.13a) \\ \tfrac{1}{2}\varphi_L, & L \geq L_c \qquad (5.13b) \end{cases}$$

Equation (5.13b) implies that where the incident particle LET $L \geq L_c$, all such flux particle tracks cause SEU, as their tracks are longer than necessary with respect to their energy deposition capabilities. However, for $L < L_c$, (5.13a), the track incidence angles θ must satisfy the critical angle criterion [viz. $\theta = \theta_c = \cos^{-1}(L/L_c)$] to cause an SEU. This also implies that for both parallelepiped faces, the corresponding $\Omega_c = 2\,(2\pi) \int_{\theta_c}^{\pi/2} \sin\theta\, d\theta = 4\pi\cos\theta_c = 4\pi(L/L_c)$, which is the solid angle outside the cones in Fig. 5.2a and which diminishes for decreasing L/L_c. Hence, in that case, fewer particles are available to produce SEU, as given by (5.13a). Therefore, from (5.13a), and (5.13b), the integral flux $\Phi_e(L_c)$ is given by

$$\Phi_e(L_c) = \int_{L_{\min}}^{L_{\max}} \varphi_e(L)\,dL = \begin{cases} \tfrac{1}{2}\int_{L_{\min}}^{L_c} \varphi_L(L)\left(\dfrac{L}{L_c}\right)^2 dL + \tfrac{1}{2}\int_{L_c}^{L_{\text{cutoff}}} \varphi_L(L)\,dL, & 0 \leq L_c \leq L_{\text{cutoff}} \\ \tfrac{1}{2}\int_{L_{\min}}^{L_{\text{cutoff}}} \varphi_L(L)\,dL, & L_c > L_{\text{cutoff}} \end{cases}$$
$$(5.14)$$

where the integration limits have been partitioned as before, and L_{cutoff} marks the high LET end of the LET spectrum.

For an average SEU cross section $\bar{\sigma}_L$, the estimated SEU error rate is proportional to

$$E_r \sim \bar{\sigma}_L \phi_e(L_c) \qquad (5.15)$$

To evaluate the integrals in (5.14), a differential flux LET spectrum fit is given by [7]

$$\varphi_L(L) = \begin{cases} \dfrac{65.1}{L^2}, & 0.07 \leq L \leq 1.0 \text{ MeV cm}^2\text{ mg}^{-1} \\ \dfrac{434}{L^{3.8}}, & 1.0 < L \leq L_{\text{cutoff}} \text{ MeV cm}^2\text{ mg}^{-1} \end{cases} \qquad (5.16)$$

This ϕ_L corresponds to the flux maximum condition that occurs during the sunspot minimum. For $L_{cutoff} = 40$ MeV cm^2 mg^{-1}, inserting (5.16) into (5.14) and using (5.15) gives the SEU error rate in dimensions of errors per cell per day:

$$E_r \cong \begin{cases} 3 \times 10^{-10} \dfrac{\bar{\sigma}_L}{L_c^2}, & 1 \leq L_c \leq 40 \text{ MeV cm}^2 \text{ mg}^{-1} \\ 2.87 \times 10^{-10} \dfrac{\bar{\sigma}_L}{L_c^2}, & L_c > 40 \text{ MeV cm}^2 \text{ mg}^{-1} \end{cases} \quad (5.17)$$

In place of $\bar{\sigma}_L/L_c^2$, abc^2/Q_c^2 can be used above, where a, b, and c dimensions are in microns and Q_c dimensions are in picocoulombs, similar to (5.8). For a variable cross section $\sigma_L(L)$, (5.15) can be written as [6]

$$E_r \sim \int_{L_{th}}^{\infty} \Phi_e(L) \, d\sigma_L(L) \quad (5.18)$$

This is similar to the development as given in Section 1.3.

The heavy-ion SEU cross section corresponding to the parallelepiped SEU-sensitive volume can also be derived from first principles [8]. The resulting SEU error rate is given as the classical integral of the cross section–incident flux product; that is, a differential cross section $d\sigma_{SEU}(\theta, \phi, \Delta E_c)$ for the SEU-sensitive volume shown in Fig. 5.2' can be defined as an incremental lamina of the parallelepiped that is projected normal to the incident flux [8]. The integration of $d\sigma_{SEU}$ to obtain $\sigma_{SEU}(\theta, \phi, L(s), \Delta E_c)$, the integral cross section, is straight forward [48]. It results in a quadratic polynomial in s_{min}, the minimum SEU chord length, [i.e., $s_{min} = (\Delta E_c/L - s_f) \geq 0$, where s_f is the funnel length]; θ and ϕ are the polar and azimuthal angles shown in Fig. 5.2'. The integration over the y' coordinate yields [8]

$$\sigma_{SEU}(\theta, \phi, L, \Delta E_c) = A_p - g \equiv A_p(\theta, \phi) - g_1(\theta, \phi) s_{min} + g_2(\theta, \phi) s_{min}^2 \quad L \geq L_{min} \quad (5.19)$$

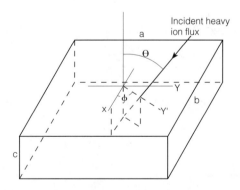

Figure 5.2'. Rectangular parallelepiped sensitive volume showing incident flux and coordinates for computing heavy-ion SEU cross section. From Ref. 8.

where $g = g_1 s_{min} - g_2 s_{min}^2$, $g_1 = [(c \sin 2\phi) \sin \theta + 2 (b \cos \phi + a \sin \phi) \cos \theta] \sin \theta$, and $g_2 = (3/4) [(\sin 2\phi)(\sin 2\theta) \sin \theta]$. g is called the "removal" cross section because it is that part of σ_{SEU} that corresponds to chords shorter than s_{min} in the sensitive volume. They supply insufficient ionization charge to produce SEU. σ_{SEU} can now be substituted in the classical integral expression for the SEU error rate for heavy ions, where the corresponding differential flux φ is not necessarily isotropic. For example, the φ of preponderant concern could be that due to the effects of the solar wind protons coming from the direction of the sun.

The SEU error rate is given by

$$E_r = \int_{4\pi} d\Omega \int_{L_{min}(s)}^{L_{max}} dL' \, \varphi(\theta, \phi, L') \, \sigma_{SEU}(\theta, \phi, L', \Delta E_c) \quad (5.20)$$

Inserting the isotropic differential flux approximation for φ [viz. $\phi_{fm} = 6 \times 10^8 /L^3$ as done in (5.6)], in the integral yields the error rate whose first term is equivalent to the "figure of merit" error rate (5.8). This can be seen from the resulting integration of (5.20) over the L variable to give

$$E_r = 6 \times 10^8 \left(\frac{c_2}{2L_c^2} - \frac{c_3}{3L_c^3} + \frac{c_4}{4L_c^4} \right) \quad (5.21)$$

where the integration is taken to L_c, with the remainder (L_c to L_{max}) deemed negligible due to the rapid $1/L^3$ drop in the flux approximation. Also, the funnel length s_f is not included, and the following integrations over solid angle are confined to the first octant as in (1.32). So that

$$c_4 = (\Delta E_c)^2 \int g_2(\theta, \phi) d\Omega, \quad c_3 = (\Delta E_c) \int g_1(\theta, \phi) d\Omega, \quad c_2 = \int A_p d\Omega = \frac{\pi \bar{A}_p}{2} \cong \frac{\pi \sigma_{sat}}{8} \quad (5.22)$$

where $\sigma_{sat} = ab$, the area of the parallelepiped face and $a, b \gg c$. Note that the first term of E_r in (5.21) becomes $6 \times 10^8 \pi ab/L_c^2$, which yields the figure of merit expression for the error rate within a constant. The second and third terms are corrections for the corresponding "removal" cross sections. The integrations result in $c_4 = (3/8)(\Delta E_c)^2$ and $c_3 = (2/3)(\Delta E_c)(a + b + c)$. Graphical comparisons between (5.21) and the figure of merit error rate (5.8) essentially show coincidence [8], as in Fig. 5.2''. This also implies that the SEU cross section and the cumulative chord distribution function $C(s)$ are indeed equivalent. This is seen from the geometric probability methods used to obtain $C(s)$.

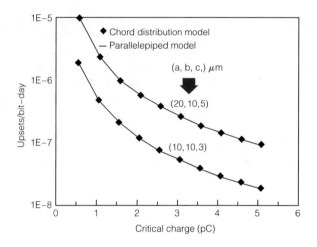

Figure 5.2''. Comparison of the parallelepiped cross section model (5.20) and chord distribution model (5.2) for error rate prediction. The curves describe parallelepiped model results; the diamonds correspond to the chord distribution model. From Ref. 8.

It is written in four terms, each dependent on the magnitude of s_{min}; that is, for $0 < s_{min} \leq c$,

$$C(s_{min}) = \left(\frac{8}{\pi S}\right)\left(ab + \frac{c(a+b)\pi}{4} - \frac{2}{3}(a+b+c)s_{min} + \frac{3}{8}s_{min}^2\right) \quad (5.23)$$

[9] which is seen to be a quadratic polynomial in s_{min} as is its equivalent σ_{SEU} in (5.19). For $c \ll s_{min} \leq a$,

$$C(s_{min}) = \left(\frac{8}{\pi S}\right)(a+b)c^2\left[\left(\frac{(\pi/4)-1/2}{c} - \frac{1}{4(a+b)}\right) - \frac{1}{2s_{min}}\right.$$
$$\left. + \frac{\frac{1}{4}[c + \pi ab/(a+b)]}{s_{min}^2} - \frac{c^2}{12s_{min}^3}\right] \quad (5.24)$$

[9] which is a cubic polynomial in s_{min}^{-1}. The corresponding two expressions for $C(s_{min})$ over the remaining two ranges $a < s_{min} \leq b$ and $b < s_{min} \leq (a^2+b^2+c^2)^{1/2}$ are available [9]. However, they contribute only 3% to the overall $C(s_{min})$ for $c < a/3, b/3$, which is the case for "domino" parallelepiped sensitive regions of interest.

An SEU error rate model also can be derived in terms of the incident particle energy instead of its LET, to account for the nonlinear dependencies between range and energy. One example of this occurs during the investigation of the effect of incident-particle-induced spallation on the SEU error rate in spacecraft avionics [10].

Referring to Fig. 5.2' showing a spherical coordinate system atop the parallelepiped sensitive volume of face area A, it is seen that dN, the rate of particles in an energy increment incident on it, is within a constant given by

$$dN(E) = \left[\int \varphi(E) A \cos\theta \, d\Omega\right] dE = A\varphi(E) \int_0^{2\pi} d\phi \int_0^{\pi/2} \cos\theta \sin\theta \, d\theta \, dE = \pi A \varphi(E) \, dE \tag{5.25}$$

$A\cos\theta$ is the projection of the sensitive volume face in the direction of incident flux at that angle θ. However, $\varphi(E)$ is assumed isotropic herein. Because the distribution of chord lengths s is given by $f(s)$, dn_i, the rate of incident type-i particles whose chord lengths lie in ds about s, is from (5.25),

$$dn_i(E) = \pi A f(s) \, ds \, \varphi_i(E) \, dE \tag{5.26}$$

Let $R_i(E)$ be the range of type-i particles of energy E and let E_{los} be the energy lost by these particles along their chords in the sensitive volume. Then, the residual range of these particles after their chord traversal therein is given by $R_i(E - E_{\text{los}})$, where $E - E_{\text{los}}$ is the corresponding residual (remaining) energy. As discussed in Sections 7.6 and 7.7, it is seen that

$$R_i(E - E_{\text{los}}) = R_i(E) - s \tag{5.27}$$

This simply asserts that the residual range is given by the range at incidence prior to traversing the chord minus the range in the chord (i.e., the chord length per se). Considering R_i as an operator with an inverse R_i^{-1} such that $R_i R_i^{-1} = R_i^{-1} R_i = I$, where I is the identity operator, (5.27) yields

$$E_{\text{los}} = E - R_i^{-1}[R_i(E) - s] \tag{5.28}$$

Note that neither R_i or R_i^{-1} has the distributive property; that is,

$$R_i(E - E_{\text{los}}) \neq R_i(E) - R_i(E_{\text{los}}), \qquad R_i^{-1}[R_i(E) - s] \neq E - R_i^{-1}(s) \tag{5.29}$$

$R_i^{-1}(r)$ can be construed as the residual energy of a particle possessing a residual range r (Section 3.2). For an SEU to occur, there must be a critical chord length s_c and corresponding critical energy loss E_{clos} in the sensitive volume. The latter is given from (5.28), analogously as [10]

$$E_{\text{clos}} = E - R_i^{-1}[R_i(E) - s_c(E)] \tag{5.30}$$

Extracting s_c from (5.30) yields

$$s_c(E) = R_i(E) - R_i[E - E_{\text{clos}}] \quad (5.31)$$

Integrating (5.26) over all feasible chord lengths results in the error rate E_{ri} due to type-i incident particles, namely

$$E_{ri} = \pi A \varphi_i(E) \, dE \int_{s_c}^{s_{\max}} f(s) \, ds = \pi A \varphi_i(E) \, dE \, C(s_c(E)) \quad (5.32)$$

where $C(s)$ is the cumulative chord length distribution function. The upper limit, $s_{\max}$, is the sensitive volume parallelepiped main diagonal, which corresponds to the cosmic ray particle spectrum high-energy cutoff. At high cosmic ray energies, only those incident particles that deposit energy $E \geq E_{\text{clos}}$ after traversing chords whose lengths are limited by $s_{\max}$ will produce SEUs. At low cosmic ray particle energy, only incident particles of energy E will cause SEU if they traverse chord lengths $s \geq s_c$, hence the lower limit in (5.32). To generalize to all cosmic ray particles for all energies, (5.32) yields the grand SEU error rate as

$$E_r = \pi A \sum_i \int_0^\infty C(s_c(E)) \varphi_i(E) \, dE \quad (5.33)$$

Again, this form of the SEU error rate expression is employed especially to include fragmentation (spallation) processes [10]. They occur mainly for cosmic ray particles whose energies are at the low end of the cosmic ray energy spectrum, usually corresponding to those in the low-earth-orbit (LEO) environment [10]. The variable change in (5.33) from energy to LET (i.e., $L = E_{\text{clos}}/s$) yields (5.8′) as shown, which is the SEU error rate expression form used in the CREME code discussed earlier.

5.3. Proton-Induced SEU

Proton-induced SEU is of importance because its major source environments include the Van Allen belts, where an increase in the number of orbits are planned for future spacecraft. Many present spacecraft orbits pass through the Van Allen belts but few, if any, lie wholly within them. The main Van Allen belt particles consist of protons and electrons, with some heavy ions trapped therein as well. The detailed structure and flux levels of the belts are discussed in Section 2.6 [13, 14]. The source of at least the inner belt is due to cosmic-ray-induced neutron production reactions with the rarified extraterrestrial atmosphere at those altitudes. The neutrons, with a half-life of about 12

Table 5.3. Proton-Induced Spallation Reactions in Device Silicon

	Reaction	L_{max}		L_{max}		L_{max}
(1)	$^{28}_{14}Si + ^{1}_{1}P = ^{1}_{1}P$	0.5	$+ ^{4}_{2}\alpha$	1.5	$+ ^{24}_{12}Na$	11
(2)	$^{28}_{14}Si + ^{1}_{1}P = ^{1}_{1}P$	0.5	$+ ^{12}_{6}C$	6	$+ ^{16}_{8}O$	7.5
(3)	$^{28}_{14}Si + ^{1}_{1}P = ^{1}_{1}P$	0.5	$+ 2(^{14}_{7}N)$	6.5		
*(4)	$^{28}_{14}Si + ^{1}_{1}P = ^{1}_{1}P$	0.5	$+ ^{4}_{2}\alpha$	1.5	$+ ^{24}_{12}Mg$	12
(5)	$^{28}_{14}Si + ^{1}_{1}P = $ —	—	$+ ^{4}_{2}\alpha$	1.5	$+ ^{25}_{13}Al$	12
(6)	$^{28}_{14}Si + ^{1}_{1}P = ^{1}_{1}P$	0.5	$+ ^{28}_{14}Si$	14		

* Ex (4): $^{28}_{14}Si + ^{1}_{1}P = ^{1}_{1}P + ^{4}_{2}\alpha + ^{24}_{12}Mg$

Note: L_{max} measured in MeV cm^2/mg

min, then decay to form protons, electrons, and neutrinos. The neutrinos, which are chargeless and virtually massless, disappear into the cosmos, whereas the electrons normally do not have sufficient stopping power to produce SEU, although they, as well as protons, can cause ionizing damage. The second major source of protons are those emanating from the sun in the form of the solar wind, and especially the tremendous increase in their flux level during solar flares. With the advent of submicron feature size devices, protons will have sufficient LET to cause SEU directly as well as indirectly. The latter mode is through nuclear reactions with the device silicon to produce energetic heavy ions that can deposit enough energy to cause SEU. This is shown in Table 5.3 where virtually all of the reaction products can cause SEU, as seen from their L_{max} values in the table.

A very important aspect of proton-induced SEU is that for submicron ICs. The total proton cross section in silicon consists of a sum of cross sections corresponding to a number of proton–silicon reactions. One such reaction that does not produce SEU in present-day "large" ICs is proton-induced direct ionization in silicon. On an individual track basis, energy output from this reaction is insufficient to cause enough ionization charge to lead to SEU, even though its cross section is extremely large for this process at proton energies of interest, such as in the inner Van Allen belt and in the South Atlantic Anomaly where the inner belt can dip to very low altitudes, [51].

Diffuse collective ionization does not usually produce SEU due to the lack of charge concentration onto a sensitive node, as from an ionization track near to or striking a node. The charge collected to cause SEU from proton-induced ionization in current ICs corresponds to energy deposited due to reactions other than direct ionization, such as spallation reactions. These produce relatively heavy ions such as magnesium and aluminum (Table 5.3) which will cause SEU. They yield ionization energy that dwarfs that from direct ionization, even though the latter cross section is extremely large, as already mentioned.

Future ICs will have feature sizes in the deep submicron range. For these, the relatively small energy available from the direct ionization process will never-

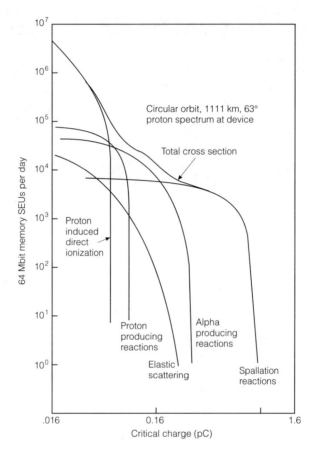

Figure 5.2'''. SEU rate for 64M-bit spacecraft memory as a function of critical charge Q_c. For sufficiently small feature size (Q_c), the SEU rate increases rapidly due to proton-induced direct ionization as seen in the figure [49].

theless be sufficient to induce SEU in them because of their miniscule Q_c. Other reactions mentioned above will still produce SEU with their energy output, but with their lesser cross sections governing their contribution to the SEU rate.

When the submicron feature size reaches approximately 0.3 μm, direct ionization by protons will be able to create SEU. Only about 1 proton in about 10^5 produces a nuclear reaction which leads to a SEU in contemporary devices (indirect ionization). However, when the feature size drops below about 0.3–0.5 μm, the SEU rate will rapidly increase to overwhelm that from nuclear reactions to about three to five orders of magnitude over the latter due to direct ionization.

SEU calculations as shown in Fig. 5.2''' reveal the trend in the very rapidly increasing SEU rate due to direct ionization as the critical charge decreases

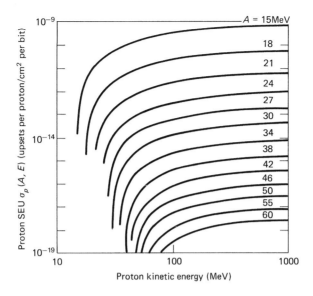

Figure 5.3. Semiempirical one-parameter proton SEU cross section. From Ref. 12.

toward increasing device SEU sensitivity (i.e., toward decreasing feature size in the submicron region) [49].

Parenthetically, it is seen from Fig. 3.1 that proton energy loss varies from about 1 to 3 fC μm^{-1} as the Bragg peak is approached. As the feature size decreases to deep submicron levels, the SEU rate will first increase slightly because several protons can penetrate the device sensitive region within an integration time following the strike in the affected node to result in a single SEU [50]. This is the obverse of a single particle penetrating an IC to cause multiple SEUs. Several particles per single strike is particularly applicable in an anomolously large solar flare environment.

The reactions given in Table 5.3, where the reaction products are sizable nuclei, are called spallation reactions, which are similar to fission. As seen in the table, except for reaction No. (5), the proton acts as a catalyst since it causes the reaction and appears following the reaction seemingly having taken no part in it.

As already mentioned for proton-induced SEU, there is no analogous error rate expression as there is for heavy-ion geosynchronous orbit environments. This is primarily because protons mainly cause SEU indirectly through various complex nuclear reactions that produce SEU-inducing ions as in Table 5.3. However, there are a few methods available to calculate proton-induced SEU using accelerator data, as will be discussed in the following [12, 15]. The method for obtaining SEU interaction rates is to integrate the product of the cross section with the environmental flux (Fig. 5.3) over an appropriate range of LET or energy values. A quite accurate semiempirical one-parameter proton SEU cross section expression has been constructed for this purpose. It is given by [12, 15]

$$\sigma_p(A, E)(10^{-12}\text{ cm}^2) = \begin{cases} (24/A)^{14}\{1-\exp-(0.18)(18/A)^{1/4}(E-A)^{1/2}\}^4, & E > A \\ 0, & E \leq A \end{cases}$$
(5.34)

where E is the energy of the incident proton, and the single parameter A (MeV) is the proton SEU cross-section energy threshold. A is normally unique to the microcircuit of interest, and details for obtaining A are given later.

With the known proton differential flux $\Phi_p(E)$ at environment altitudes that include the Van Allen belts [13, 14], and containing a built-in attenuation factor to account for the spacecraft material shielding thickness, the corresponding proton-induced SEU error rate is given by [15]

$$E_{\text{pr}}(A) = \int_A^\infty \sigma_p(A, E)\Phi_p(E)\,dE$$
(5.35)

Equation (5.35) is numerically integrated to yield E_{pr}, and various results are depicted in Figs. 5.4–5.7. Figure 5.5 compares E_{pr} with that due to heavy ions at geosynchronous altitudes, where it is seen that the proton SEU peaks exceed the latter SEU levels in the inner Van Allen belt region.

As an aside, Figs, 5.4 and 5.5 show a maximum in the error rate, implying a Van Allen belt proton flux peak at about 1500 nmi for a 60° inclination

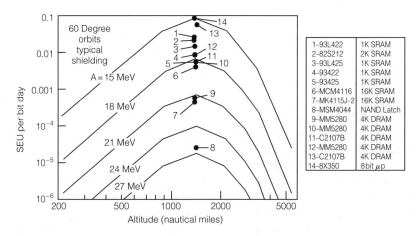

Figure 5.4. Proton-induced SEU versus altitude for a 60° spacecraft orbit using the one-parameter cross section fit for various semiconductor microcircuits. Same parts with more than one point on graph correspond to different vendors. From Ref. 15.

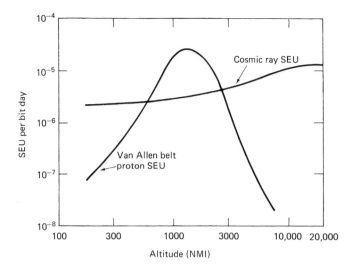

Figure 5.5. Comparison of geosynchronous cosmic ray SEU with that of proton-induced SEU for a 60° spacecraft orbit for a moderately sensitive ($A = 25$) microcircuit. From Ref. 11.

angle orbit. Compare this peak with those in Fig. 2.17 which shows proton peaks at 1.5 (6000 mi) and 3 (12,000 mi) earth radii, for equatorial orbits (0° inclination angle). This shows the large differences in the proton peak altitudes with inclination angle; this is due to the earth's magnetic field drop, for a given altitude, near the magnetic pole singularities (60° degree inclination) which allows the proton peak flux level to drop in altitude by a factor of 4–8 as well.

Figures 5.4, 5.6, and 5.7 depict E_{pr} as a function of spacecraft altitude for various devices, orbit inclination angles, and solar minima and maxima. It is seen from Figs. 5.6 and 5.7 that, for a given altitude, E_{pr} does not change much as a function of orbit inclination angle, except for the 30° inclination during a solar maximum at low altitudes. Such relatively low inclination angles correspond to near-equatorial orbits, and are therefore, influenced maximally by the interaction of the earth's magnetic field with the solar wind. This is especially the case at low earth orbit (LEO) altitudes, as the earth's magnetic field strength increases as the orbit drops well below the Van Allen belts. Again, this is in contrast to the higher inclination orbits, which are influenced by their being in the vicinity of the magnetic poles.

Further, for high proton fluxes, and other low-atomic-number (Z) particles such as high fluxes of deuterons and tritons, a multiple-particle coherent SEU can occur. This is in contradistinction to a single particle-induced multiple-bit SEU. The former is due to the coherent action of two or more low-Z, high-flux particles to produce one SEU, thereby enhancing the corresponding (coherent) SEU cross section [16, 17].

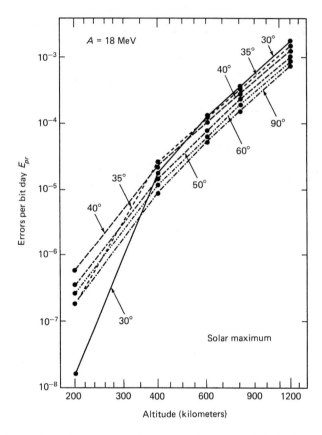

Figure 5.6. Proton solar maximum SEU error rates versus altitude. From Ref. 15.

A refined approach to the preceding semiempirical proton-induced SEU cross section is obtained from a two-parameter model [18], namely,

$$\sigma_p(A, B; E) = (B/A)^{14}\{1 - \exp[-(0.18)](18/A)^{1/4}(E - A)^{1/2}\}^4, \quad E > A \tag{5.36}$$

in the same units as $\sigma_p(A, E)$ in (5.34). Comparing (5.34) and (5.36), it is seen that the second parameter B is related to the two cross sections as

$$\sigma_p(A, B; E) = (B/24)^{14}\sigma_p(A, E) \tag{5.37}$$

For the newer micron and submicron feature size, small-geometry devices, the two-parameter cross section gives a better fit to measured cross sections [19]. Its A parameter correlates with the apparent SEU cross-section energy thresh-

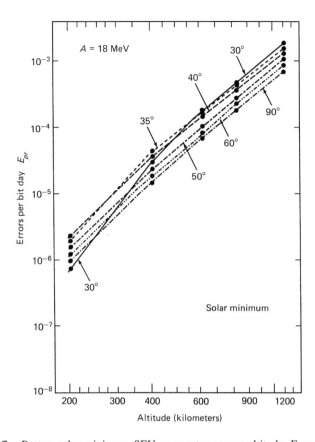

Figure 5.7. Proton solar minimum SEU error rates versus altitude. From Ref. 15.

old, whereas the B parameter aids in fitting the asymptotic (large E) cross-section values, because $\sigma_p(A, B; \infty) \sim (B/A)^{14}$ and that near the energy threshold A, $\sigma_p(A, B; E) \sim (E - A)^2$.

As discussed, the proton-induced SEU error rate E_{pr} is given by an integral as in (5.35) of the product of the proton SEU cross section $\sigma_p(A, E)$ and the proton flux $\Phi_p(E)$. The proton flux, besides being a function of incident proton energy, depends on the environment as expressed in terms that include the orbit inclination angle (with respect to the equator), the altitude, the South Atlantic Anomaly discussed in Section 6.7, as well as the two extremes of solar activity reflected in the solar wind.

The proton SEU cross section is a function of the incident proton energy and the cross-section threshold energy A; that is, for $E < A$, there is insufficient proton energy to initiate the proton SEU-inducing reactions. To compute the SEU error rate for a given part in a spacecraft, whose orbit lies in or near the

Table 5.4. Proton Energy Threshold A for Various Devices

Ser.	Part Name	Mfr.	Function	Energy thresh. A	Technology
1	93L422	FSC	1 K SRAM	9.88	TTL
2	7164	IDT	64 K SRAM	29.85	NMOS
3	93ZL422	FSC	1 K SRAM	10.35	TTL
4	2901B	AMD	4Bit Slice	14.16	TTL
5	8X350	SIG	8Bit Microp	14.96	LSTTL
6	2909A	AMD	4Bit Microp	15.60	TTL
7	2909	AMD	4Bit Microp	16.39	TTL
8	93L422	AMD	1 K SRAM	16.42	TTL
9	9407	FSC	9Bit Reg	16.58	IIL
10	C2107B	INTEL	4 K DRAM	16.77	NMOS
11	93L425	AMD	1 K SRAM	17.04	TTL
12	93422	FSC	1 K SRAM	17.66	TTL
13	93425A	FSC	1 K SRAM	18.00	TTL
14	MCM4116-20	MOT	16K DRAM	18.68	NMOS
15	MM5280	NSC	4 K DRAM	18.98	NMOS
16	9407	FSC	9Bit Reg	20.56	TTL
17	54LS395	SIG	4Bit Sh Reg	>21.4	LSTTL
18	54LS374	TI	Octal F/F	>21.4	LSTTL
19	MK4116J-2	MOST	16K DRAM	21.66	NMOS
20	54LS169	AMD	4Bit Cntr	22.80	LSTTL
21	MSM4044	SSS	4Bit NAND	26.40	CMOS
22	HM6514	HA	64K DRAM	$>50.$	CMOS

Source: Ref. 21 and 22.

inner Van Allen belt neighborhood, the first step is to obtain the corresponding energy threshold parameter A. Ideally, this can be done by obtaining a single accurate measurement of the proton cross section for the part of interest in the energy range 10–100 MeV. Then A is immediate by using that data point to identify the corresponding curve in Fig. 5.3, a one-parameter family of curves for A. Alternatively, A can be obtained from the now known $\sigma_p(A, E)$ by correlating it with its analytic expression in (5.34) or (5.36). A is elusive if the above cross-section measurement is unavailable for the part of interest. In this case, Table 5.4 can be used for part types that are alike or similar in part technology or part function by interpolating to yield a credible value of A. In this regard, recall that the proton SEU susceptibility decreases with increasing A. As discussed, in the next section, there is a correlation between proton and geosynchronous cosmic ray susceptibilities and A values. If the geosynchronous SEU value exists for this device, it may be very cautiously used to extrapolate an appropriate A value for protons. Such interpolations and extrapolations usually yield only upper or lower bounds for A. This can be a robust approach, as there are extensive data bases of device SEU parameters in the geosynchronous environment as discussed in Chapter 8.

The second step is to obtain the appropriate proton flux spectrum [i.e., $\Phi_p(E)$] for a particular orbit altitude, inclination angle, and sunspot activity extrema epoch. This is usually available from certain computer orbit environment programs, such as CREME [23]. The third step is to obtain the distribution of shielding materials surrounding the device of interest in the form of attenuation factors with which to multiply $\Phi_p(E)$ to obtain the flux incident on the part per se. The last step is to combine the above factors into (5.35) to realize the SEU error rate by integration.

However, once the energy threshold A is obtained, Tables 5.5 and 5.6 can be used to get the proton-induced SEU error rate $E_{pr}(A)$ directly for various orbit parameters. For example, assume an orbit with the following parameters: 800 km altitude, 35° inclination angle, and a solar minimum epoch. The example device is the MK4116J-2 16K NMOS DRAM. From Table 5.4 it is seen that the corresponding parameter $A = 21.66$. To use Tables 5.5 and 5.6 it is necessary to correct for this A value, as tabulations are for $A = 18$ only. This is done by using the empiricsm [15]

$$E_{pr}(A) = E_{pr}(18) \frac{(18/A)^{14}(18+B)}{A+B} \quad (5.38)$$

where B is the correction factor obtained from the last columns in Tables 5.5 and 5.6. From the former table for 800 km altitude, inclination angle of 35° and a solar minimum condition, $E_{pr}(18) = 4.69 \times 10^{-4}$ with a corresponding $B = 23.5$. Inserting this B and A into (5.38) yields a resultant $E_{pr}(21.66) = 3.23 \times 10^{-5}$ errors per bit-day. For parameter values other than those listed in the tables, linear interpolation can be used to obtain them when applicable.

5.3.1. Approximate Methods for Proton-Induced SEU

For protons at normal incidence to the ab face of the parallelepiped SEU sensitive region of thickness c, a model for the corresponding cross section $\sigma_{pseu}(E)$ is given by

$$\sigma_{pseu}(E) = ab\{1 - \exp[-\mu_p(E)]c\}f(E), \qquad \sigma_{pseu}(\infty) = ab \quad (5.39)$$

where $\mu_p(E) = N_{si}\sigma_p(E)$. $\sigma_p(E)$ is the microscopic cross section for protons interacting with silicon to produce SEU-inducing ions in the sensitive volume $V = abc$ as in Table 5.3. N_{si} is the number of silicon nuclei per cubic centimeter. $\mu_p = 1/\lambda_p$ is also the linear attenuation coefficient, where λ_p is the corresponding mean free path in silicon. Equation (5.39) is a reasonable assumption for $\sigma_{pseu}(E)$ because, first, the asymptotic cross section $\sigma_{pseu}(\infty) = ab$. Second,

Table 5.5. Proton SEU Error Rates in Circular Earth Orbits in Van Allen Proton Belts (1980)

Orbit Inclination	Upsets per bit-day for $A = 18$ MeV		B (MeV)	
	Solar min	Solar max	Solar min	Solar max
Altitude: 200 km				
30°	8.20-7	1.58-8	17.5	32.1
35°	2.13-6	1.84-7	12.3	15.4
40°	2.65-6	5.42-7	11.2	12.9
50°	1.68-6	3.17-7	9.3	12.4
60°	1.23-6	2.40-7	10.5	13.5
90°	1.02-6	1.80-7	10.5	12.7
Altitude: 400 km				
30°	3.43-5	1.68-5	22.1	27.5
35°	4.39-5	2.28-5	20.7	23.5
40°	3.97-5	2.13-5	19.0	22.0
50°	2.50-5	1.35-5	18.2	21.4
60°	2.22-5	1.16-5	19.5	22.0
90°	1.66-5	8.58-6	19.6	22.3
Altitude: 600 km				
30°	1.80-4	1.13-4	23.9	25.6
35°	1.81-4	1.15-4	22.3	23.7
40°	1.57-4	1.03-4	21.7	22.7
50°	1.11-4	7.11-5	21.4	22.6
60°	9.13-5	5.73-5	21.8	23.1
90°	7.99-5	5.10-5	22.3	23.4
Altitude: 800 km				
30°	5.04-4	3.59-4	24.6	25.4
35°	4.69-4	3.33-4	23.5	24.5
40°	4.09-4	2.89-4	23.0	23.8
50°	3.03-4	2.13-4	23.1	23.8
60°	2.54-4	1.78-4	23.6	24.5
90°	2.14-4	1.49-4	23.8	24.8
Altitude: 1200 km				
30°	2.08-3	1.62-3	24.1	24.5
35°	1.84-3	1.43-3	23.7	24.1
40°	1.58-3	1.24-3	23.4	23.8
50°	1.23-3	9.56-4	23.6	24.0
60°	1.06-3	8.17-4	23.9	24.3
90°	8.95-4	6.93-4	24.0	24.4

Note: $X.YZ - W \equiv X.YZ \cdot 10^{-W}$
Source: Ref. 21.

Table 5.6. Proton SEU Error Rates for Selected Orbits in High Van Allen Proton Belts

Orbit Inclination	Altitude (km)	Upsets per bit-day for $A = 18$ MeV	B (MeV)
Solar max.; time = 1989.5			
60°	1667	3.03-3	22.6
60°	2593	6.69-3	19.1
60°	3889	3.18-3	11.5
60°	5186	8.19-4	5.0
60°	6389	2.10-4	(0.0)
60°	10371	1.36-6	
Solar max; time = 1981.8			
63°	1111	5.69-4	24.4
Solar min; time = 1985.8			
63°	1111	7.91-4	23.9
Solar max; time = 1989.0			
63°	1111	6.59-4	24.2

Source: Ref. 21.

$\exp(-\mu_p)c = \exp(-c/\lambda_p)$ is the probability that the proton penetrates the device thickness without interacting in the material, as discussed in Section 1.1. Hence, $1 - \exp(-\mu_p)c$ is the interaction probability within c. Third, $f(E)$ is a factor to account for proton interactions that occur inside or in the immediate neighborhood outside the sensitive region to still produce SEU [24]. For energetic protons of energy $E \gg A$, $f(E) > 1$. For protons of energy $E \gtrsim A$ greater than but near the A threshold, $f(E) \ll 1$. It is also claimed that $f(E)$ decreases with increasing LET threshold [24].

Because the proton mean free path λ_p is normally much larger than the thickness c, $\exp(-c/\lambda_p) \cong 1 - c/\lambda_p$. Inserting in (5.39) gives

$$\sigma_{\text{pseu}}(E) = ab\mu_p(E)cf(E) \tag{5.40}$$

or

$$\sigma_{\text{pseu}}/\sigma_{\text{hseu}}c = \mu_p(E)f(E) \tag{5.41}$$

where the heavy-ion cross section $\sigma_{\text{hseu}} = ab$, as the ion path is normal to ab.

The proton SEU cross section $\sigma_p(E, A)$ in (5.34) can be rewritten in units of 10^{-12} cm² as

$$\sigma_p(E, A) = 2.1 \times 10^7 A^{-14}\{1 - \exp-(0.18)(18/A)^{1/4}(E-A)^{1/2}\}^4, \quad E > A \tag{5.42}$$

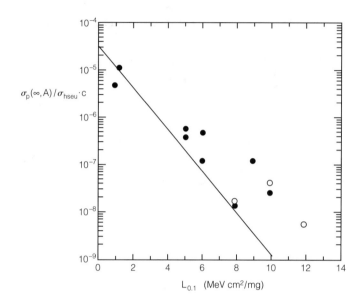

Figure 5.8. Ratio of asymptotic proton SEU cross section to the product of the heavy-ion cross section with device thickness of the parallelepiped sensitive volume. From Ref. 20.

so that its asymptotic form σ_{pasy} is

$$\sigma_{\text{pasy}} = \sigma_p(\infty, A) = 2.1 \times 10^7 A^{-14} \quad (5.43)$$

From this point, guidelines using experimental heavy-ion data are available to obtain the proton-induced SEU error rate [24] and are discussed in Sections 8.3 and 8.7. There is a straightforward simple algorithm [20] by which the proton SEU error rates can ultimately be obtained from heavy-ion LET data to determine the sought-after parameter A, and thus to compute E_{pr} from (5.35). The algorithm is obtained by first fitting the plot of $[\sigma_p(\infty, A)/\sigma_{\text{hseu}}] c$ versus $L_{0.1}$ [LET value at 10% of $\sigma_p(\infty, A)$] in Fig. 5.8 with

$$\left(\frac{\sigma_p(\infty, A)}{\sigma_{\text{hseu}} \cdot c} \right) = 3 \times 10^{-5} \exp(-cL_{0.1}) \quad (5.44)$$

$L_{0.1}$ is also representative of L_c of a median cell in an IC memory array [16]. Then, eliminating $\sigma_p(\infty, A)$ between (5.43) and (5.44) yields A as

$$A \cong 7 \exp\left(\frac{cL_{0.1} - \ln c\, \sigma_{\text{hseu}}}{14} \right) \quad (5.45)$$

Each factor in the exponent is sufficiently small that its exponential can be approximated by the first two terms in its series expansion. This gives

$$A \cong 7\left(1 + \frac{cL_{0.1}}{14}\right)\left(1 - \frac{\ln c\sigma_{\text{hseu}}}{14}\right) \quad (5.46)$$

For the range of σ_{hseu} and parallelepiped dimensions abc of interest, $\sigma_{\text{hseu}} \simeq 8 \times 10^{-3}$ cm^2 and $c = 1\,\mu$m; then, the second term in (5.46), $1 - (1/14)\ln(\sigma_{\text{hseu}}c) \approx 2$. So, (5.46) yields [20]

$$A \cong 14 + L_{0.1} \quad \text{(in units of } c = 1\,\mu\text{m)} \quad (5.47)$$

This simple expression provides the bridge between the device heavy-ion LET threshold ($L_{0.1}$) and the proton SEU energy threshold parameter A to use in computing the corresponding proton-induced error rate for the same device.

It is now seen that devices whose heavy-ion LET thresholds are very low can also suffer proton-induced SEU, for protons whose energies are as low as 14–15 MeV. Recall that the Van Allen belt protons have energies much higher. Actually, a plot of A versus $L_{0.1}$ as in Fig. 5.9 yields a better fit [20] under the

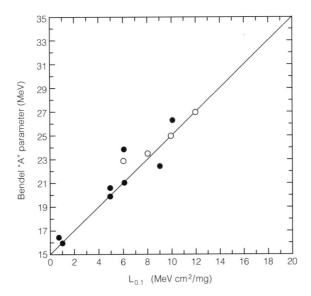

Figure 5.9. The Bendel "A" parameter plotted as a function of the LET threshold $L_{0.1}$. From microcircuit data it is found that the line can be fitted by $A = L_{0.1} + 15$. From Ref. 20.

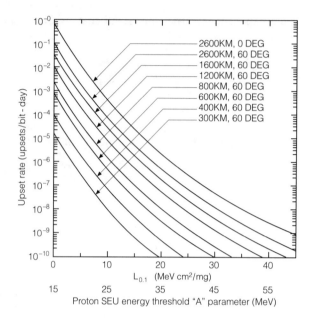

Figure 5.10. Proton upset rate versus LET threshold for various altitudes and orbit inclinations using the Bendel "A" parameter. From Ref. 20.

same assumptions, giving

$$A \cong 15 + L_{0.1} \qquad (5.48)$$

Hence, using the one-parameter proton-induced SEU cross-section model with A determined above, proton-induced SEU error rates can be computed using (5.35) or using the guidelines discussed in Section 8.3. Equation (5.48) is to be construed as resulting in an *estimation* of either proton- or heavy-ion-induced SEU, when corresponding $L_{0.1}$ experimental data are available. Figure 5.10 shows proton-induced SEU error rates versus $L_{0.1}$ threshold LET for various spacecraft altitudes and inclination angles using A from (5.48). SEU calculations based on these methods show that for certain orbits in the radiation belts, microcircuits with $L_{0.1}$ thresholds from 10–45 MeV cm^2 mg^{-1} could be subject to unacceptable SEU error rates. Proton-induced SEU error rates for devices whose LET thresholds are greater than 10 MeV cm^2 mg^{-1} are probably due to relatively rare (very low cross section) proton-induced spallation reactions caused by very high-energy protons [20].

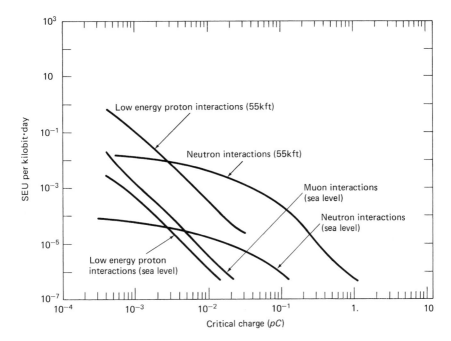

Figure 5.11. SEU rates at high latitudes versus critical charge, due to protons, neutrons, and muons at high-flying-aircraft altitudes and at sea level. From Ref. 25.

5.4. Neutron-Induced SEU

Below the altitudes of nominal levels of Van Allen belt particle densities in the far-stratospheric regions, cosmic ray reactions with these atmospheric particles produce substantial numbers of neutrons. The neutrons can penetrate to ground level to cause SEU in hypersensitive electronics and to cause SEU in high-flying aircraft such as the Concorde airliner [26], as shown in Fig. 5.11. Other neutron source environments for inducing SEU include fission and fusion reactors and radioisotope thermoelectric generators (RTGs). Neutron-induced SEUs also have been demonstrated using accelerator beam particles incident on neutron-rich targets [27–29], as well as Pu–Be (α, n) reactions [30].

Realizing that ionization produces spurious charge (and current) to cause SEU in sensitive node regions of microcircuits, neutrons can produce such ionization charge through (a) neutron collisions that excite nuclei, which deexcite by emitting ionizing gamma rays; (b) neutron collisions that produce ionizing recoil atoms or ions from lattice displacements if they are sufficiently energetic; (c) inelastic collisions by which the neutron is absorbed in the nucleus which then emits a charged particle, as in (n, α) and (n, p) reactions where the α and p-product particles can ionize; and (d) neutron-induced fission

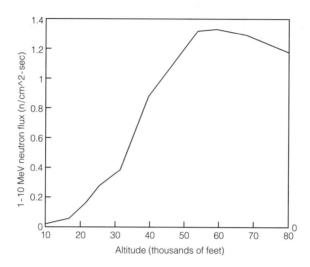

Figure 5.12. Neutron flux versus altitude showing peak at about 60,000 ft. From Ref. 32.

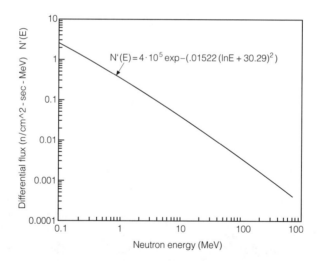

Figure 5.13. Normalized differential neutron flux in the atmosphere versus neutron energy; based on measurements by NASA–Ames at 40,000 ft and 45°–50° north geomagnetic latitude. Flux normalized to 0.85 n cm^{-2} s^{-1} at 1–10 MeV. From Ref. 32.

of trace amounts of natural uranium and thorium found in the microcircuit package material to produce ionizing fission fragments.

In the atmosphere, the maximum number of neutron-induced SEU error rates occur at approximate altitudes of 60,000 ft [31-32]. Figures 5.12 and 5.13

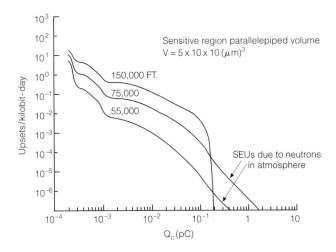

Figure 5.14a. SEU error rate versus critical charge at different altitudes. From Ref. 35.

depict neutron flux versus altitude, and neutron flux spectrum in the atmosphere, respectively. Above 60,000 ft, cosmic ray heavy ions become the dominant source of SEU. Below 25,000 ft, the atmosphere attenuates the neutron flux to sea level amounts about 10^{-3} times those at high altitudes. Atmospheric neutron flux levels are also a function of latitude, being greatest at high latitudes near the magnetic poles [31, 32]. Over the last two decades, a methodology for studying and computing neutron-induced SEU error rates has been formulated. It is given in the following with an example of a computation for a high-flying aircraft.

Figure 5.14a depicts single event upset (SEU) error rates versus critical charge for various altitudes. Collaterally, Figure 5.14b depicts cumulative SEU at 29 kft. Note that in Figure 5.14a for the larger Q_c, the curves intersect because of the neutrons at these altitudes, whereas higher altitudes are relatively neutron free. The computational approach for neutron-induced SEU error rate is somewhat different than that for geosynchronous or proton SEU. Neutrons produce ionization and thus electron charge indirectly, as discussed. In this case, the corresponding neutron reactions are typified by neutron-induced production of low-energy silicon lattice atom displacement primaries in the device. These become silicon recoil ions that upon deceleration lose their energy principally by ionization (as opposed to producing atomic displacement cascades of neighboring silicon atoms) in the form of tracks of electron ionization. The electron charge causes SEU when it aggregates to a critical charge at a susceptible node.

The Mott–Rutherford differential cross section for such recoil ions of energy E to produce ionization electrons, where E_n is the energy of the incident

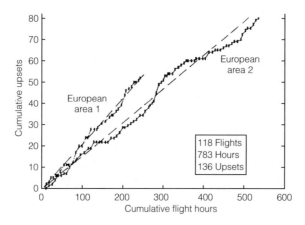

Figure 5.14b. Accumulated avionics upsets at 29,000 ft for Nov. 1990–1992 from 1560 each 64K SRAMs. From Ref. 32.

neutron [viz. $d\sigma/d\Omega\,(\theta, E; E_n)$], provides an appropriate model for this phenomenon [33, 34]. θ is the angle of interaction in the center-of-mass coordinates of the silicon recoil atom and the incident neutron. Then, the cross section for producing recoil ions of energies $E \geq E_c$, where Q_c (pC) = E_c(MeV)/22.5 in silicon, that yield sufficient ionization electrons to provide a critical charge Q_c in the sensitive volume is $\sigma_{th}\,(E_c, E_n) = \int_{E_c}^{\infty} dE \int_{\Omega_c} (d\sigma/d\Omega\,(\theta, E; E_n))\,d\Omega$. Integration over the solid angle that only includes angles corresponding to ion tracks of sufficient length to produce Q_c is denoted by Ω_c.

Now, a burst generation rate (BGR) of SEU-producing silicon recoils is [33, 34] $B_g \equiv \mathrm{BGR}\,(E_c, E_n) = N\sigma_{th}\,(E_c, E_n)$. N is the number of silicon atoms per cubic microns of sensitive region volume. BGR has units of cm^2 μm^{-3}. Commonly in units of cm^{-1}, it is often termed the macroscopic cross section, or linear attenuation coefficient μ, and thus can be seen to be the probability per centimeter of the production rate of SEU recoils of energy $E \geq E_c$ per unit neutron flux and per unit volume of the device sensitive region. Its reciprocal $(N\sigma_{th})^{-1}$ is the corresponding mean free path λ, long with respect to any device dimension. Here, any SEU recoil ion with energy $E \geq E_c$ is meant as one that by losing its energy through ionization will yield sufficient electron charge within the device sensitive volume to produce Q_c and thus SEU with probability unity. BGR being a very small number (Fig. 5.15) allows the probabilistic interpretation of a one-to-one correspondence between the production rate probability of SEU recoils whose energy $E \geq E_c$ and the production rate probability of SEU errors. Hence, the following computation of the SEU recoil ion production rate in the sensitive volume directly yields the SEU error rate probability. Figure 5.15 shows the BGR versus incident neutron energy as a one-parameter family of curves, with the parameter E_c expressed on the curves in terms of Q_c.

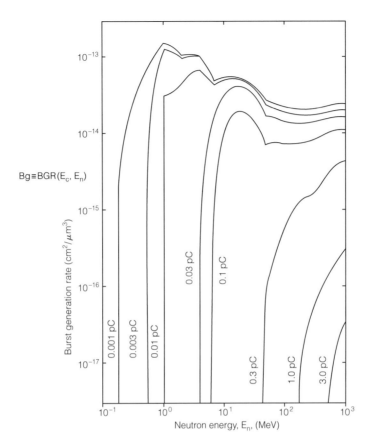

Figure 5.15. Burst generation rate, (BGR), versus neutron energy for various E_c expressed in units of Q_c. From Ref. 31.

Continuing, the rate of neutron-induced production of SEU recoil ions in the sensitive volume is essentially $V\int_{1\,\text{MeV}}^{\infty} \varphi_n(E_n) B_g(E_c, E_n) dE_n$. V is the sensitive node volume (μm^3). $\varphi_n(E_n) = \varphi_0 N'(E_n)$ is the incident neutron flux (cm^{-2} s^{-1}), and $N'(E_n)$ is the neutron spectrum normalized to unity; that is, $\int_{1\,\text{MeV}}^{\infty} N'(E_n) dE_n = 1$, with 1 MeV as a sensible lower limit, so the integral flux level $\varphi_0 = \int_{1\,\text{MeV}}^{\infty} \varphi_n(E_n) dE_n$. $N'(E_n)$ is depicted in Fig. 5.16 for five pertinent neutron spectra. Then the SEU error rate is $E_r = \epsilon V \varphi_0 \int_{1\,\text{MeV}}^{\infty} N'(E_n) B_g(E_c, E_n) dE_n$. ϵ is the collection efficiency parameter of the sensitive region volume as not all the ionization electrons in V contribute to SEU. For example, some will leak from the sensitive region [30, 36] without participating in the SEU process.

For the computation of neutron-induced SEU in high-flying-aircraft avionics, the preceding has been incorporated into a graphical format for ease of calculation [23]. First, define a neutron-induced SEU error function $NIE(Q_c)$,

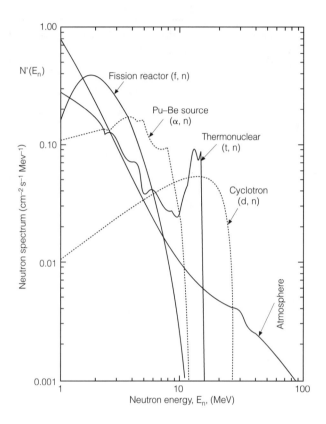

Figure 5.16. Neutron energy spectra for various environments normalized to unit flux. From Ref. 31.

which depends on the critical charge through E_c. It is given by [31] $\text{NIE}(Q_c) = \int_{1\,\text{MeV}}^{\infty} N'(E_n) B_g(E_c, E_n) dE_n = E_r/\epsilon V \varphi_0$ and is plotted in Fig. 5.17. So, the neutron-induced error rate is $E_r = \epsilon V \varphi_0 (\text{NIE})$ (errors per bit second).

As a numerical example [37] consider an actual SRAM which is part of a system memory with the following characteristics in a specified high-altitude neutron environment. The avionics SRAM device of $16\text{K} \times 1$ (NMOS/CMOS) memory cells, as part of a 2.4-Mbit memory array has the following parameters:

Critical LET L_c: 3 MeV cm^2 mg^{-1}

Critical Charge Q_c: 0.13 pC

Feature Size: 0.9 μm

Device cross section: $\sigma_d = 0.2$ cm^2 (16K cells)

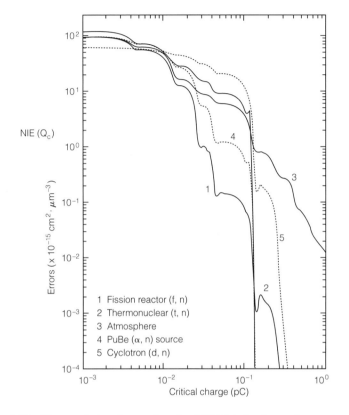

Figure 5.17. Neutron-induced error function NIE(Q_c) for the environments shown in Fig. 5.16 versus critical charge. NIE(Q_c) is the SEU error rate per unit neutron flux per unit volume. From Ref. 31.

Cell depth: $c = 2\mu m$

Latitude: 40° N

Collection efficiency: 0.3

Neutron flux: 0.1 cm^{-2} s^{-1}

Altitude: 25,000 ft

System memory: 2.4×10^6 bits

From Fig. 5.17, $Q_c = 0.13$ pC yields from the "atmosphere" spectrum plot No. 3, NIE $= 0.9 \times 10^{-15}$ cm^2 μm^{-3}. The cell bit sensitive volume $V = \sigma_d c$ from the above tabulation is $V = (0.2/16 \times 10^3)(10^8)(2) = 2500$ μm^3. The neutron flux at these altitudes is about 0.1 cm^{-2} s^{-1}. Thus, $E_r = (0.3)(2.5 \times 10^3) \cdot (0.1)(9 \times 10^{-16})(86400$ s day$^{-1}) = (5.83 \times 10^{-9})(2.4 \times 10^6) = 0.014$ SEU per system day, or $(1/0.014)(24) = 1700$ h between system SEU errors. The measured value in the actual avionics was about 1 SEU error every 1460 h [37].

5.5. Alpha-Particle-Induced SEU

The alpha particle, or $_2^4He^{2+}$ ion, can cause SEU by virtue of its mass, charge, and stopping power (LET). Of interest herein, there are three main sources of alpha particles. The first is that alpha particles are the decay products of naturally occurring radioactive heavy actinide nuclei, such as uranium and thorium and their daughter nuclei which comprise their decay chains. The actinides make up a very small portion of the earth's crust due to their synthesis, along with other nuclei, within a relatively few stars whose subsequent nova dispersed them to eventually form other stars and planets, including our solar system. They are found in trace impurities in device materials such as Si, SiO_2, Si_3N_4, and so forth that are used to fabricate integrated circuits and their packages. For example, ^{238}U (half-life: 4.5×10^9 years) decays ultimately to ^{206}Pb through its daughter nuclei emitting eight alpha particles in its transmutation to lead. Similarly, ^{230}Th (half-life: 1.4×10^{10} years) decays to ^{208}Pb, discarding six alphas in its decay chain to end at a different isotope of lead. Table 5.7 depicts the results of a quantitative analysis of device package materials that contain uranium and thorium nuclei and the corresponding alpha flux.

The second source of alpha particles are products of (a) Van Allen belt proton flux that interacts with the silicon devices in a spacecraft, as discussed in Section 5.3, which reactions produce alpha particles, as seen in Table 5.3: also, (b) cosmic-ray-induced neutrons can produce alpha particles in reactions

Table 5.7. Uranium and Thorium Traces in DRAM Package Materials

Ser.	Material	U (ppm)	Th (ppm)	Φ_α (cm^{-2} h^{-1})	ZrO_2 (weight %)
(1)	Alumina (Al_2O_3)–A	2.5	0.6	0.6	1.0
(2)	Alumina (Al_2O_3)–B	—	—	0.3	—
(3)	Alumina (Al_2O_3)–C	—	—	0.5	—
(4)	Glass–A (sealing)	12.	6.	29.	17.5
(5)	Glass–B (sealing)	2.5	3.	5.2	3.
(6)	Glass–C (sealing)	17.	6.	45.	25.
(7)	Glass–D (sealing)	12.	6.	18.	6.
(8)	Glass–E (sealing)	—	—	32.	20.
(9)	Epoxy	—	—	1.7	—
(10)	Silicone	—	—	1.3	—
(11)	Au-plated lids	—	—	0.04–1	—

Source: Ref. 38.

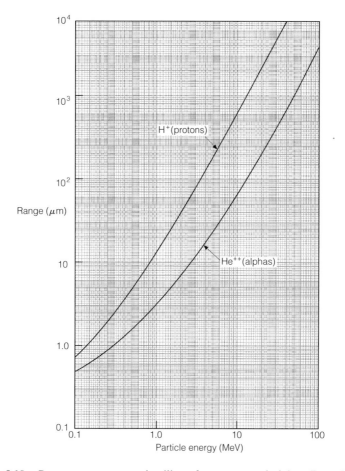

Figure 5.18. Range–energy curves in silicon for protons and alphas. From Ref. 41.

with spacecraft silicon, as discussed in Section 5.4. For low-energy neutrons (~ 3 MeV), an important reaction is $^{28}_{14}\text{Si} + ^{1}_{0}\text{n} = ^{4}_{2}\alpha + ^{25}_{12}\text{Mg}$, with a maximum cross section of 0.20 barns [33]. For high-energy neutrons (≥ 14 MeV), the predominant reaction is $^{28}_{14}\text{Si} + ^{1}_{0}\text{n} = ^{4}_{2}\alpha + ^{1}_{0}\text{n} + ^{24}_{12}\text{Mg}$ with a corresponding cross section of 0.24 barns. In both of these reactions, as well as the proton-induced reactions given in Table 5.3, the reaction product alphas are always accompanied by much heavier ion isotope products, such as Mg, Al, and Na, with correspondingly much larger LETs than the alphas. However, these alpha particles can still produce SEU but with perhaps lesser intensity than the heavier nuclei. Figures 3.2 and 5.18 depict stopping power and range–energy curves, respectively, for protons and alpha particles. From Fig. 5.18, it is seen that the proton range is roughly an order of magnitude larger than that of the alpha particles, except at the lower energies. This is in concurrence with the well-known fact that alphas are stopped relatively easily, compared to most other particles. For example, a 3-MeV alpha particle would be stopped by the paper thickness of this page.

The SEU error rate expressions for protons and neutrons, as in Sections 5.3 and 5.4, respectively, are constructed in a manner that they either include alpha-induced SEU by implication or render their contribution to SEU negligible by comparison. For these reasons then, the effects of alpha particle SEU are focused on alpha-induced SEU in devices and packaging, where the alphas are the decay products of heavy actinides inside the material, as discussed, and suffer virtually no competition from heavier SEU-inducing nuclei, certainly during manufacture, storage, and installation. Table 5.7 shows that ZrO_2 (zirconia), a known alpha source, in the package sealing glasses is the main source of alpha-particle-induced SEU in the chip package by virtue of its relative abundance (column 5) as compared to uranium and thorium.

The third source of alpha particles is, like protons, the solar wind, where the alpha flux is second only to that of the protons. Again, the alphas are incident together with protons, so that, as discussed previously, protons produce relatively heavy spallation reactions products whose LET is much greater than that of the alpha particles. The accompanying alphas are treated as already mentioned.

An ancillary source of alpha particles are those from thermal neutrons ($< 1\,eV$) incident on boron p-type dopant in a microcircuit transistor. The reaction has a relatively high cross section of 3850 barns and is: $^{10}_{5}B + ^{1}_{0}n \rightarrow ^{11}_{5}B \rightarrow ^{4}_{2}He + ^{7}_{3}Li^*$. The $^7Li^*$ denotes a lithium nucleus in an excited state, from which it quickly stabilizes by $^{7}_{3}Li^* \rightarrow ^{7}_{3}Li + (0.478\,MeV)$ gamma. Besides the probability of ionizing damage occurring from the gamma ray, the lithium and alpha particle can produce recoil displacement damage as well as SEU in the device [39].

The NMOS dynamic RAM (DRAM) is a representative part type susceptible to SEU and is used here as a model. It stores data in the form of the presence or absence of charge on storage capacitors which are part of the memory bit cell. The stored information is maintained by periodic refresh cycles. The individual cell capacitors, each representing one bit, are charged with electrons. The capacitors are made up of potential wells in the p-type substrate under a positively charged gate, as in Fig. 5.19. The value of the charges range from about 50 to 500 fC, corresponding to $3 \times 10^5 - 3 \times 10^6$ electrons [40]. In this context, the critical charge Q_c can be defined as the number of electrons that differentiates between a zero and a one of a nominal memory cell. A single alpha particle can cause an SEU in some 4K and 16K DRAMs. This is appreciated by realizing that, for example, a 5-MeV alpha particle that deposits all of its energy into a memory cell sensitive region yields $5 \times 10^6 \div 3.6\,eV$ (required energy to produce one electron–hole pair in silicon) to yield 1.4×10^6 electrons.

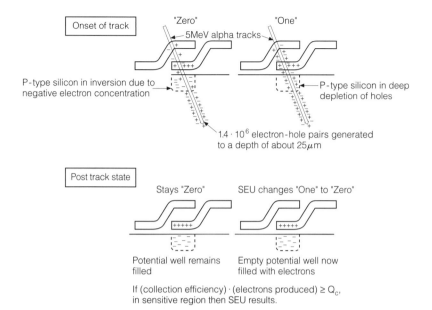

Figure 5.19. Effects of a 5-MeV alpha particle on a 16K RAM memory cell. From Ref. 40.

The range of the alpha particle, shown in Fig. 5.18, in most materials suffers very little statistical deviation (straggling), because usually $\Delta R/R \simeq 0.02$. Alphas are usually emitted with discrete energies. However, when absorbed in thick materials (thickness greater than its range), they will produce a continuous energy spectrum from their maximum energy to zero. The electron–hole pair generation takes place within a radial distance of about 2–3 μm from the alpha particle track [40]. Fast electrons (beta particles) and gammas normally cannot produce SEU directly in silicon because of their very low stopping power. Certain cosmic ray primary and secondary particles which include muons and neutrons can cause SEU at ground level. For example, 64K charge-coupled devices (CCDs) suffer SEU from the heavy cosmic ray ions incident at ground level (Section 5.6).

On the other hand, 90% of SEUs detected in microcircuit vendor's inventories are caused by alpha particles. One very strong argument for the fact that it is indeed the alpha particles that induce SEU is that measured SEU error rates are proportional to the incident alpha flux over eight orders of magnitude [40]. However, some high alpha fluxes can cause multiple-bit SEU errors (i.e., one particle produces many SEUs in a microcircuit). In some devices, electrons collected by floating bit nodes in other parts of the microcircuit can contribute to SEU. The bit sense lines can collect enough charge to result in an

SEU—either in the "ones" or "zeros" stored in the cells [40]. The remedy is, of course, to eliminate floating bit lines. Actually, nothing should be left floating in any part of a microcircuit system.

Figure 5.19 depicts two individual 16K DRAM memory cell capacitors struck by alpha particle tracks. The leftmost capacitor stores a "zero," as this particular memory convention asserts that "zero" corresponds to a concentration of electrons in its p-type silicon potential well in the inversion state (electron population is inverted from its normal hole population, because it is p-type silicon). Prior to the track onset, the rightmost capacitor potential well is essentially empty of both electrons and holes, which corresponds to storing a "one." Following the track onset, the leftmost capacitor is still in the same state, as the alpha particle track only added electrons to its store. However, the rightmost capacitor potential is now filled with electrons produced by the alpha particle track ionization electrons from the silicon material. This transforms it to a filled electron state like the leftmost capacitor, thus changing its information state from "one" to "zero," which constitutes an SEU.

Figure 5.20a depicts a DRAM fragment of four cells. The horizontal lines are the word or select lines, and the vertical lines are called bit or data lines. Peripheral memory buffer circuits translate requests to read from, or write into specific memory cells, signals for the corresponding words and bit lines. A particular cell at the intersection of a word and bit line consists of one MOS access transistor plus one storage capacitor. The storage capacitor is built into the transistor and is comprised of a thin layer of silicon dioxide as its dielectric. One capacitor plate is one end of the transistor source or drain elongated to extend some distance under the SiO_2, as in Fig. 5.20b. The other plate is the layer of conducting polysilicon extending over the dielectric as shown in the figure.

To write, the normally off access transistor at the particular location is momentarily turned on by a voltage pulse on its gate from the corresponding

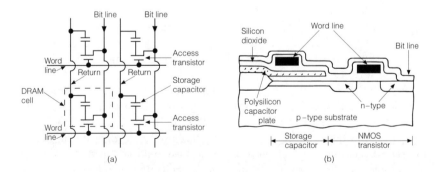

Figure 5.20. MOSFET n-Channel DRAM cell: (a) DRAM schematic fragment of four DRAM cells; (b) vertical section through a single DRAM cell. From Ref. 42.

word line. Almost simultaneously, the data voltage pulse reaches and charges only the appropriate storage capacitor through the corresponding bit line and the on-access transistor. After these pulses, the access transistor reverts to its off-state.

To read, the access transistor is again turned on as above. This then connects the appropriate cell by way of the bit line to a sense amplifier, which sends the signal to the input/output circuitry. The sense amplifier refreshes the cell by restoring the bit line to its original voltage (charge state) written into the cell. The incident ions, including alpha particles, degrade the information in the cell as discussed previously.

5.5.1. Alpha-Particle-Induced Error Rate

For alpha particles, the error rate computations are basically like those for particles previously discussed. The usual flux–cross section product integral is used to obtain the interaction (SEU) rates, but with certain variations unique to the incident particles. As an example [40], consider a 4K × 1 DRAM with the memory cells divided equally between ones and zeros and each cell's sensitive region area is 675 μm^2. As seen in Fig. 5.21 (No. 3), the alpha flux

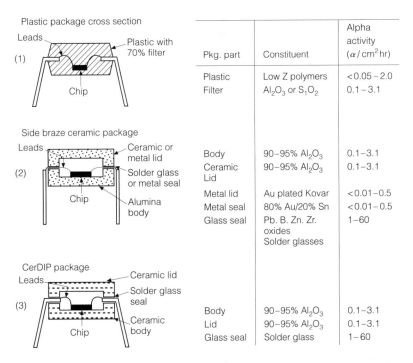

Figure 5.21. Package design and radioactivity levels. From Ref. 43.

emanating from the package lid of this glass-sealed ceramic package is 0.2 cm^{-2} h^{-1}. However, the major portion of the alpha flux comes from the glass seal flux incident on the chip (die) of 3.8 alphas cm^{-2} h^{-1}. From Fig. 5.21, it is evident that the major amount of the flux onto the chip comes from the glass seal, that is, at low angles with respect to the vertical (about 15°).

Generally, alpha particles that strike the chip at nearly grazing (low) angles disperse their energy over too large an area to normally provide the critical charge Q_c. The mean range of these alphas is about 20 μm, and that the chip circuitry per se lies in the top 4 μm implies that the ionization track lies well below the circuitry in the chip substrate for near-vertical incidence angles. The optimum angles for which the die suffers the maximum energy deposition are between 45° and 60° [44].

Returning to the example, a sensitivity factor S, defined as the number of SEUs induced per alpha particle, is given by [40]

$$S = \left(\frac{A_{\text{stor}}}{A_{\text{cell}}}\right) \int_0^\infty \sigma_\alpha(E; Q_c) N(E) \, dE \tag{5.49}$$

where $\sigma_\alpha(E; Q_c)$ is the probability (cross section) that an alpha particle of energy E causes an SEU. $N(E)$ is the alpha flux energy spectrum. Φ_α is the alpha flux level [i.e., $\Phi_\alpha(E) = \Phi_{0\alpha} N(E)$, where $\Phi_{0\alpha} = \int_0^\infty \Phi_\alpha(E) \, dE$ because $\int_0^\infty N(E) \, dE = 1$]. $A_{\text{stor}}/A_{\text{cell}}$ = (cell SEU sensitive area)/(total cell area) is the fraction of alphas that impinge on the cell sensitive area region. Here, S was estimated [40] using a Monte Carlo code with known $N(E)$ and Q_c. The result is $S = 0.015$ SEUs per alpha particle. Then, the corresponding SEU error rate E_r is given by

$$E_{r\alpha} = A_{\text{cell}} N_{\text{cell}} \Phi_{0\alpha} S \tag{5.50}$$

where N_{cell} is the number of memory cells, all assumed to experience the same alpha particle level. $A_{\text{cell}} N_{\text{cell}} = (6.75 \times 10^{-6})(4000) = 0.027$ cm^2. For this 4K DRAM, $\Phi_{0\alpha} = 3.8$ cm^{-2} h^{-1} and, with S computed from (5.49) gives

$$E_{r\alpha} = (0.027)(3.8)(0.015) = 0.0015 \text{ SEU h}^{-1} \tag{5.51}$$

To obtain estimates of desired parameters, monoenergetic particles corresponding to an average energy E_0 are often assumed. Then the spectrum $N(E) \sim \delta(E - E_0)$, where $\delta(x)$ is the well-known delta function with the property that $\int f(x) \delta(x - x_0) \, dx = f(x_0)$. Inserting $N(E) = \delta(E - E_0)$ into (5.49) yields $S \sim \sigma_\alpha(E_0, Q_c)$ with the approximate SEU error rate given by

$$E_r \sim \sigma_\alpha(E_0, Q_c) \Phi_\alpha(E_0) \tag{5.52}$$

thus reducing the cross section–flux product integral to a product as above.

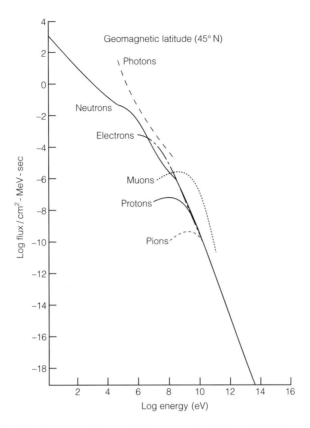

Figure 5.22. Cosmic rays and secondaries at sea level. From Ref. 33.

5.6. Ground-Level SEU

It has been established conclusively for incident cosmic rays that as a result of cosmic ray interactions, their secondary product particles also cause SEU in microcircuits at ground/sea level [33, 45]. Mesons, neutrons, and protons are included in the secondaries as shown in Fig. 5.22 [33]. The high-energy particles are principally incident at ground level within a solid angle of about 1 sr about the vertical, whereas the lower-energy particles are more nearly isotropic due to their many more interactions with the atmosphere. This aspect could influence the orientation of devices relative to the vertical direction at sea level [33].

For pions and muons, their maximum stopping power (LET) is about 0.5 MeV cm^2 mg^{-1}, whereas, by comparison, it is ~ 2 MeV cm^2 mg^{-1} for alpha particles [46]. Until recently, pions were largely ignored because of the relatively small fraction of the ground-level particles they represent, as noted in

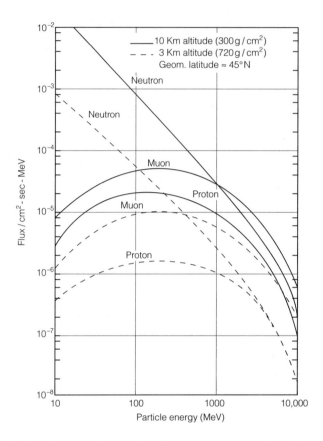

Figure 5.23. Cosmic ray and secondary flux versus energy at 3 and 10 km altitude. From Ref. 33.

Fig. 5.22. However, it has been shown that pions are relatively more effective in producing SEU than muons, as can be seen in Table 5.8 [46, 47]. Even acknowledging their lower abundance, pions are apparently as important as neutrons in producing SEU at ground level [46].

The relative abundance of cosmic rays and their secondaries at ground level is shown in Fig. 5.22. However, large fluctuations in their fluxes are attributed to many factors, including magnetic latitude, time of day, season, solar cycle, angle of incidence, and others [33]. The geomagnetic latitude of 45° was chosen because of the extensive amount of data available for this orientation [33]. These fluxes also increase with altitude from sea level, as in Fig. 5.23 where they are compared at 3 km and 10 km altitudes. In this figure, it is seen that their energies maximize (neutrons excepted) at about 300 MeV, and their altitude maxima is about 10 km, as in Fig. 2.7.

Examination of Tables 5.8–5.10 reveals that the SEU error rates are quite low. This could lead to an initial conclusion that they are inconsequential at sea level, except for CCDs. CCDs are used mainly in flat panel imaging systems. For them, the image observer can usually glean the information from the visual context. Most other current devices are relatively insensitive to SEU at ground level, and perhaps at very low altitudes. However, if the investigator uses multipliers to compute SEU at the ground system level, from that available herein at the bit level, the relative insensitivity is seen to begin to blur. For example, a commercial airline fleet can accumulate millions of flight hours in a few months. Roughly, the computed SEU rate for a nominal SRAM amounts to about one SEU per microcircuit per aircraft over this time [46]. This may or may not be tolerable currently, but for future VLSI and ULSI devices, including CCDs, the SEU error rate will rapidly increase with chip packing density, even with fault-tolerant architecture and EDAC methods.

Table 5.8. Flux, Cross Section, and SEU Error Rates of a 4K SRAM For Six SEU-Inducing Particle Types at Sea Level

Particle	Energy (MeV)	SEU cross section (cm^2)	Particle flux (cm^{-2} s^{-1})	SEU error rate Per bit second	SEU error rate Per bit day
π^-	90	3.7×10^{-8}	1.1×10^{-4}	4.1×10^{-12}	3.5×10^{-7}
π^-	164	1.6×10^{-7}	1.1×10^{-4}	1.7×10^{-11}	1.5×10^{-6}
π^+	164	8.1×10^{-8}	1.1×10^{-4}	8.8×10^{-12}	7.6×10^{-7}
μ^+	164	1.2×10^{-9}	2.4×10^{-6}	3.0×10^{-15}	2.6×10^{-10}
μ^-	164	1.2×10^{-9}	2.4×10^{-6}	3.0×10^{-15}	2.6×10^{-10}
p	>100	5.0×10^{-8}	2.2×10^{-4}	1.1×10^{-11}	9.5×10^{-7}
n	>10	1.4×10^{-9}	4.0×10^{-3}	5.6×10^{-12}	4.8×10^{-7}
Totals:				4.7×10^{-11}	4.1×10^{-6}

Sorce: Refs. 46–48

Table 5.9. SEU Error Rates of a 4K SRAM For Six SEU-Inducing Particle Types at 10 km Altitude

Particle	SEU error rate Per bit second	SEU error rate Per bit day
π^+, π^-	1.7×10^{-10}	1.5×10^{-5}
μ^+, μ^-	1.7×10^{-12}	1.5×10^{-7}
p	4.3×10^{-10}	3.7×10^{-5}
n	1.1×10^{-9}	9.5×10^{-5}
Totals:	1.7×10^{-9}	1.5×10^{-4}

Source: Ref. 47.

Table 5.10. SEU Error Rates For 64K DRAM, 64K CCD, and 256K CCD at Ground Level

Interaction	Chip error rate (events per 10^6 hours)		
	64K DRAM	64K CCD	256K CCD
(1) e^- ionization wake	0	0	0
(2) $e^- \rightarrow$ Si recoils (EM)[a]	0	<0	<1
(3) p^+ ionization wake	0	140	1300
(4) $p^+ +$ Si $\rightarrow$ HN recoils[b]	<1	<1	<1
(5) $p^+ +$ Si $\rightarrow$ He	<1	<1	<1
(6) $n +$ Si $\rightarrow$ HN recoils	1	100	250
(7) $n +$ Si $\rightarrow$ He	6	22	20
(8) Muon ionization wake	0	330	1700
(9) Muon $\rightarrow$ silicon recoils (EM)[a]	<1	3	4
(10) μ^- capture $\rightarrow$ HN recoils[b]	<1	7	8
Total: Chip Error Rate per 10^6 h	~7	~600	~3000
Chip Error Rate per bit day	2.63×10^{-9}	2.3×10^{-7}	2.8×10^{-7}

[a] "Si recoils (EM)" indicates close electromagnetic (EM) collisions that induce energetic recoiling silicon nuclei.

[b] "HN recoils" indicates a summation over all recoiling heavy nuclei (HN) from the nuclear reaction indicated.

Source: Ref. 33.

Hence, how well will these systems tolerate SEU with future mitigation technology at ground levels, even generally, is a query still to be addressed.

The following notes refer to the consecutive rows in Table 5.10 to explain the major cosmic-ray-produced particle reactions with silicon at ground level [33].

(1) Electrons interact with silicon to produce an ionization wake of hole–electron pairs but too weak to cause SEU.

(2) Electrons elastically scattered from silicon produce recoiling silicon ions that ionize. However, no SEU or permanent damage due to these electrons has been detected in the devices mentioned in the caption of Table 5.10.

(3) Protons also interact with silicon to produce a wake of hole–electron pairs, but presently too weak to cause SEU directly on current technology, except for CCDs. Protons cause SEU indirectly, as discussed, but for future submicron technologies, protons will certainly be able to cause SEU directly.

(4) Protons also scatter from heavy ions to produce heavy-ion recoils that ionize.

(5) Protons can produce alpha particles in nuclear reactions, as in Table 5.3.

(6) Neutrons scatter elastically and inelastically from silicon to produce silicon recoils that ionize to, in turn, produce SEU. This occurs mainly in CCDs.

(7) Neutrons can produce alphas, like protons, in nuclear reactions which can cause SEU, as discussed in Section 5.5.

(8) Mesons also can interact with silicon to produce an ionization wake, but of low density, too weak to cause a significant SEU problem at the present time.

(9) Muons electromagnetically scatter from silicon to produce weak silicon recoils of little SEU consequence.

(10) Muons can be captured by the silicon nucleus, which can cause it to recoil to ionize, again producing little SEU except for CCDs.

For large land-based systems, shielding methods may be practical. Figure 5.24 depicts the SEU error rate as a function of aircraft-type altitudes down to sea level for 64K CCDs, 256K CCDs, and a nominal 64K DRAM, plus

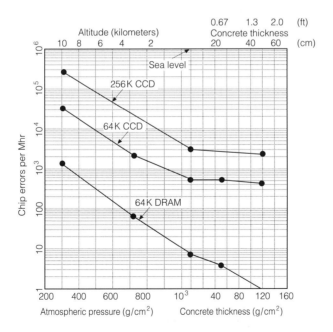

Figure 5.24. SEU error rate for 64K SRAM, 64K CCD, and 256K CCD versus altitude from 10 km to ground plus 2 ft of concrete. From Ref. 33.

their attenuation due to concrete shielding [33]. The 64K DRAM is affected at sea level principally by neutrons, as witnessed by its SEU drop off in concrete. However, the figure shows that concrete shielding has a negligible effect on the CCDs. This is because the CCD SEU is mainly affected by the sea-level mesons. Metallic shielding, such as that already in place in the system for electronics electromagnetic shielding, could be buttressed to further attenuate the mesons, because one of their fast decay modes transforms muons into electrons [48] for which shielding is effective. It should be noted that the preceding refers to current devices. Future reduction in feature size will cause a dramatic increase in the SEU problem, as discussed in Section 5.3. Currently, 1M DRAMs are being marketed that are found to have much higher SEU rates in the lower atmosphere, by virtue of their reduced feature size as compared to today's devices of like complexity.

Problems

5.1 For a particular 1K SRAM, destructive physical analysis (DPA) methods reveal that the individual SEU-sensitive region dimensions are approximately $3 \times 10 \times 10 \ \mu m^3$ and can be modeled as 0.05-pF flat-plate capacitors. The manufacturer's specification sheets for this SRAM assert that the cell V_{OH} (output voltage HI) = 5.5 V and V_{OL} (output voltage LO) = 2.5 V. Compute (a) the SEU error rate per cell-day and (b) the corresponding SEU error rate per device-day.

5.2 An individual galactic cosmic ray heavy ion can enter the spacecraft avionics ICs at any angle, to proceed along a chord to produce electron–hole pairs by ionization. The particular IC chip cell has a sensitive volume, assumed to be a parallelepiped, with dimensions a, b, and thickness c. (a) Show that for this sensitive volume, the mean chord length $\bar{s} = 2c$. (b) Compute the corresponding L_c in units of MeV cm^2 mg^{-1}, if the critical charge $Q_c = 10^6$ electrons and thickness $c = 1 \ \mu m$.

5.3 SEU is usually associated with spacecraft avionics. Name at least three nonspacecraft systems for which SEU could cause difficulty, and why.

5.4 A monoenergetic heavy-ion beam of flux equal to 1.6×10^8 ions cm^{-2} s^{-1} caused 10 SEUs before being interrupted after 10 s of beam time because the accelerator was shutdown.

   (a) What was the corresponding SEU cross section?
   (b) After the accelerator was brought up and running again after some adjustments, its flux was now only found to be 10^5 ions

cm^{-2} day^{-1}. What is the SEU error rate per bit per day using the above cross section?

(c) Can this SEU cross section be used to compute SEU error rates for other heavy-ion species?

5.5 Why was the SEU phenomenon of funneling postulated, and later proved?

5.6 It has been established that many SEU-sensitive regions, of volume abc, are comprised of depletion layers near a junction. Using the "figure of merit" SEU error rate $E_p = 5 \times 10^{-10} abc^2/Q_c$, show that E_r is proportional to c^4.

5.7 As the answer in Problem 5.6 yields $E_r \sim c^4$, does this mean that the thinner (i.e., the smaller) the device, the smaller the SEU error rate will be?

5.8 For proton-induced SEU, (e.g., from the Van Allen proton belts): (a) Why is there no analogous expression for SEU error rate as in the "figure of merit" formula for galactic cosmic rays? (b) How is the proton-induced error rate handled?

5.9 In Figure 5.14a, which plots SEU rate versus critical charge Q_c at the three altitudes 55, 75, and 150 kft for a device-sensitive region volume of dimensions $5 \times 10 \times 10 \, \mu m^3$. Explain why the 150-kft curve intersects the other two.

5.10 The autopilot controller avionics of, say, a Concorde class aircraft contains 16×1 SRAMs comprising its 10MB memory array. At cruising altitude, the neutron flux $\phi_n \cong 0.5$ cm^{-2} s^{-1}. The SRAM cell LETs = 5 MeV cm^2 mg^{-1} with SEU-sensitive regions of thickness $c = 2 \, \mu m$, and collection efficiency $\epsilon = 0.3$. The SRAM device asymptotic SEU cross section $\sigma_d = 0.2$ cm^2 (all 16K cells). What is (a) the SEU error rate and (b) the mean number of hours between SEUs?

5.11 To check on the alpha-induced SEU in the 13-bit cache memory of a certain part, it was exposed to an 8-μCi alpha source at a distance of 1 mm from the part for 20 min. It experienced only one SEU in that time. What is the corresponding alpha-induced SEU cross section for this device?

5.12 When calculating the SEU error rate using the integral of the product of the differential flux $\Phi'(L) dL$ and the cross section $\sigma_u(L)$, how is N, the number of targets per cm^3, to be construed?

References

1. D. Chlouber, P. O'Neill, and J. Pollack, "General Upper Bound on Single Event Upset Rate," *IEEE Trans. Nucl. Sci.*, **NS-37** (2), 1065–1071 (1990).

2. E.L. Petersen, J.B. Langworthy, and S.E. Diehl, "Suggested SEU Figure of Merit," *IEEE Trans. Nucl. Sci.*, **NS-30** (6), 4533–4539 (1983).

3. E.L. Petersen, P. Shapiro, J.H. Adams, Jr., and E.A. Burke, "Calculation of Cosmic Ray Induced Soft Upsets and Scaling in VLSI Devices," *IEEE Trans. Nucl. Sci.*, **NS-29** (6), 2055–2063 (1982).

4. M.D. Petroff, in J.C. Pickel, and J.T. Blandford, "Cosmic Ray Induced Errors in MOS RAMs," *IEEE Trans. Nucl. Sci.*, **NS-27** (2), 1006–1015 (1980), Appendix I. [Errata: **NS-27** (6), 1221 (1992).]

5. J.H. Adams, Jr., "Cosmic Ray Effects on Microelectronics, Part IV, Naval Research Labs. NRL Memo Report 5901, 1986.

6. D.L. Chenette, "Petersen Multipliers for Several SEU Environment Models," Aerospace Corp. Report AF Systems Command POB 92960, LA 90009, 1986.

7. D. Binder, "Analytical SEU Rate Calculation Compared with Space Data," *IEEE Trans. Nucl. Sci.*, **NS-35** (6), 1570–1572, (1988).

8. L.W. Connell, P.J. McDaniel, A.K. Prinja, and F.W. Sexton, "Modeling the Heavy Ion Upset Cross Section," *IEEE Trans. Nucl. Sci.*, **NS-42** (2), 73–82 (1995).

9. J.N. Bradford, "A Distribution Function for Ion Track Lengths in Rectangular Volumes," *J. Appl. Phys.*, **50** (6), 3799–3801 (1979).

10. J.L. Shinn, F.A. Cucinotta, J.W. Wilson, G.D Badhwar, P.M. O'Neill, and F.F. Badavi, "Effects of Target Fragmentation on Evaluation of LET Spectra from Space Radiation in Low Earth Orbit (LEO) Environment: Impact on SEU Predictions," *IEEE Trans Nucl. Sci.*, **NS-42** (6), 2017–2025 (1995).

11. E.L. Petersen, and E.L. Marshall, "Single Event Phenomena in the Space and SDI Areas," *J. Radiat. Effects Res. Eng.* (1989).

12. W.L. Bendel, and E.L. Petersen, "Proton Upsets in Orbit," *IEEE Trans. Nucl. Sci.*, **NS-30** (6) 4481 (1983).

13. E.G. Stassinopoulos and J.M. Barth, "Non-Equatorial Terrestrial Low Altitude Charged Particle Radiation Environment," NASA, Goddard SFC Report. No. X-601-82-9, 1982.

14. E.G. Stassinopoulos, "Orbital Radiation Study for Inclined Circular Trajectories," NASA Goddard SFC, Report No. X-601-81-28, 1981.

15. W.L. Bendel, and E.L. Petersen, "Predicting Single Event Upsets in the Earth's Proton Belts," *IEEE Trans. Nucl. Sci.*, **NS-31** (6), 1201–1206 (1984).

16. G.C. Messenger, "Single Event Upset Considerations for Multiple Particles," Hardened Electronics and Research and Technology (HEART) Conf., 1985.

17. T.C. May, "Soft Errors in VLSI: Present and Future," *IEEE Trans.* **CHMT-2** (4), 377–387 (1979).

18. W.J. Stapor, J.P. Meyers, J.B. Langworthy, and E.L. Petersen, "Two Parameter Bendel Model Calculations for Predicting Proton-Induced Upsets," *IEEE Trans. Nucl. Sci.*, **NS-37** (6), 1966–1973 (1990).

19. F.W. Sexton, "Measurement of Single Event Phenomena in Devices and ICs," IEEE–1992 NSRE Conf. Short Course Syllabus (Chap. 3), 1992.

20. E.L. Petersen, "The Relationship of Proton and Heavy Ion Upset Thresholds," *IEEE Trans. Nucl. Sci.*, **NS-39** (6), 1600–1604 (1992).

21. W.L Bendel, and E.L. Petersen, "Proton Upsets in Orbit," *IEEE Trans. Nucl. Sci.*, **NS-30** (6), 4481–4485 (1983).

22. D.K. Nichols, W.E. Price, L.S. Smith, G.A. Soli, "The Single Event Upset (SEU) Response to 590 MeV Protons," *IEEE Trans. Nucl. Sci.*, **NS-31** (6), 1565–1567 (1984).

23. J.H. Adams, Jr., C.H. Tsao et. al., "CREME—Cosmic Ray Effects on Microelectronics," Part I NRL Memo 4506, 1981; Part II NRL Memo 5099, 1983; Part III NRL Memo 5402, 1984.

24. J.G. Rollins, "Estimates of Proton Upsets Rates from Heavy Ion Test Data," *IEEE Trans. Nucl. Sci.*, **NS-37** (6), 1961–1965 (1990).

25. R. Silberberg, C.H. Tsao, and J.R. Letaw, "Neutron Generated Single Event Upsets in the Atmosphere," *IEEE Trans. Nucl. Sci.*, **NS-31** (6), 1183–1185 (1984).

26. C.S. Dyer, J. Farren, A.J. Sims, J. Stephan, and C. Underwood, "Comparative Measurements of Single Event Upsets and Total Dose Environments Using the CREAM Instruments," *IEEE Trans. Nucl. Sci.*, **NS-39** (3), 413–417 (1992).

27. C.S. Guenzer, E.A. Wolicki, and R.G. Allas, "Single Event Upset of Dynamic RAMs by Neutrons and Protons," *IEEE Trans. Nucl. Sci.*, **NS-26** (6), 5048–5055 (1979).

28. C.A. Gossett, B.W. Hughlock, M Katoozi, G.S. LaRue, and S.A. Wender, "Single Event Phenomena in Atmospheric Neutron Environments," *IEEE Trans. Nucl. Sci.*, **NS-40** (6), 1845–1852 (1993).

29. S. Wender, and A. Gavron, "High Altitude Neutron Simulation," Executive Summary, H-803 Group P-17, Los Alamos Scientific Laboratory, 1993.

30. E. Normand, J.L. Wert, W.E. Doherty, D.L. Oberg, P.R. Measel, and T.L. Criswell, "Use of Pu–Be Sources to Simulate Neutron Induced Single Event Upset Rates in Static RAMs," *IEEE Trans. Nucl. Sci.*, **NS-35** (6), 1523–1528 (1988).

31. J.R. Letaw and E. Normand, "Guidelines for Predicting Upsets in Neutron Environments," *IEEE Trans. Nucl. Sci.*, **NS-38** (6), 1500–1506 (1991).

32. A. Tabor and E. Normand, "Single Event Upsets in Avionics," *IEEE Trans. Nucl. Sci.*, **NS-40** (2), 120 (1993).

33. J.F. Ziegler and W.A. Lanford, "Effects of Cosmic Rays on Computer Memories," *Science*, **206**, 776–788 (1979).

34. C.H. Tsao, R. Silberberg, and J.R. Letaw, "A Comparison of Neutron Induced SEU Rates in Silicon and Gallium Arsenide Devices," *IEEE Trans. Nucl. Sci.*, **NS-35** (6), 1634–1637 (1988).

35. C.H. Tsao, R. Silberberg, J.H. Adams, Jr., and J.R. Letaw, "Cosmic Ray Effects on Microelectronics," Part III, Naval Research Laboratory, Memo Report No. 5402, 1984.

36. E. Normand and W.R. Doherty, "Incorporation of ENDF-V Neutron Cross Section Data for Calculating Neutron Induced Single Event Upsets," *IEEE Trans. Nucl. Sci.*, **NS-36** (6), 2349–2355 (1989).

37. C.P. Capps and J.P. Raymond, "Aircraft Single Event Upset Phenomena," Eighth Annual IEEE SEU Symposium, 1992.

38. T.C. May and M.H. Woods, "A New Physical Mechanism for Soft Errors in Dynamic RAMs," *IEEE Reliability Physics (IRPS) Symposium*, 1978.

39. T.R. Oldham and J.M. McGarrity, "Worst Case Prediction of Single Particle-Induced Permanent Failures in Microelectronics," Harry Diamond Labs. HDL-TR-1966, 1981.

40. T.C. May and M.H. Woods, "Alpha Particle Induced Soft Errors in Dynamic Memories," *IEEE Trans. Electron. Dev.*, **ED-26** (1), 2–9 (1979).

41. A.H. Seidle and L. Adams, *Handbook of Radiation Effects*, Oxford University Press, Oxford 1993, Appendix B.

42. G.C. Messenger and M.S. Ash, *The Effects of Radiation on Electronic Systems*, 2nd ed., Van Nostrand Reinhold, New York, 1992, Section 3.2.

43. E.S. Meieran, P.R. Engel, and T.C. May, Proc. IRPS 1979, 1979, pp. 13–22.

44. R.J. Redman, R. M. Sega, and R. Joseph, "Alpha Particle Induced Soft Errors in Microelectronics Devices I," *Military Electronic Countermeasures*, 42–47 (1980).

45. T.J. O'Gorman, "The Effect of Cosmic Rays on the Soft Error Rate of a DRAM at Ground Level," *IEEE Trans. Electron. Dev.*, **ED-41** (4), 553–557, (1994).

46. J.F. Dicello, "Microelectronics and Microdosimetry," *Nucl. Instrum. Methods in Phys. Res.*, **B24/25**, 1044–1049 (1987).

47. J.F. Dicello, M. Paciotti, and M.E. Shillaci, "An Estimate of Error Rates in Integrated Circuits at Aircraft Altitudes and Sea Level," *Nucl. Instrum. Methods in Phys. Res.*, **B40/41**, 1295–1299 (1989).

48. J.F. Dicello, M.E. Shillaci, C.W. McCabe, J.D. Doss, M. Paciotti, and P. Berardo, "Meson Interactions in NMOS and CMOS Static RAMs," *IEEE Trans. Nucl. Sci.*, **NS-32** (6), 4201–4205 (1985).

49. E. Petersen, "Soft Errors Due to Protons in the Radiation Belt," *IEEE Trans. Nucl. Sci.*, **NS-28** (6), 3981–3986 (1981).

50. G.C. Messenger, "Single Event Considerations for Multiple Particles," 1985 HEART Conf., 1985.

51. E.L. Petersen, "Approaches to Proton Single-Event Rate Calculations," *IEEE Trans. Nucl. Sci.*, **NS-43** (2), 496–504 (1996).

6

Single Event Phenomena I

6.1. Introduction

This chapter and the next provide brief descriptions of selected important single event phenomena (SEP) that it is felt the reader should be aware of and understand the basic concepts thereof. This chapter first discusses the process of funneling, which is of prime importance in the assessment of additional charge collected at the sensitive nodes of vulnerable devices. As can be appreciated, the charge collection dependencies of the device SEU-sensitive node areas is a major factor in the determination of whether SEU and/or single event latchup (SEL) will occur in the device.

Section 6.3 discusses the dynamics of the transport of the incident ion that forms the charge track in the semiconductor material. The kinematics of the transport behavior are germane with regard to understanding the SEU-connected effects of the time history of the track ionization response. The fact that the life of the track is usually less than 1 ns, and always less than 1 μs, influences the semiconductor response, its circuit function, and therefore its SEU propensities.

The final sections include discussion of the South Atlantic Anomaly and its role in enhancing SEU production, especially for low-earth-altitude-orbiting spacecraft. Also discussed are the effects of incident heavy cosmic ray ions on certain power MOSFETs due to their unique structure. Other single event phenomenon in MOSFETs is their susceptibility to gate rupture and SEU-induced burnout. Discussions of the burnout effects in this context in bipolar power transistors is also given.

In most of the preceding, the phenomenology of the particular effect is described together with the physical models from which computations regarding its SEU or SEL consequences can be ascertained. A good share of the models are currently tentative, and efforts to refine them are ongoing as part of SEU investigative studies that include experimental data by teams in this field.

6.2. Funneling

When a heavy high-energy ion penetrates a semiconductor device through its junctions and depletion layers, it produces a track of ionization, composed of electrons and holes from the semiconductor material atoms. The presence of the track temporarily collapses the depletion layers local to the track, distorting the equipotential surfaces of the depletion layer electric fields in the track vicinity. This distortion results in a nesting of funnel-shaped equipotential surfaces that can extend into the substrate bulk of the device, as shown in Fig. 6.1. The funneling of the equipotentials produces a large potential gradient (i.e., a large electric field in the funnel that propels the ionization electrons up the funnel into SEU-sensitive node regions). This produces an enhanced charge

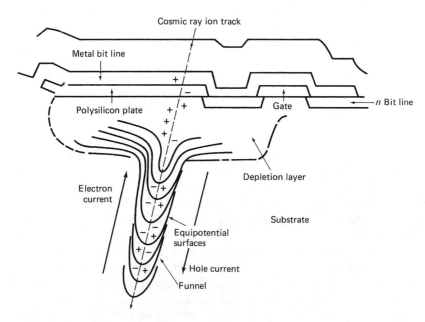

Figure 6.1. Collapse of DRAM cell depletion layer equipotential surfaces along ion-track-producing funnel in device bulk.

concentration in the latter. This originally puzzling enhanced charge density alerted researchers to the possibility of the funnel existence. There is a corresponding hole current directed down the funnel into the semiconductor substrate, as in the figure. As mentioned, there is a large increase in charge density collected, over that were there no funnel, thus increasing the probability that a critical charge is collected at a device information node.

If the incident ion track impales or passes very near the information node, most of the charge in the funnel is collected by the node depletion layers in fractions of a nanosecond following the track penetration. If the track lies remote from the node, the charge collection time is much longer, reckoned in nanoseconds and microseconds, as the charge transport is principally by diffusion. After collection of the charge, plus other modes of track charge depletion such as ambipolar diffusion in the radial direction, the charge density falls to a level comparable to the substrate dopant density, and the disturbed depletion layer field relaxes back to its state prior to the track onset.

It should be noted that when the major portion of charge collected is due to funneling, the parallelepiped model of the sensitive node is well taken. However, when most of the charge collected is by diffusion, the parallelepiped model is somewhat vitiated, due in part to the much longer times involved in the diffusional phenomena, as mentioned above. It should be appreciated that the expressions for SEU error rate discussed in Chapter 5 do not include corrections for the additional charge collected due to funneling. However, some inroads have been made to include this charge by attributing portions of it to internal bipolar action [1]. Also, this charge is usually included by extending the depth c of the parallelepiped volume, as used in obtaining SEU error rates for geosynchronous environments given in Section 5.1. Details are given in the ensuing discussions on funneling models. Certain of these models center about increasing c by a multiplicative factor $\{$e.g., $[1 + (\mu_n/\mu_p)^k]^n$, where k and n are chosen empirically$\}$.

The following detailed qualitative description of the funneling process is based on a number of computer studies and corresponding experimental verifications [2–5].

(a) The initial phase is the passage of the incident heavy ions through the device material, which ionize it to form a track consisting of a wake of ionized particle plasma. This plasma track loses energy to its material surround, coming into thermal equilibrium with it (thermalization) within picoseconds following track onset. The track radius is initially less than 0.1 μm, with an initial charge density of about 10^{18}–10^{21} particles cm^{-3}, orders of magnitude greater than the nominal substrate dopant concentration. The funnel in its traversal of the junction depletion layers is saturated with charge along its immediate neighborhood.

(b) Following thermalization of the track, its cylindrical shape begins to expand radially by ambipolar diffusion, as the charge density is still extremely high. Briefly, ambipolar diffusion asserts that electrons and holes tend to diffuse in train and that a quasineutrality holds (i.e., $p - n \cong p_0 - n_0$, where the subscript zero implies initial conditions). Strong electrostatic restoring fields in the track maintain this charge state.

(c) During this epoch of track expansion, holes and electrons begin to separate on the track periphery where the track carrier densities approach that of the material dopant densities. The holes are drawn away from the track by vertical components of the electric field and driven down into the substrate. In contrast, the track electrons (opposite charge) are driven up the track and most of them are collected at an SEU-sensitive node.

(d) With the reduction in track charge density, the depletion layers begin to revert to their original configuration in the vicinity of the track—first at the track periphery, and inward toward the track center. Charge separation begins to occur during this phase because of the track's electron and hole restoring junction fields. The field distortions begin to decrease rapidly, and with the dissolution of the track, the junction fields relax back to their original state.

6.2.1. Funneling Models

A series of funneling model descriptions begin with simple but effective ones, proceeding to the detailed physico-electrical models that more accurately describe the important ion-track details from track onset to its demise.

(1) One model involves a time-dependent funnel length D_c created following a heavy-ion strike through a junction and its depletion layers. The initial (prompt) charge Q_c collected due to the funnel electric field (drift), as opposed to the delayed charge collected by diffusion, is given by [6]

$$Q_c = e \int_0^{D_c} N_0(x)\, dx = e\bar{N}_0 D_c \tag{6.1}$$

where $\bar{N}_0(x)$ is the initial line charge density along the track (funnel) produced by ionization caused by the incident heavy ion. $\bar{N}_0$ is an average line charge density over the funnel length D_c. Q_c includes charge generated both in the depletion layer and in the funnel. Let D_c be given by

$$D_c = \bar{v}_d \tau_c + x_d \tag{6.2}$$

where

τ_c = charge collection time
x_d = initial funnel depth
V_0 = potential between the sensitive node and the substrate
$\bar{v}_d$ = mean charge upward velocity
E_L = mean longitudinal funnel field = V_0/D_c

μ_n is the field-dependent mobility for electrons. With $\bar{v}_d = \mu_n E_L = \mu_n V_0/D_c$, inserting this into (6.2) yields a quadratic equation for D_c, whose pertinent root is

$$D_c = \tfrac{1}{2}[x_d + (x_d^2 + 4\mu_n \tau_c V_0)^{1/2}] \quad (6.3)$$

For a shallow depletion layer ($x_d \ll 4\mu_n \tau_c V_0$), $D_c \cong \sqrt{(\mu_n \tau_c V_0)}$. For very large E_L, the drift velocity $\bar{v}_d$ is asymptotic to v_{sat} (i.e., $\bar{v}_d \sim v_{sat} \cong 10^7$ cm s^{-1} in silicon). Then, (6.2) becomes

$$D_c \sim v_{sat} \tau_c + x_d \quad (6.4)$$

It was realized that a track plasma screening factor would aid in refining this model [8, 9], that is, because the track ionization plasma is made up of charged particles, they tend to screen or shield the track core from influences due to external fields—not unlike the "skin-effect" attenuation in high-frequency electronic systems. For example, one type of screening factor is of the form $\exp(-x/\delta)$; $\delta = \sqrt{\tau/\pi\mu_0\sigma_0}$, where μ_0 is the free-space permeability, σ_0 is the material conductivity prior to the track onset, and $\tau = L/v_{av}$, where v_{av} corresponds to an average field $E_{av} = v_{av}/\mu_n = V_0/2L$. L is the unscreened funnel field penetration depth. Then, from (6.1) and (6.2),

$$Q_c \cong e\bar{N}_0 \left[x_d + \bar{v}_d \tau_c \exp\left(-\frac{L-x_d}{\delta}\right) \right] \quad (6.5)$$

As already mentioned, the track initially begins to expand radially by ambipolar diffusion. Its radial charge density profile decreases from a maximum at the central axis to the periphery, where it is of the order of $n \simeq p \cong N_A$, the background dopant concentration. It is assumed that the time-dependent ambipolar root mean square diffusion length $r_{rms}(t)$ is given by

$$r_{rms}(t) = \sqrt{\langle r^2(t) \rangle_{av}} = 2\hat{\beta}(Dt)^{1/2} \quad (6.6)$$

with $D \simeq 25$ cm^2 s^{-1} in silicon [10, 13]. $\hat{\beta}$ is a time averaged scaling factor which accounts for high-density tracks due to ions of high stopping power. It relates the radius of the expanding track column to the ambipolar diffusion length L_D, defined from

$$r_{\text{rms}}(t) = \hat{\beta} L_D = 2\hat{\beta}(Dt)^{1/2} \tag{6.7}$$

$\hat{\beta}$ is obtained by assuming that the track density radial profile has a gaussian shape whose level falls to N_A at the track periphery. Thus, the track charge density n_D is,

$$n_D(r, t) = \left(\frac{N_D(0)}{4\pi Dt}\right) \exp\left(-\frac{r^2}{4Dt}\right) \tag{6.8}$$

At the track periphery, $n_D = N_A$, and as $r_{\text{rms}}(t) = 2\hat{\beta}(Dt)^{1/2}$,

$$\hat{\beta} = \left\langle \left[\ln\left(\frac{N_D(0)}{4\pi N_A Dt}\right)\right]^{1/2} \right\rangle_{\text{av}} \tag{6.9}$$

The subscript av indicates a time average over the duration of the funnel.

In the initial phase, charge neutrality holds strictly. In the second charge drift phase, the track particles begin to separate with electrons ultimately collected by the sensitive node. However, charge neutrality in the track is maintained even as the electron collection rate up the funnel equals the hole escape rate down the funnel, as the track begins to dissemble. The radial reduction rate of the track charge density N_D satisfies

$$\dot{N}_D(t) = -2\pi r_{\text{rms}}(t) J_p(t) = -4\pi N_A \hat{\beta} v_p (Dt)^{1/2} \tag{6.10}$$

because the escaping hole current $J_p = N_A v_p(t)$. The corresponding integral is given by

$$N_D(t) = N_D(0)\left[1 - \left(\frac{t}{\tau_c}\right)^{3/2}\right] \tag{6.11}$$

where τ_c is the time it takes to deplete the track charge density near the junction interface, as in Fig. 6.1. τ_c can be identified from (6.10) as

$$\tau_c = \left(\frac{3N_D(0)}{8\pi\hat{\beta}N_A v_p D^{1/2}}\right)^2 \tag{6.12}$$

Inserting (6.12) into the collected charge expression for a shallow depletion layer following (6.3) yields for the prompt charge produced, including both the funnel and shallow depletion layer using (6.1),

$$Q_c = \frac{3e\bar{N}_0 N_D(0)(\mu_n V_0)^{1/2}}{8\pi\hat{\beta}N_A v_p D^{1/2}} \tag{6.13}$$

It should be noted that whereas $\bar{N}_0$ is the mean track line charge density averaged over the length D_c, $N_D(0)$ is the initial track line charge density near the junction interface. The difference between the two is significant only for relatively long track lengths (large D_c). The preceding implies positive applied bias on a p-type substrate. For the opposite case of a negative bias on an n-type substrate, the roles of holes and electrons are reversed. Also, as the above Q_c (p-type) ~ $\mu_n^{1/2}$ that is, the charge funnel is longer in a p-type silicon substrate, where electrons are the carriers collected at the sensitive node, than vice versa by the factor $(\mu_n/\mu_p)^{1/2} \sim 3$ in silicon. As depicted in Figs. 6.2. and 6.3, this model is fairly well confirmed by experiments using heavy ions up to oxygen, as well as for alpha particles [7]. Again, strong evidence for the existence of the funnel phenomenon is confirmed by experiment, in that the collected charge is much greater than that from depletion layers alone [10, 11]. Previous discussions on

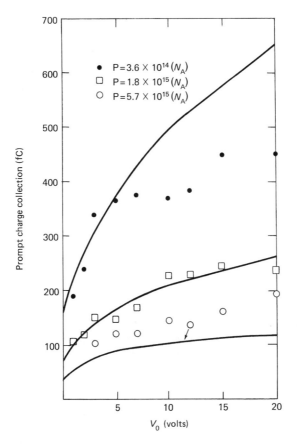

Figure 6.2. Charge collection measurements and model fit for Be ions versus applied bias. From Ref. 7.

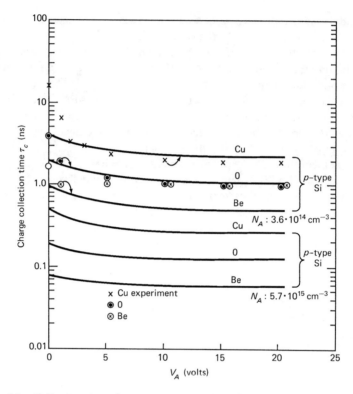

Figure 6.3. Collection times for two p-type samples of three ions versus applied bias. From Ref. 7.

further modification of the model includes a plasma screening factor based on the skin effect, which asserts that more accurate charge collection levels due to funneling can be obtained.

(2) This model approaches funnel charge creation and collection from a different vantage point, one which allows the inclusion of the ion's angle of incidence as a prime parameter [10]. This model was originally applied to alpha particles, but it is easily generalized to heavier ions. As seen in Fig. 6.4, the n^+p junction is struck by an ion, incident at an angle θ with respect to the device normal. The electric field due to an applied voltage V_0 drives the ion-track-ionization-produced holes down into the p-type substrate. The hole current density J_p creates an internal electric field $E(r, t)$ given by $J_p = \sigma E(r, t)$ in the depletion layer and in the funnel. The corresponding hole current I_p satisfies

$$I_p = g_1 e \mu_p E \tag{6.14}$$

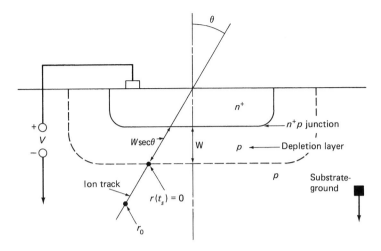

Figure 6.4. n^+p junction during incident ion strike. From Ref. 10.

where

g_1 = number of electron–hole pairs per centimeter generated by the ion

μ_p = hole mobility in the p-type silicon material

e = electron charge ($1.6 \times 10^{-19}C$)

The columnar ion-track charge thermalizes in picoseconds yielding a track with a radius of about 0.1 μm. The linear charge density $g_1 = 4 \times 10^8$ hep (hole-electron pairs) per centimeter, for alpha particles in silicon. This is equivalent to a track volumetric density $g_1/\pi r^2 \sim 10^{18}$ hep cm^{-3}. Due to ambipolar diffusion, the density is reduced to about 10^{16} hep cm^{-3} after about 100 ps.

Simultaneously with the preceding hole current, the electric field E drives an ion-track electron current I_n up the funnel toward the $r = 0$ coordinate. In Fig. 6.4, $r = 0$ is shown at the lower boundary of the depletion layer. These ion-track electrons possess a velocity v given by, using (6.14),

$$v = -\dot{r}(t) = \mu_n E(r, t) = \frac{\mu_n I_p(t)}{e\mu_p g_1} \qquad (6.15)$$

The negative sign multiplying $\dot{r}$ asserts that the electron current is in a direction toward smaller r as it moves under the influence of E. Consider an electron positioned at an extremity r_0 of the ion track at zero time, and that it is ultimately collected. Then integrating (6.15) gives the electron position coordinate $r(t)$ along its track for $t > 0$; that is,

$$r_0 - r(t) = \left(\frac{\mu_n}{eg_1\mu_p}\right) \int_0^t I_p(t')\,dt' \qquad (6.16)$$

An indicator of the ending of the charge collection process and the demise of the funnel occurs when no holes remain in the depletion layer. Let t_s be this time, with the hole charge $Q_p(t_s) = \int_0^{t_s} I_p(t')\,dt'$ which was contained in the depletion layer but is now collected in the substrate. Further, at this time all the track electrons originally between r_0 and $r(t_s) = 0$, the lower boundary of the depletion layer, have also been collected into the node regions, simply because of charge neutrality equilibrium conditions. Then at time t_s, (6.16) yields, because $r(t_s) = 0$,

$$r_0 = \left(\frac{\mu_n}{e\mu_p g_1}\right) Q_p(t_s) = \left(\frac{\mu_n}{e\mu_p g_1}\right)(eg_1 W \sec\theta) = \left(\frac{\mu_n}{\mu_p}\right) W \sec\theta \qquad (6.17)$$

as $Q_p(t_s) = eg_1 W \sec\theta$ is the collected hole charge that was contained in the depletion layer segment of the ion track, where W is the effective depletion layer width. It is assumed that the hole charge in the depletion layer on the n^+ side of the n^+-p junction is negligible. For the track electrons whose positions $r \leq r_0$ (i.e., within $r_0 = (\mu_n/\mu_p) W \sec\theta$ distance to the lower boundary of the depletion layer) are also collected by drift as already discussed. Therefore, the total funnel collection length is simply r_0 plus the length of the ion track in the depletion layer viz. $W \sec\theta$. Letting $\mathcal{L}$ be the funnel length and its vertical component depth $\mathcal{D} = \mathcal{L} \cos\theta$, so that,

$$\mathcal{L} = \left(1 + \frac{\mu_n}{\mu_p}\right) W \sec\theta, \qquad \mathcal{D} = \left(1 + \frac{\mu_n}{\mu_p}\right) W \qquad (6.18)$$

Because the spreading resistance plays a role in the development of funneling processes, as well as in many parts of semiconductor physics, a digression is apropos to explain it at this point. In most electro-physical structures where currents are exiting a confined volume and emerging in a flaring or spreading configuration, a spreading resistance is manifest, connecting this current and the applied voltage through Ohm's Law. There are a number of types of spreading resistance. One is termed base spreading resistance in a cylindrical bipolar transistor, where a portion of the current entering the base from the emitter above (Fig. 6.5a) is diverted in a transverse direction

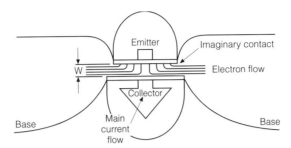

Figure 6.5a. Base current flow transverse to main emitter–collector flow in a bipolar cylindrical transistor. From Ref. 14.

into the base instead of proceeding vertically into the collector. Most of the spreading resistance is equivalent to the resistance model of a conductive disc (transistor body) where current enters from the top (emitter) and exits from the disc periphery (base), with none out of the bottom (collector). The first term in the base spreading resistance is $r_{sp} \cong 1/8\pi\sigma W$, where σ is the material conductivity and W is the disc thickness [12].

In the present case, the spreading resistance is due to the funnel contacting the substrate in Fig. 6.5b. The current exiting the funnel into the substrate spreads in a hemispherical-like shape under the depletion layer junction as in the figure. The funnel–substrate contact surface is elliptical with semidiameters a and $(a^2 - b^2)^{1/2}$. Axes are chosen so that the xy plane contains the junction contact surface [13], where the long dimension a is on the x axis. Transforming to ellipsoidal coordinates provides for an easy integration of Laplace's equation for the potential V from which the spreading resistance is derived. Recall that in the funnel, strict charge neutrality holds, so that any net charge vanishes, even at the contact surface, corresponding to the zero right-hand side of

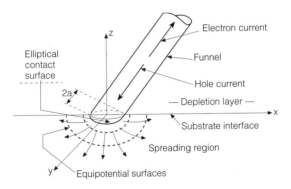

Figure 6.5b. Funnel contact at depletion layer–substrate interface.

Laplace's equation ($\nabla^2 V = 0$). The three transformation equations used from rectangular (x, y, z) to ellipsoidal (ξ, η, ζ) coordinates are

$$\frac{x^2}{u_i^2 - a^2} + \frac{y^2}{u_i^2 - b^2} + \frac{z^2}{u_i^2} = 1, \quad i = 1, 2, 3, \ a \geq b \geq 0 \quad (6.19)$$

where $u_1 = \xi$, $u_2 = \eta$, and $u_3 = \zeta$; or explicitly for $\xi > a > \eta > b > \zeta > 0$,

$$x = \left(\frac{(\xi^2 - a^2)(\eta^2 - a^2)(\zeta^2 - a^2)}{a^2(a^2 - b^2)}\right)^{1/2} \quad (6.19a)$$

$$y = \left(\frac{(\xi^2 - b^2)(\eta^2 - b^2)(\zeta^2 - b^2)}{b^2(b^2 - a^2)}\right)^{1/2} \quad (6.19b)$$

$$z = \frac{\xi \eta \zeta}{ab} \quad (6.19c)$$

In these coordinates, Laplace's equation for the potential V becomes separable into

$$\frac{d}{d\xi}\left(\sqrt{(\xi^2 - a^2)(\xi^2 - b^2)}\,\frac{dV}{d\xi}\right) = 0, \quad V(\infty) = 0, \ V(a) = V_0 \quad (6.20)$$

with the same holding for the other variables η and ζ. V_0 is the voltage with respect to infinity on the ellipse whose equation is $\xi = a$ in ellipsoidal coordinates. Integrating (6.20) twice gives

$$V = V_0 \frac{\int_\xi^\infty d\xi' \,[(\xi'^2 - a^2)(\xi'^2 - b^2)]^{-1/2}}{\int_a^\infty d\xi' \,[(\xi'^2 - a^2)(\xi'^2 - b^2)]^{-1/2}} \quad (6.21)$$

Letting $\xi' = a/\sin u$ in the above integrals yields

$$V = V_0 \frac{\int_0^{\sin^{-1}(a/\xi)} du\,[(1 - k^2 \sin^2 u)]^{-1/2}}{\int_0^{\pi/2} du\,[(1 - k^2 \sin^2 u)]^{-1/2}}, \quad k^2 = \left(\frac{b}{a}\right)^2 \quad (6.22)$$

where the contact ellipse semiaxes are a and $(a^2 - b^2)^{1/2}$. In terms of elliptic integrals, the potential V is, from (6.22),

$$V = V_0 \frac{\mathrm{sn}^{-1}(a/\xi, k)}{\mathrm{sn}^{-1}(1, k)} \quad (6.23)$$

where

$$F(\phi, k) \equiv \text{sn}^{-1}(\sin \phi, k) = \int_0^\phi \frac{dx}{(1 - k^2 \sin^2 x)^{1/2}},$$

$$F(\pi/2, k) \equiv \text{sn}^{-1}(1, k) = \int_0^{\pi/2} \frac{dx}{(1 - k^2 \sin^2 x)^{1/2}} \quad (6.24)$$

are respectively the elliptic integral of the first kind and the complete elliptic integral of the first kind.

Current density, conductivity, and electric field are related by $J = \sigma E$, where $E = -\partial V/\partial r$ and $J \approx I/2\pi r^2$ for the current density. Using (6.23) and (6.19) to transform back to spherical coordinates results in

$$I = 2\pi\sigma \left| r^2 \frac{\partial V}{\partial r} \right| \sec \theta = \frac{2\pi\sigma a V_0 \sec \theta}{\text{sn}^{-1}(1, k)} \quad (6.25)$$

where $\sec \theta$ accounts for the angle of incidence. Then the spreading resistance is readily identified from $I = V_0/r_{sp}$ as

$$r_{sp} = \frac{\text{sn}^{-1}(1, k)}{2\pi\sigma a \sec \theta} = \frac{\text{sn}^{-1}(1, k)}{2\pi a e \mu_p N_A \sec \theta} \quad (6.26)$$

Then the conductivity $\sigma = e\mu_p N_A$ is a consequence of the fact that the track hole current, which is the current descending from the funnel, has a current density $J = pev_p = \sigma E$. From the definition of hole mobility $v_p = \mu_p E$ and the hole density $p \approx N_A$, the acceptor dopant density, the identification that $\sigma = e\mu_p N_A$ in the acceptor-doped material, can be made. For a normally incident funnel, $\theta = 0$ so that the contact surface is circular of radius a and thus $b = 0$. Then $k = 0$ and $\text{sn}^{-1}(1, 0) = \pi/2$. Then from (6.26) the spreading resistance is

$$r_{sp} = \frac{1}{4\sigma a} = \frac{1}{4ae\mu_p N_A} \quad (6.27)$$

If the funnel contact is at a grazing angle, where b approaches a, then $k = 1 - \epsilon$, where $\epsilon \ll 1$, giving $\text{sn}^{-1}(1, k) \sim \ln(4a/b)$. For the corresponding contact surface $r_{sp} \sim [\ln(4a/b)](2\pi\sigma a)^{-1}$.

For high substrate dopant density greater than about 10^{17} cm^{-3}, the currents begin to be marginally contained within the funnel ion track, especially in the later phases of the funnel existence. This is because the ratio of funnel charge density to background dopant density is much less than for lightly doped material. As a result, the current begins to spread into the substrate beyond the depletion region. For a track of circular cross section with $r_{sp} = 1/4\sigma a$, the electric field

$$E = \frac{J}{\sigma} = \frac{I}{2\pi a^2 \sigma} = \left(\frac{2r_{sp}}{\pi a}\right) I \quad (6.28)$$

Using this electric field in (6.15) and the derivation following it yields the corresponding collection depth for high dopant density substrates

$$\mathcal{L}_r = \left[1 + \left(\frac{2r_{sp}eg_1}{\pi a}\right)\left(\frac{\mu_n}{\mu_p}\right)\right] W \sec\theta \tag{6.29}$$

When $\mathcal{L}$ and $\mathcal{L}_r$ are evaluated numerically, they yield similar results, but both yield an approximate $N_A^{-1/2}$ dependence [10]. This suggests that the models are not sensitive to assumptions regarding current flow patterns. For n-type substrates μ_n/μ_p is replaced by μ_p/μ_n. There is a corresponding funnel on the n^+ side of the n^+p junction, but evaluation of $\mathcal{L}_r$ results in a negligible length, as alluded to earlier.

After the ion strike, the voltage drop in the quasineutral region is given by the product of the carrier current I_0 and the following sum: (a) the track resistance in the depletion layer plus (b) the spreading resistance at the depletion layer where the track enters the substrate. The carrier current I_0 corresponds to the applied voltage V_0 plus the junction diode drop (0.7 V in silicon) across the quasineutral region; that is, I_0 is given by

$$I_0 = (V_0 + 0.7)G \tag{6.30}$$

where the conductance G is obtained from

$$G^{-1} = \frac{\mathcal{L}'}{eg_1(\mu_n + \mu_p)} + \frac{1}{4ae\mu_p N_A} \tag{6.31}$$

The first term is the resistance of the ion-track segment that spans the depletion layer, where $\mathcal{L}'$ is the length of that track segment. The denominator is its resistance per unit length. The second term is the spreading resistance reckoned from the depletion layer end of the track down into the substrate [5].

The depletion layer width $W(t)$ varies during the funneling process. From (6.17) for $t \leq t_s$,

$$W(t)\sec\theta = \int_0^t \frac{I_p(t')\,dt'}{eg_1} = \int_0^t \frac{I(t')\,dt'}{(1+\mu_n/\mu_p)eg_1} \tag{6.32}$$

where $I = I_p + I_n$ and $p \simeq n$ in the depletion region. It is well known that the voltage drop across the depletion layer is approximately $eN_AW^2/2\epsilon\epsilon_0$, where ϵ is the dielectric constant and ϵ_0 is the permittivity of free space. Then the voltage drop $V_{qn}(t)$ in the quasineutral region is

$$V_{qn}(t) = V_0 + 0.7 - \frac{eN_AW^2}{2\epsilon\epsilon_0} \tag{6.33}$$

Then,

$$I(t) = V_{qn}(t)G \tag{6.34}$$

where the conductance G is time independent. Eliminating W and V_{qn} in (6.32)–(6.34) results in a differential equation for the collected charge $Q(t) = \int_0^t I(t')\,dt'$, namely

$$\dot{Q}(t) = I_0 - GmM^2 Q^2(t), \qquad Q(0) = 0 \tag{6.35}$$

where $m = eN_A/2\epsilon\epsilon_0$ and $M = \cos\theta[(1 + \mu_n/\mu_p)eg_1]^{-1}$. Its integral is

$$Q = \left(\frac{I_0/G_m}{M}\right)^{1/2} \tanh[(I_0 G_m)^{1/2} Mt] \tag{6.36}$$

and the total collected charge is $Q(\infty) = (I_0/Gm)^{1/2} M^{-1}$. The corresponding current $I = \dot{Q}$ is

$$I = I_0 \operatorname{sech}^2[(I_0 Gm)^{1/2} Mt] \tag{6.37}$$

which is depicted in Fig. 6.6 for two different background dopant densities. It should be noted that $I(t)$ in the figure has an initial jump discontinuity because $I(0-) = 0$ and $I(0+) = I_0$, which is an artifact of the model. Actually, $I(t)$ has a finite initial slope increasing to a maximum at a very early time [10].

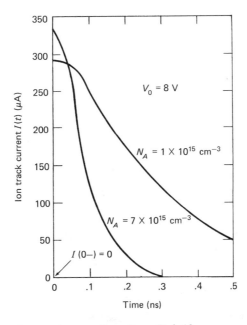

Figure 6.6. Ion-track current versus time. From Ref. 10.

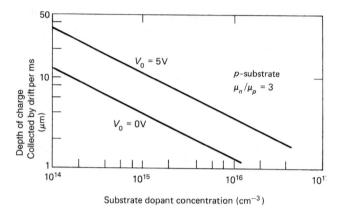

Figure 6.7. Funnel depth versus substrate dopant density. From Ref. 10.

The collected charge $Q(t)$ (Fig. 6.7) and funnel depth (Fig. 6.8) as given above have been verified experimentally, at least for alpha particles as the SEU-inducing ions [10].

Besides providing detailed descriptions of the funneling process, the two major parameters that these and other funneling models should provide are (a) the additional prompt charge collected due to the funnel and (b) the effective funnel length or depth. Both (a) and (b) are of immediate use in the computation of critical charge for the device of interest.

The preceding two funneling models provide similar results for the additional charge collected due to the funnel. Differences lie in the computation of the charge collection time. The latter model derives this parameter assuming that the longitudinal (along the track) charge separation is operative throughout the funnel life existence. The former model derives the charge collection time only from charge separation in the radial direction (i.e., normal to the track). This is because in established computer simulations [5], longitudinal charge separation is not manifest until the late phase of the funneling process, when the ion-track densities approach the background dopant densities. Both models are in reasonable agreement as to charge collection in the p-type substrate junction example used, except for low values of bias ($V_0 \leq 5$ V). For the n-type substrate case, the latter model apparently underestimates the charge collected and the funnel depth as validated by expermment. For p-type substrates, the derived funnel lengths of both models are in sensible agreement, the funnel length is about three times the depletion layer width in silicon.

6.3. Track Transport

To understand the kinematics of SEU-inducing ions within and without their ionization tracks, it is necessary to investigate their detailed transport proper-

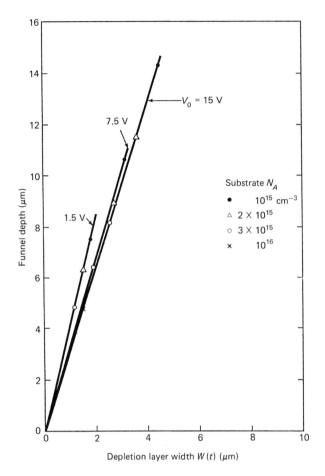

Figure 6.8. Funnel depth versus depletion layer width compared with experimental data. From Ref. 10.

ties. By this is meant the ion–material interactions as well as ion–ion interactions, insofar as they affect the ultimate production of single event upset errors either hard or soft. A typical ionization track contains about 4×10^{18} hole–electrons pairs per cubic centimeter generated in the material by a single incident ion. For an assumed cylindrical track, it is generally agreed that the initial track radius is of the order of 0.1 μm [5, 15]. The time required to create the initial track of ionized charge plasma is less than 10 ps from the time of arrival of the incident ion. The very high density of initial charge density in the track implies ambipolar transport, as mentioned earlier. This means that the electrons and holes in the track are in a "lockstep" which causes them on average to diffuse in train. This corresponds to an effective diffusion coefficient whose value lies between that of the slower and faster particles in the track. The ambipolar state also implies, at least, quasineutrality (i.e., $p - n \cong p_0 - n_0$, where the zero subscripts imply initial values). The corresponding ambipolar transport equation [16] for the excess hole concentration δp is given by

$$\frac{\partial}{\partial t}(\delta p) = D^* \nabla^2 (\delta p) - \mu^* \mathbf{E} \cdot \nabla (\delta p) - \frac{\delta p}{\tau} \quad (6.38)$$

where D^* is the ambipolar diffusion coefficient, for both holes and electrons because they move together, given by

$$D^* = \frac{(n_0 + p_0 + 2\delta p) D_n D_p}{(n_0 + \delta n) D_n + (p_0 + \delta p) D_p} \quad (6.39)$$

with D_n and D_p being the diffusion coefficients of electrons and holes, respectively. δn and δp are excess particle concentrations over their respective equilibrium values n_0 and p_0. The ambipolar mobility μ^* is given by

$$\mu^* = \frac{(n_0 - p_0) \mu_n \mu_p}{(n_0 + \delta n) \mu_n + (p_0 + \delta p) \mu_p} \quad (6.40)$$

$\mathbf{E}$ is the electric field within the track and τ is the excess ambipolar carrier lifetime defined implicitly by

$$\frac{\delta n}{\tau} = \frac{n_0 + \delta n}{\tau_n} - \frac{n_0}{\tau_{n0}} = \frac{\delta p}{\tau} = \frac{(p_0 + \delta p)}{\tau_p} - \frac{p_0}{\tau_{p0}} \quad (6.41)$$

where the subscripts n and p refer to electrons and holes, respectively. Equation (6.38) holds for δn as well, with the same definitions for the pertinent parameters.

The corresponding initial and boundary conditions for (6.38) are given as

1. An initial condition on the form of the charge generation function $\delta p(r, 0)$ in that it is radially symmetric, [e.g.,

$$\delta p(r, 0) = \left(\frac{N}{\pi b^2}\right) \exp\left[-\left(\frac{r}{b}\right)\right]^2 \quad (6.42)$$

where N, the track lineal density, bounds are approximately $10^7 \leq N \leq 10^{11}$ hep/cm^{-1}, and b is an assumed initial track radius. Such

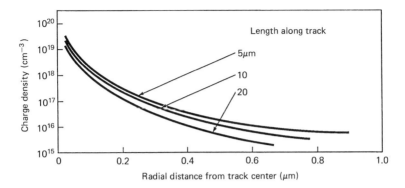

Figure 6.9. Radial charge density in silicon along 20-GeV iron ion track. From Ref. 17.

track density profiles are depicted in Figure 6.9 obtained from Monte Carlo simulations of their radial densities [17]. These are of such high density and their radial gradients are so extremely large compared to their axial ones that the ambipolar mobility $\mu^* \cong 0$. Equation (6.40) yields $n_0 - p_0 \cong n - p \cong 0$ and charge neutrality is strictly held by the initial electric field, and the track particles are now essentially frozen in place with zero net motion, except for ambipolar diffusion, by which the track now begins to expand in the radial direction.

2. For relatively long times after track formation, its carrier density is reduced in that D^* and μ^* revert to their prestrike values D and μ, respectively.

In rectangular coordinates, the solution to (6.38) for a constant electric field E_c is

$$\delta p(x, y, t) = \left(\frac{N}{4\pi Dt}\right) \exp\left(-\frac{(x - \mu_i E_c t)^2 + y^2}{4Dt + t/\tau}\right) \quad (6.43)$$

with a modified mobility defined by

$$\mu_i = \frac{n_i \mu_n \mu_p}{n_i \mu_n + (\mu_n + \mu_p) \delta p} \quad (6.44)$$

From (6.43), it is seen that for early times, the track charge is diffusing radially in an ambipolar fashion, and with no drift due to the electric field. However, for relatively long times of about 500 ps after track formation, the charge begins to drift due to the electric field as the track charge density falls to levels comparable with the background material dopant density. Therefore, in the early diffusional phases, the track only expands radially, as mentioned. This is followed by a second phase characterized by charge motion along the track, driven by the electric field.

When the SEU-inducing ion penetrates the junction at normal incidence to traverse the junction depletion layer, the governing equation connecting excess hole density δp and its current $\mathbf{J}_p$ is

$$\frac{\partial}{\partial t}(\delta p) = \frac{\nabla \cdot \mathbf{J}_p}{e} - \frac{\delta p}{\tau} \tag{6.45}$$

In the initial phase, not only is drift parallel to the track negligible but so is the corresponding diffusion term $-eD_p\nabla(\delta p)$; that is, $-eD_p\partial/\partial x(\delta p)$ in the equation complementary to (6.45), namely

$$\mathbf{J}_p = e\delta p\mu_p \mathbf{E} - eD_p\nabla p. \tag{6.46}$$

Thus, now from (6.46), $J_p \cong e\mu_p \delta pE$, and differentiating yields $\partial J_p/\partial x = e\mu_p\delta pE'(x)$. Inserting this $\partial J_p/\partial x$ into (6.45) yields, for the one-dimensional case,

$$\frac{\partial}{\partial t}(\delta p) = -\left(\mu_p E'(x) + \frac{1}{\tau}\right)\delta p, \quad \lim_{t \to 0}(4\pi Dt\delta p) = N \tag{6.47}$$

Integrating with respect to time gives

$$\delta p = \left(\frac{N}{4\pi Dt}\right)\exp\left[-\left(\mu_p E' + \frac{1}{\tau}\right)t\right] \tag{6.48}$$

Note that the time-dependent area through which the current flows is $x^2(t)$, where $x^2(t) = 4Dt$, a mean square diffusion distance. $E'(x)$ is readily available from the corresponding Poisson's equation $\nabla \cdot \mathbf{E} = E'(x) = \rho/\epsilon\epsilon_0$, where ρ is the included charge density; that is, $E'(x) = -eN_D/\epsilon\epsilon_0$, because $\rho = -eN_D$.

The total δp can be constructed empirically by adding a term to (6.48), representing the high-density electrons from the track. It is characterized by a time constant $\beta^{-1} = t_0$. After a relatively long time, the resultant expression is given by [5]

$$\delta p \sim \left(\frac{N}{4\pi Dt}\right)[\exp(-\mu_p E't) - \exp(-\beta t)] \tag{6.49}$$

Using the time-varying area above, the corresponding current I_p can be written by using (6.46), by neglecting its diffusion term, to give

$$I_p = J_p(4\pi Dt) = 4\pi Dt e \delta p \mu_p E_0 = -e\mu_p N E_0 [\exp(-\mu_p E't) - \exp(-\beta t)] \tag{6.50}$$

where the maximum value of the junction depletion layer electric field $E_{max} \equiv E_0$, where E_0 is

$$E_0 = \left(\frac{2eN_D(-V_0 + \phi_0)}{\epsilon\epsilon_0} \right)^{1/2} \tag{6.51}$$

a well-known quantity [16]. ϕ_0 is the contact potential of the junction.

The total current due to both δp and δn is obtained by summing an expression similar to (6.49) for δn with (6.50) to yield

$$I(t) = -e\bar{\mu}NE_0[\exp(-\bar{\mu}E't) - \exp(-\beta t)] \tag{6.52}$$

with $\bar{\mu} = \frac{1}{2}(\mu_n + \mu_p)F(E)$. $F(E)$ supplies a mobility correction factor in the transition from the ambipolar phase to asymptotically long time values [5]. Hence, the SEU current pulse $I(t)$ is given by the difference of two exponential terms. The track decay time constant $(\bar{\mu}E')^{-1} = \epsilon\epsilon_0/e\mu_p N_D$ is the "RC time constant" of the junction field region [5].

The total SEU charge collected, neglecting the diffusion component, is obtained from (6.52) as

$$\int_0^\infty I(t)\,dt = \begin{cases} -eN\ell, & \ell < X_j, \\ -eNX_j, & \ell \geq X_j \end{cases} \tag{6.53}$$

where ℓ is the track distance into the depletion layer and X_j is the junction width [5].

6.4. SEU Cross-Section Morphology

If all the cells in an IC memory were physically and electronically identical, the curve of SEU cross section (σ_{seu}) versus L_{eff} would consist of a step function at the threshold linear energy transfer (LET) (L_{th}), with a constant σ_{seu} for increasing $L_{eff} > L_{th}$ as in Fig. 6.10 [18]. The step at L_{th} would correspond to the critical charge Q_c. However, it is well known that the experimentally determined σ_{seu} is a convex downward curve with a critical threshold LET as seen in the preceding figure. Also, an asymptotic value of σ_{seu} (σ_{asy} or σ_{sat}) can be

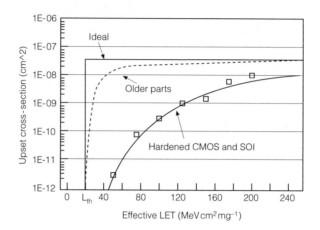

Figure 6.10. Representative SEU cross sections. The step function depicts the ideal where all IC memory cells are identical. From Ref. 18.

discerned in the limit of large L_{eff}, especially for proton-induced SEU cross sections. One reason for the actual behavior of σ_{seu} is that memory cells are not identical due to manufacturing vagaries—some are more sensitive than others. Thus, individual cells can upset over a range of L_{eff} about a threshold LET. Other reasons [19] have to do with accelerator beam experimental difficulties, and the applicability of LET as a simple descriptor for SEU processes is still open to question (Section 3.9).

A model follows to yield σ_{seu} versus L_{eff} for heavy ions [19], as from accelerator beams. It attempts to accommodate key considerations regarding σ_{seu} discussed throughout this treatise while providing additional insights into the SEU phenomenon [19].

Assume that the cell volume consists of a rectangular parallelepiped with dimensions a and b and thickness c. Using the methods of Section 1.4, a probability density distribution of parallelepiped chord lengths ℓ is obtained, where $p(\ell)\,d\ell$ is the probability density that an incremental chord length lies between ℓ and $\ell + d\ell$. For the accelerator beam, consider parallel chords as shown in Fig. 6.11. The determination of the chord length distribution $p(\ell)$ in this case devolves to a two-dimensional problem [19, 20]. $p(\ell)$ can be decomposed into two terms $p_1(\ell)$ and $p_2(\ell)$. $p_2(\ell)$ corresponds to the SEU occurrence at the incident angle $\theta = \theta_c$ with a probability of unity, where the corresponding chord length $\ell \geq c \sec \theta$. Thus,

$$p_2(\ell) = c_2 \delta(\ell - c \sec \theta_c), \quad \ell \geq c \sec \theta_c \tag{6.54}$$

c_2 is a constant to be determined. The delta function is chosen because its integral, the cumulative probability, $P(\ell) = \int_\ell^\infty p(\ell')\,d\ell' = 1$, by virtue of the

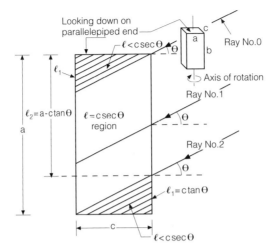

Figure 6.11. End view of sensitive volume parallelepiped showing incident ion rays from accelerator ion beam. Note that the subscript on θ_c are dropped.

delta function property that $\int_a^b \delta(x)\,dx = 1$ for $a \leq x \leq b$. Other properties are discussed in Problem 1.11.

From Fig. 6.11, it is seen that there are chords whose lengths are less than $c \sec \theta_c$ (i.e., for $0 \leq \ell < c \sec \theta_c$). From geometric probability considerations discussed in Section 1.4, these chord lengths occur with equal probability $p_1(\ell)$. So, in this region,

$$p_1(\ell) = c_1, \quad 0 \leq \ell < c \sec \theta_c \tag{6.55}$$

where c_1 is a second constant to be determined. Then a total probability $p(\ell) = p_1(\ell) + p_2(\ell)$, where p_1 and p_2 appear according to their chord length range of validity. c_1 and c_2 are obtained from the following two conditions:

(a) Because the cumulative probability $P = \int_0^\infty p(\ell)\,d\ell = 1$ then using (6.54) and (6.55), where θ_c subscripts on θ are now dropped,

$$P_1 + P_2 = 1 = \int_0^{c \sec \theta_c} p_1(\ell)\,d\ell + \int_{c \sec \theta_c}^\infty p_2(\ell)\,d\ell = c_1 c \sec \theta + c_2 \tag{6.56}$$

where the cumulative probabilities $P_1 = c_1 c \sec \theta$ and $P_2 = c_2$.

(b) The second condition is found by computing the ratio P_1/P_2. From Fig. 6.11 it can be appreciated that the probabilities P_1 and P_2 are each proportional to the lengths of their respective chord type, which are, in turn, proportional to their projections onto the parallelepiped edges i.e., proportional to ℓ_1 and ℓ_2 in the figure. The two chord types are those whose chord lengths are less than $c \sec \theta$, and those whose lengths equal $c \sec \theta$. Thus, with (6.56)

$$P_1/P_2 = \frac{2\ell_1}{\ell_2} = \frac{2c \tan \theta}{a - c \tan \theta} \tag{6.57}$$

where the factor of 2 in the numerator accounts for the (smaller) chord lengths at both ends of the parallelepiped.

Then, using (6.56) and (6.57) yields

$$c_1 = \frac{2 \sin \theta}{a + c \tan \theta}, \quad c_2 = \frac{a - c \tan \theta}{a + c \tan \theta} \tag{6.58}$$

Then the respective probabilities can be written for their corresponding ranges as

$$p_1 = \frac{2 \sin \theta}{a + c \tan \theta}, \quad 0 \leq \ell < c \sec \theta, \quad p_2 = \left(\frac{a - c \tan \theta}{a + c \tan \theta}\right) \delta(\ell - c \sec \theta), \quad \ell \geq c \sec \theta \tag{6.59}$$

To obtain σ_{seu} as a function of L_{eff}, the above geometric probabilities are converted to charge deposition probabilities i.e., to $p_{\text{cd}}(Q) \, dQ$, which is defined as the probability that a charge between Q and $Q + dQ$ is deposited by an ion into the memory cell. The relationship between $p(\ell)$ and p_{cd} is simply given by

$$p_{\text{cd}}(Q) Q_{\text{av}} = p(\ell) c \sec \theta \tag{6.60}$$

and Q_{av} is the mean charge deposited along the chord length $c \sec \theta$. Therefore, from (6.59) and (6.60) for the various ranges of ℓ values,

$$p_{\text{cd}} = \left(\frac{2c \tan \theta}{a + c \tan \theta}\right) Q_{\text{av}}, \quad 0 \leq \ell < c \sec \theta; \quad p_{\text{cd}} = \left(\frac{a - c \tan \theta}{a + c \tan \theta}\right) p_G, \quad \ell \geq c \sec \theta \tag{6.61}$$

where p_G is a gaussian approximation replacement for the computationally untenable delta function [19]; that is,

$$p_G = \left(\frac{1}{\sqrt{2\pi s_Q}}\right) \exp\left[-\frac{1}{2}\left(\frac{Q - Q_{\text{av}}}{s_Q}\right)^2\right] \tag{6.62}$$

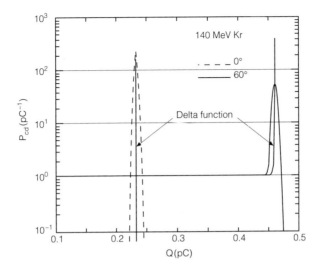

Figure 6.12. Calculation of charge deposition probabilities for krypton ions incident on a CMOS/SOS memory cell with dimensions $a = 3\,\mu m$, and $c = 0.55\,\mu m$. From Ref. 19.

and Q_{av} can be computed from $L_{eff}c \cong 96.6 Q_{av}$, where $[Q_{av}] \doteq pC$ and $[L_{eff}] \doteq$ MeV cm^2 mg^{-1}. The standard deviation is given by [21] $s_Q = (4.46 \times 10^{-5} Q_{av} \overline{\delta^2}/\overline{\delta})^{1/2}$ and $\overline{\delta^2}/\overline{\delta}$ is the ratio of the second moment to the first moment of the energy deposited per collision by the incident ion [21]. It is approximated by [21] $\overline{\delta^2}/\overline{\delta} \cong 0.4443 t_{ion}^{0.655}$, where t_{ion} is the incident ion energy in units of MeV per nucleon (Section 3.9). Figure 6.12 depicts p_{cd} with and without the delta function approximation for two θ values.

A function $P_R = 1 - \exp[-\beta(Q/Q_c)^\alpha]$, where β and α are empirical data-fitting parameters [21], is introduced as the mean SEU probability of a cell when a charge Q is deposited therein. Equivalently, P_R provides the sensitive volume fraction that is susceptible to SEU for a given deposited charge Q. P_R is needed to yield charge deposition probabilities in the cell, as the exact response of a given cell to an incident ion is essentially not calculable [21]. This is, in part, because the precise shape of the sensitive volume as well as the corresponding chord distribution function are generally not available.

With the above $p_{cd}(Q)$ and $P_R(Q)$, the SEU cross section can be written as

$$\sigma_{seu} = \sigma_{asy} \int_{Q_c}^{\infty} p_{cd}(Q) P_R(Q) \, dQ \qquad (6.63)$$

that is, $p_{cd}(Q) \, dQ$ is the probability of depositing charge between Q and $Q + dQ$ in the memory cell. P_R is the average probability that the memory cell experi-

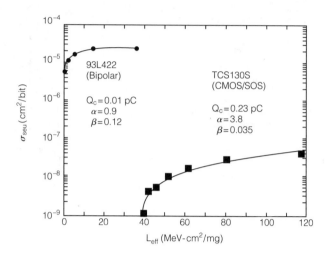

Figure 6.13. Comparison of model calculations with experimental cross-section measurements for two memory technologies. From Ref. 21.

ences an SEU when this charge is deposited therein. Thus, the product $P_R p_{cd} dQ$ is the probability that an SEU will be caused by the deposition of charge dQ about Q. Hence, the integral over Q is proportional to σ_{seu}, as in (6.63).

Figure 6.13 depicts the results of calculations of σ_{seu} from (6.63) compared to experimental cross section data for the 93L422 bipolar SRAM and the TCS-130S CMOS/SOS SRAM. The model calculations of σ_{seu} compare favorably with the experimental data as seen in the figure [21].

6.5. Device SEU Scaling

To appreciate how single event phenomena (SEP) impact device and circuit design parameters, certain scaling relations between them have been identified. They provide estimates of SEU susceptibility to aid in extrapolation of present microcircuit design and development to future product versions. They are pertinent for planning new process design rules as well as for future system development.

To gain insight into the SEP problems facing suppliers and users, Fig. 6.14 illustrates the evolution of microelectronics integrated-circuit chip sizes and circuit packing densities. It is seen in the figure that the number of bits per chip as well as circuits per chip are increasing at an almost exponential rate with time, whereas individual device feature size is decreasing at a similar rate. This bodes ill in terms of SEU mitigation for these devices because of the greatly reduced critical charge Q_c required to cause SEU, as individual device dimensions are becoming so small.

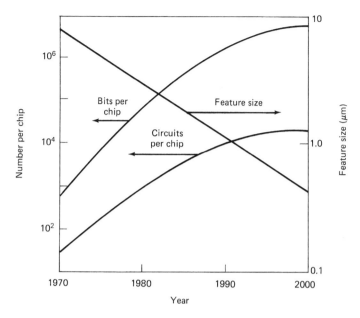

Figure 6.14. Microcircuit density and dimensions versus production year. From Ref. 14.

Feature size is usually considered as the smallest dimension on a lithographic processing mask for a particular microcircuit. It is usually the width of a stripe (circuit connection, line, trace, via, lead) and is normally much smaller than the size of an integrated-circuit device element. For example, a microcircuit bipolar transistor emitter area might be $10 \times 15 \, \mu m^2$, whereas its corresponding stripe width could be only $2 \, \mu m$. For MOSFETs, the feature size is also the length of the channel between the source and drain.

Another parameter that measures microcircuit evolution in terms of information handling capability is the functional throughput rate (FTR) [22]. For a particular digital family or part type, it is defined as the product of the number of gates and its clock rate. For constant FTR, the number of gates and clock rate are then hyperbolically related and shown as iso-FTR lines with a common slope in log–log coordinates in Fig. 6.15. For increasing FTR. the required switching energy per bit decreases to maintain the power rating of the part at levels commensurate with the associated circuitry.

A method for scaling of MOSFET devices [23] is discussed, it is a forerunner of that to be used for SEP purposes. The scaling rules are developed around the criterion that the electric fields remain invariant through the scaling transformation equations; that is, the scaling rules derived maintain constant electric fields for varying dimensions and other device parameters. Specifi-

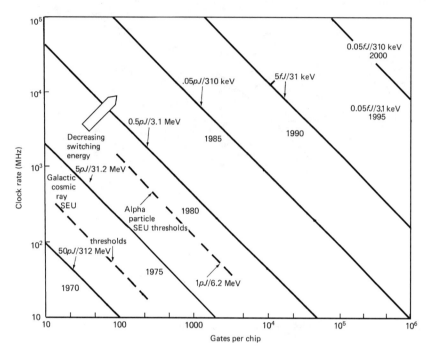

Figure 6.15. Clock rate versus gate number per chip showing iso-FTR lines labeled with switching energy in picojoules and MeV. From Ref. 22.

cally, to design a device smaller than a nominal prototype, its dimensions, voltages, and dopant concentrations are scaled by a constant factor to reproduce the same performance. The scaling rules are derived by noting that depletion layer widths are reduced in proportion to the device dimensions, due to the scaled-down voltages and scaled-up dopant densities within the device; that is, the depletion layer width ℓ can be scaled down to a smaller width ℓ', for $N_A \ll N_D$, where the well-known expression for depletion layer width is given by

$$\ell' = \left(\frac{2\epsilon\epsilon_0(\phi_0 + |V_0|/k)}{ekN_A} \right)^{1/2} \cong \frac{\ell}{k} \quad (6.64)$$

and k is the scale factor. The built-in contact potential $\phi_0 \ll |V_0|/k$ is the usual case, so that this scaling rule is a very good approximation. Similarly for the decrease in MOSFET gate threshold voltage V_T, for n^+ polysilicon gate termination and $|V_{\text{sub}}| \ll 2V_F$,

$$V_T' \cong \left(\frac{t_{\text{ox}}}{k\epsilon_0\epsilon_{\text{ox}}} \right) \left[-Q_{\text{ss}} + \left(\frac{2\epsilon_0\epsilon_{\text{si}}kN_A|V_{\text{sub}}|}{k} \right)^{1/2} \right] = \frac{V_T}{k} \quad (6.65)$$

where

t_{ox} = gate oxide thickness
ϵ_{ox} = dielectric constant of SiO_2
ϵ_{si} = dielectric constant of silicon
Q_{ss} = gate insulator (SiO_2) charge density
V_{sub} = substrate voltage
N_A = acceptor dopant density
V_F = fermi voltage

because (6.64) and (6.65) imply a proportional reduction in all voltages and dimensions, it is seen that the corresponding electric field remains invariant (i.e., $V'/x' = V/x$). These scaling rules have implications for the critical charge of a sensitive node in the device. Consider the node sensitive volume to be a parallelepiped so that it can be modeled as a parallel-plate bit storage capacitor $C = \epsilon\epsilon_0 A_r/t$, where t is its thickness, A_r is the plate area, and V is the voltage across C. Then the preceding scaling rules yield for the scaled-down critical charge Q_c'

$$Q_c' = C'V' = \left(\frac{\epsilon\epsilon_0 A_r/k^2}{t/k}\right)\left(\frac{V}{k}\right) = \frac{Q_c}{k^2} \tag{6.66}$$

so that the scaled-down critical charge is inversely proportional to k^2.

In terms of part-type family technologies, the critical charge increases exponentially with feature size, as seen in Fig. 6.16. Interestingly, the critical charge seems independent of the semiconductor material, whether silicon or gallium arsenide, or others, or device technology, as seen in the figure.

With respect to the funnel length $\mathcal{L}$, its vertical projection $\mathcal{D}$, and the charge it collects Q_{coll} as a result of scaling, the Hu [10] funnel model asserts that there are two cases:

(a) Low dopant density ($N_A \ll 10^{16}$ cm^{-3})

$$\mathcal{L}_{low}' = \left(1 + \frac{\mu_n}{\mu_p}\right)\left(\frac{1}{k}\right)\sec\theta = \frac{\mathcal{L}}{k}, \quad \mathcal{D}_{low}' = \left(1 + \frac{\mu_n}{\mu_p}\right)\left(\frac{1}{k}\right) = \frac{\mathcal{D}}{k} \tag{6.67}$$

(b) High dopant density ($N_A \gg 10^{16}$ cm^{-3})

$$\mathcal{L}_{high}' = (kN_A)^{-1/2} = \frac{\mathcal{L}}{k^{1/2}}, \quad \mathcal{D}_{high}' = \frac{\mathcal{D}_{high}}{k^{1/2}}, \quad Q_{coll}' = \frac{Q_{coll}}{k} \tag{6.68}$$

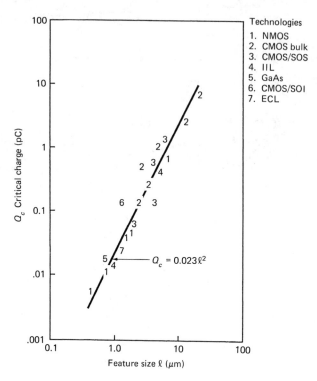

Figure 6.16. SEU critical charge versus feature size. From Ref. 24.

The McLean–Oldham funnel model [6, 7, 9] does not address a dopant concentration difference but yields

$$\mathscr{L}' \sim \left[\left(\frac{V_0}{k}\right) N_A^{-2/3} k^{-2/3}\right]^{1/2} = \frac{\mathscr{L}}{k^{5/6}}, \qquad Q'_{\text{coll}} = Q_{\text{coll}} k^{5/6} \qquad (6.69)$$

Q_{coll} includes the charge collected in the depletion layer, as well as that in the funnel proper.

To investigate how scaling affects SEU error rate, E_r, from (5.4) is again used and the steps of the "approximate" calculation of SEU to obtain E_r as in (5.8) are retraced. However the scaling factor k is introduced as appropriate in these steps. For example, $s'_{\text{max}} = s_{\text{max}}/k$, where, recall, s_{max} is the main diagonal of the parallelepiped sensitive region and $\mathscr{L}' = \mathscr{L}/k$. The "exact" calculation consists of a numerical integration of (5.4) using the known differential cosmic ray flux φ_L, the actual integral chord distribution $C(s(L))$, and including the scale factor k as it appears in the development [25]. Fig. 6.17 depicts a

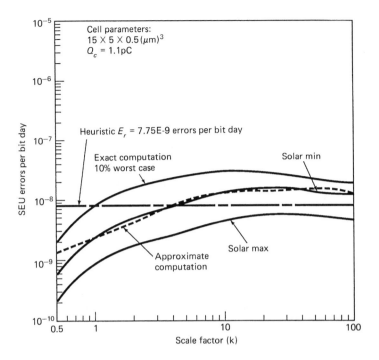

Figure 6.17. SEU error rate versus scale factor using the exact, approximate, and heuristic formulas for cell parameters given. From Ref. 25.

comparison of E_r versus scale factor for a $15 \times 5 \times 0.5$ μm^3 parallelepiped cell of critical charge 1.1 pC obtained using the approximate, exact, and heuristic computation dicussed below. It is seen in the figure that, for the three cosmic ray environments (viz. solar minimum, solar maximum, and Adams 90% worst case), E_r is roughly a linear function of the scale factor for the "exact" and "approximate" calculations. This implies that E_r increases with decreasing scale, as expected. However, as also seen in Fig. 6.17, these E_r approach an asymptotic mean constant behavior for k factors of 10 or more. The E_r corresponding to the heuristic argument is obtained simply by substituting the scale factor k directly into the "approximate" formula for E_r, as given by (5.8). From (5.8) and (6.66), the scaled heuristic SEU error rate E_r' is,

$$E_r' = \frac{a'b'c'^2}{Q_c'^2} = \frac{(a/k)(b/k)(c/k)^2}{(Q_c/k^2)^2} = \frac{abc^2}{Q_c^2} \qquad (6.70)$$

Hence, to this approximation, the SEU error rate is independent of scale change.

For a value of $k \leq 10$, the "exact" and "approximate" formulas for E_r show the proper behavior. However, for $k > 10$, all of these E_r are or have become roughly constant. This implies that this scaling is beginning to break down for these relatively high values of k. In actuality, E_r increases with decreasing device parameters, even for $k > 10$, because of other factors related to the actual reduction in device size but not modeled herein. One major factor is that the SEU charge collection sensitive volume is not merely a parallelepiped but is a topologically complex volume shape, which is circuit dependent with or without including the funnel. One extreme that can be appreciated is that if this phenomenon behaved as described by (6.70) (i.e., E_r were independent of scale), the SEU problem would be trivial. Essentially, simple scaling as developed in the preceding degenerates for SEU error rates corresponding to relatively large parameter changes (i.e., large k). E_r is a complicated function of the electronic and material device parameters. Nevertheless, these scaling rules are valid for derived quantities, such as the critical charge as a function of device dimensions, including feature size.

It is important to note that the approximation to the flux versus LET curve (viz. $\phi_L \sim 5.8 \times 10^8 / L^3$, as in Fig. 5.1) is not valid for the case of direct ionization to produce SEU charge from protons. This will occur for submicron feature size, at which juncture the above coefficient will increase by approximately two orders of magnitude in a proton environment such as the inner Van Allen belt.

6.6. Single Event Latchup

Single event phenomena are well known to include single event latchup (SEL). Latchup is also the well-known pathological state of a device which under certain conditions can operate as a parasitic silicon-controlled rectifier (SCR), often leading to device burnout if not powered-down. The SEU-induced latchup pulse is one to two orders of magnitude shorter (10–100 ps) than that for transient dose-rate-induced latchup (10–100 ns). SEU-induced latchup usually affects only a single component function such as a parasitic *npnp* chain in an inverter cell, whereas dose-rate-induced latchup can affect a complete circuit or an entire subsystem. In the latter case, a single latchup can expand to include latching the entire chip as a result of heat produced in the neighborhood of the original latchup. With the advent of submicron feature size, the incident latchup-producing particle track can encompass a goodly fraction of the areas of a group of information storage cells. Most SEU-induced latchup occurs in CMOS bulk and CMOS-epi devices [27] which are also prone to dose-rate-induced latchup. Table 6.1 provides a tabulation of SEU-induced latchup devices obtained from accelerator test data. SEU-induced latchup can also be caused by neutrons and protons [27].

Table 6.1. Devices Susceptible to Single Event Latchup (SEL) from Heavy Ions.

Ser.	Device	Vendor	Technology	Function	Latchup LET (MeV cm² mg⁻¹)	σ_u (cm²)	Remarks/beam ion	Refs: IEEE Trans. Nucl. Sci. NS-XX(Y)		
1	SP1016A	ADI	CMOS bulk	Multiplier	≪37	4E-4	Krypton	38(6)	Dec. 1991	p. 1529
2	320C25	TI	CMOS-epi	DSP	≪37	>1E-2		38(6)	Dec. 1991	p. 1529
3	SP2100ASG	ADI	CMOS-bulk	DSP	12	1E-3		38(6)	Dec. 1991	p. 1529
4	MC68882	MOT	CMOS-epi	Microprocessor	12	2E-6	32 Bit, flt. pt.	38(6)	Dec. 1991	p. 1529
5	MHS80631	HA	CMOS-epi	Microprocessor	<25	2E-6	8 Bit	38(6)	Dec. 1991	p. 1529
6	P1750A	PFS	CMOS epi	Microprocessor	<11	7E-3	16 Bit	38(6)	Dec. 1991	p. 1529
7	68020	MOT	CMOS bulk	Microprocessor	6			38(6)	Dec. 1991	p. 1529
8	72016A	IDT	CMOS bulk	FIFO	13	2E-3	512 × 9	38(6)	Dec. 1991	p. 1529
9	CS5016	CRY	CMOS bulk	ADC	≪37		16 Bit	38(6)	Dec. 1991	p. 1529
10	HM65262	HA	CMOS bulk	SRAM	≪37	>4E-3	16K × 1	38(6)	Dec. 1991	p. 1529
11	61CD16	TI	CMOS-epi	SRAM	<13	6E-3	16K	38(6)	Dec. 1991	p. 1529
12	28C256	XIC	CMOS bulk	EEPROM	≪37	1E-3	32K × 8	38(6)	Dec. 1991	p. 1529
13	54AC163	RCA	ACMOS	4-Bit Bin Cntr	37	>5E-6		38(6)	Dec. 1991	p. 1529
14	54AC163	NSC	ACMOS	4-Bit Bin Cntr	40	1E-5		38(6)	Dec. 1991	p. 1529
15	ZR34161	ZOR	CMOS-epi	Sig. Processor	17 ± 5	4E-3	12μm epi	38(6)	Dec. 1991	p. 1529
16	32C016	NSC	CMOS bulk	Microprocessor	3		Boron	34(6)	Dec. 1987	p. 1332
17	27C64	INT	CMOS bulk	PROM	14			34(6)	Dec. 1987	p. 1332
18	HM6516	HA	CMOS bulk	DRAM	<17		Krypton, argon	34(6)	Dec. 1987	p. 1332
19	HM6516	HA	CMOS-epi	DRAM	<29	3E-6	Krypton, argon	34(6)	Dec. 1987	p. 1332
20	IN1600	INM	NMOS	DRAM	2	3.1E-5	LU w/ mult. SEU	34(6)	Dec. 1987	p. 1332
21	IN1601	INM	NMOS	DRAM	<3	1.6E-6		34(6)	Dec. 1987	p. 1332
22	HS6504RRH	HA	CMOS-epi	DRAM	40		Addr latch only	34(6)	Dec. 1987	p. 1332
23	54HC165	TI	HCMOS	Shift register	<40	2.5E-5		34(6)	Dec. 1987	p. 1332
24	80C86	HA	CMOS-epi	Microprocessor	≪12	4E-3	Bromine	34(6)	Dec. 1987	p. 1332
25	3022	RWL	MNOS/SOS	EAROM	<37		Krypton	32(6)	Dec. 1985	p. 4189

Single event upset can also produce hard errors [28] in the sense that the information storage nodes are not merely overwritten by incoming data, as is the case with SEU, but that the data are at least semipermanently imprinted. These are sometimes referred to as "stuck bits." Hard errors may be often erased by powering-down that node neighborhood as in ordinary dose-rate-induced latchup, provided node burnout has not occurred. It could be difficult for a microcircuit to sense a lone single-event-induced latchup if it is a very localized situation [29], as opposed to a latchup involving a number of nodes, as in rail span collapse [14].

Single event latchup has been observed in bipolar ECL technology and in bipolar linear amplifier circuitry. Also, SEL has been seen in the modeling of advanced current mode logic (CML) devices [30]. CML is a differential nonsaturating bipolar logic, possessing several internal logic states, each separated by one diode junction voltage drop. These studies have found a CML SEL rate of about 10^{-7} latchups per cell-day [30]. From SEL model analyses it has been found that it is the relatively slower diffusion current collection occurring in fractions of a microsecond, depending on device parameters, following pulse onset, that is responsible for SEL. This in contrast to the collection current tracks into the node depletion layer (and funnel), which are only a few nanoseconds in duration, that produce SEU.

As discussed previously, ^{252}Cf fission fragments have sufficient energy and LET to cause SEU latchup in microcircuits. An interesting experiment for determining SEL characteristics [31] in CMOS devices with ^{252}Cf as the heavy-ion source uses the device under test (DUT) as part of a monostable RC oscillator circuit. When a ^{252}Cf fission fragment incident on the DUT causes a latchup within, the device is latched into the SCR state. This "SCR" is then used to trigger "on" the oscillator to produce a current pulse. As subsequent particles of the fission fragment flux repeat the process, the succession of pulses (SELs) are counted. The ^{252}Cf source is mounted directly over the de-lidded DUT and the whole placed in an evacuated (10^{-3} torr) chamber. This setup yields the SEU latchup cross section directly as

$$\sigma_{\text{sel}} = \lim_{t \to \infty} \frac{N_{\text{SEL}}(t)}{^{252}\Phi(t)} \quad (6.71)$$

where $N_{\text{SEL}}(t)$ is the number of single event latchups at time t from the onset of the incident fluence $^{252}\Phi(t) = {}^{252}\dot{\Phi}(t)\,t$. The results of devices tested in this manner have been tabulated [31].

For certain devices, recent accelerator data [32] tentatively indicate that the SEU cross section as well as the critical LET L_c depend on device temperature and supply voltage [33]. The following describes a model for the increase in SEL cross section with temperature [34] while also indicating that L_c is essen-

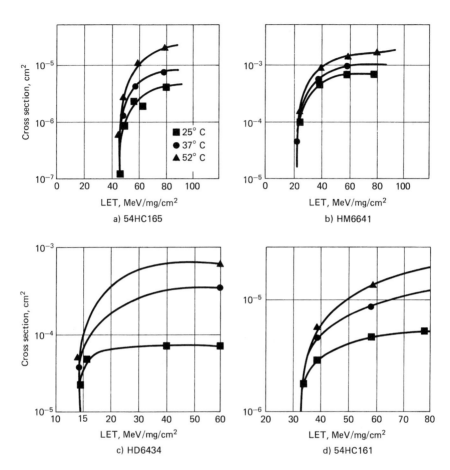

Figure 6.18. Device latchup cross sections versus LET for various temperatures. From Ref. 34.

tially independent of temperature. Figure 6.18 depicts the increase in SEL cross section versus LET for four different CMOS devices at three different ambient temperatures [34]. None of these particular devices manifests a change in L_c as a function of temperature. However, this may be due to an artifact in the device type.

Consider the common CMOS building block inverter as in Fig. 6.19. Superposed on this figure is the schematic of a classic SCR latchup, whether latchup is induced by SEL or transient dose rate photocurrent. It is seen from the figure that one latchup path consists of a lateral *pnp* parasitic transistor Q_1, directly coupled to a vertical *npn* transistor parasitic transistor Q_2, together comprising the typical *pnpn* four-layer parasitic SCR.

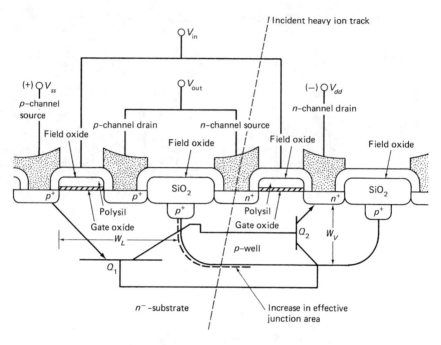

Figure 6.19. CMOS inverter showing *pnpn* latchup path transistors and incident ion track. From Ref. 34.

Single event latchup can be initiated by an incident heavy ion whose track is shown in Fig. 6.19 activating parasitic bipolar transistors Q_1 and Q_2 to a high-current low-impedance (SCR) state. The well-known latchup criterion is $\beta_1\beta_2 \simeq 1$, where β_1 and β_2 are the respective common emitter current gains of Q_1 and Q_2. Also well known is that the temperature dependence of the bipolar common emitter current gain is given approximately by [14]

$$\beta(T) = \left(\frac{\sigma_e L_e}{\sigma_b W}\right)\left(\frac{T}{T_0}\right)^2 \tag{6.72}$$

where

σ_e = parasitic transistor emitter conductivity
σ_b = parasitic transistor base conductivity
T_0 = device ambient temperature
L_e = emitter diffusion length
W = base width
T = device temperature (°K)

Inserting (6.72) into the latchup criterion $\beta_1\beta_2 = 1$ yields

$$\frac{(\sigma_{e1}L_{e1}/\sigma_{b1})(\sigma_{e2}L_{e2}/\sigma_{b2})(T/T_0)^4}{W_V W_L} \cong 1 \quad (6.73)$$

where W_V and W_L are the respective base widths of the vertical and lateral transistors. Equation (6.73) asserts that for device-elevated temperatures above T_0, the parasitic transistor current gains can be increased from their ambient values. It is seen in Fig. 6.19 that the preceding implies that the increased current gains correspond to an effective increase in the SEL cross section. Hence, σ_{SEL} increases with temperature, as shown in Fig. 6.18. The detailed electro-physical mechanisms by which the base widths extend to increase their effective area is currently being studied [32].

For the variation of L_c with temperature [34], it can be written as

$$L_c = \frac{Q_c}{W_d + W_f} \quad (6.74)$$

where W_d is the depletion layer width of the p well–n substrate junction and W_f is the corresponding funnel length. Now $Q_c = C\Delta V_{CB}$ with the depletion layer capacitance $C = \epsilon\epsilon_0 A/(W_d + W_f)$, A is the junction area and ΔV_{CB} is the voltage across the junction. Then (6.74) becomes

$$L_c = \frac{\epsilon\epsilon_0 A \Delta V_{CB}}{(W_d + W_f)^2} \quad (6.75)$$

The well-known expression for the depletion layer width, using the abrupt junction approximation, is given by

$$W_d = \left(\frac{2\epsilon\epsilon_0(\phi_0 + |\Delta V_{CB}|)}{eN_A}\right)^{1/2} \quad (6.76)$$

where the quantities are defined earlier. The funnel length W_f is essentially temperature independent, and L_c must depend on temperature only through W_d, whose temperature-dependent parameter is the contact potential $\phi_0 = (kT/e)\ln(N_A N_D/n_i^2)$, with the intrinsic carrier density $n_i = C_1 T^{3/2}/\exp(-E_g/2kT)$. With the usual values of the parameters involved, it is also known [33] that $\Delta\phi_0/\Delta T \cong 2\text{mV}/°\text{C}$. Even for a very large ambient temperature increase (e.g., $\Delta T \cong 200°\text{C}$), $\Delta\phi_0$ is only 400 mV. This implies that any change in ϕ_0 is dominated by ΔV_{CB} because it is of the order of 5–10 V and that $\phi_0 \ll \Delta V_{CB}$ is well known. Hence, this leads to the conclusion that L_c is essentially unaffected by device temperature changes, but again temperature changes can affect σ_{SEL}, as discussed in the preceding.

216 / Single Event Phenomena

6.7. South Atlantic Anomaly, Low Earth Orbits

The earth's magnetic field has a certain detailed structure that is of concern to those who calculate space radiation levels for this environment, as well as for susceptible spacecraft avionics systems. In the Southern Hemisphere, the magnetic field of the earth is offset by approximately 11 degrees from the earth's axis of rotation and displaced about 500 km toward the western Pacific [35]. This results in dips in the magnetic field over various parts of the earth, including a major dip in the vicinity of Brazil (30° S lat., 345° E long.) where magnetic field lines reenter the earth normally, as at the poles. These not only produce dipolelike singularities in the magnetic field for cosmic rays to intrude along field lines but also allows the Van Allen radiation belt particles to extend down into the earth's atmosphere. This can be seen in Figs. 6.20. and 6.21.

For spacecraft in low earth orbits (LEO), usually construed as "under" the Van Allen belts at small inclination angles, the vehicle suffers little incident radiation, except for that due to solar flares (protons), gamma ray bursts, and the South Atlantic Anomaly (SAA). For a spacecraft at an orbital altitude of about 850 km in a polar orbit of inclination angle of 99° measured clockwise from the equator, its exposure to the SAA corresponds to a fraction of a steradian of area on the globe, or about 20–25 min of polar orbit time (Fig. 6.21'). The dose-rate

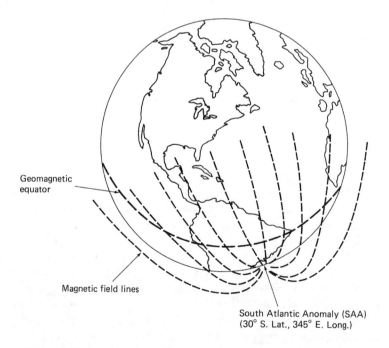

Figure 6.20. South Atlantic Anomaly.

Single Event Phenomena I / 217

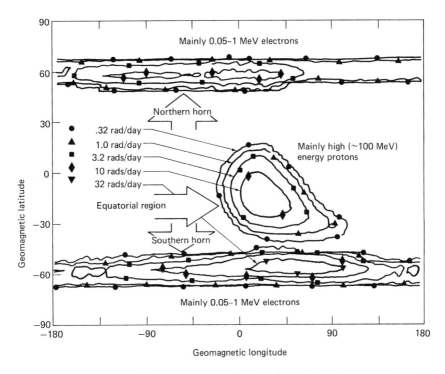

Figure 6.21. Dose rate fine structure over South Atlantic Anomaly measured within (840 km, 98.8° inclination angle) DMSP F-7 spacecraft. From Ref. 35.

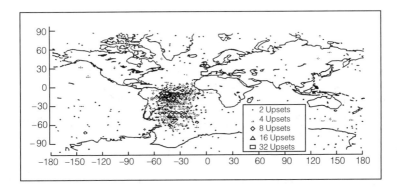

Figure 6.21'. South Atlantic Anomaly SEUs on board solar magnetosphere explorer (SAMPEX) nearly polar low-altitude spacecraft between Sept. 5, 1992 and Dec. 1, 1992. From Ref. 26.

218 / Single Event Phenomena

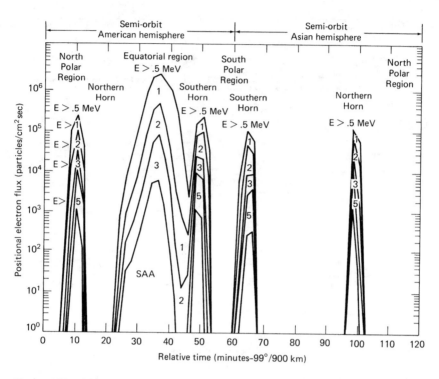

Figure 6.22. Most severe pass through the South Atlantic Anomaly showing integral omnidirectional trapped electrons. From Ref. 35.

and particle-energy fine structure of the SAA is shown in Figs. 6.21 and 6.22. These figures show that in addition to the principal equatorial singularity (SAA), there are magnetic dipole-like singularities in the Asian hemisphere, as well as northern and southern "horns" accompanying the former. These are of lesser size and are due to second-order magnetic dips into the earth's atmosphere, stronger in the south than in the north, probably due to the nearness of the antarctic regions. The above Southeast Asian anomaly evokes greater magnetic field levels, but the corresponding radiation is incident at higher altitudes [35]. It is known that certain complete orbits during the spacecraft precessional cycle around the earth are electron flux free [35]. This implies that with respect to ionizing dose levels (not SEU), extravehicular activity possibly involving astronauts could take place during these flux free orbits.

For the above low-earth-altitude spacecraft, the incident ionizing dose rate is about 2 rads per day behind 0.55 g cm^{-2} or 1 rad per day behind 5.9 g cm^{-2} [37]. The annual incident dose due to gamma ray bursts (Section 2.3) is about 1 krad (Si) per annum at that altitude, which is the same order of magnitude as the above SAA.

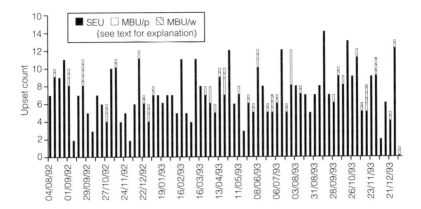

Figure 6.23. Weekly in-flight SEUs (Aug. 1, 1992–Jan. 4, 1994) for the HM65756 SRAM in the Russian MIR space station. Dates are da/mo/yr. From Ref. 39.

The corresponding SEU error rates for low-earth-orbit spacecraft varies. For a polar orbit of 90° inclination angle, at altitudes between 300 and 800 km, the SEU error rate varies between 5×10^{-7} (HM6264 CMOS SRAM) and 2×10^{-5} (TM4116 CMOS DRAM) errors per bit-day, due mainly to SAA protons [38]. These parts are all in the same spacecraft avionics.

As implied above, the major radiation component encountered is protons in the SAA singularity. There are a number of models for corresponding low-earth-orbit proton fluxes [35]. Proton energies at low earth orbits are much larger than those at geosynchronous altitudes. At these altitudes, the protons are so "soft" that they can be stopped with 50 mils of aluminum, implying that they are of little radiation significance. Protons at low earth altitudes impact the spacecraft as it passes through the SAA. Although the proton energies are higher than those at geosynchronous altitudes, their flux levels are not excessive in that usually only "spot" shielding for individual radiation-susceptible parts would be required.

One example of a part enduring SEU in a low earth orbit is in the Russian–French experiment aboard the MIR space station using the HM65756 [39]. Fig. 6.23 depicts a series of 75 successive weekly SEU rates for the HM65756 SRAM ($L_c = 6$ MeV cm^2 mg^{-1}, $\sigma_{\text{SEL}} = 3 \times 10^{-3}$ cm^2 per device). The MIR space station is in a parking orbit at 350 km altitude, 51.6° inclination angle, and 91.6 min period. In the figure, MBU/w means that two or more SEUs occurred in the same word in the same week. MBU/p means that a single incident particle produced several SEUs in adjacent bits in the device. Apparently, no significant fluctuations in these events occurred during the exposure epoch, within the large statistical uncertainties because of the relatively few events per day [39].

6.8. Single Event Gate Rupture, Device SEU Burnout

In MOSFET devices, a heavy ion piercing the gate insulator can cause damage through its perforation and consequent rupture. This is called single event gate rupture (SEGR). It usually occurs when the gate dielectric is being stressed by a high internal electric field, simultaneous with the heavy-ion strike. Such fields occur during a write or clear operation in a nonvolatile SRAM or EEPROM (electrically erasable, programmable, read-only memory) [40]. The heavy ion in its transit through the gate, produces a highly conductive plasma track through ionization of the dielectric. This forms a very low-resistance path between the gate and substrate. If sufficient energy was stored in this gate capacitor prior to the ion strike, the plasma track becomes the discharge path for the capacitor. This discharge can cause excessive heating of the dielectric—enough to melt or otherwise degrade it. Device SEGR tendencies include dependence on the LET of the incident particle, dielectric properties, corresponding electric field, and the angle of incidence of the heavy ion.

For SEU hard errors due to dielectric breakdown including SEGR, a simple hyperbolic relationship obtains between the device gate failure threshold voltage and the LET of the incident particle in the dielectric material. This is derived [40, 41] by noting that the power, P, dissipated in the track plasma in the dielectric is given by

$$P = iV_0 \qquad (6.77)$$

where V_0 is the bias voltage across the dielectric capacitor and i is the instantaneous current in the ion plasma track. The latter can be written as

$$i = nev_{sat}A \qquad (6.78)$$

n is the track electron density, A is the track cross-sectional area, and v_{sat} is the track particle saturated velocity, usually associated with a composite dielectric gate device in contrast to a single (pure) dielectric gate. Also, $v_{sat} = d_{ox}/t_r$, where d_{ox} is the dielectric thickness (track length for vertical incidence), and t_r is the average particle transit time through the ion track. Inserting this for v_{sat} above gives

$$i = \frac{ned_{ox}A}{t_r} \qquad (6.79)$$

Substituting for i in (6.77) from (6.79), integrating over an epoch Nt_p, a multiple N of the transit time that easily encompasses the track duration, yields the energy E_{tr} in the track; that is, within a constant of proportionality,

$$E_{tr} = \int_0^{Nt_r} P\,dt = \int_0^{Nt_r} \left(\frac{ned_{ox}V_0A}{t_r}\right) dt = (\text{LET}) \cdot V_0 Ned_{ox}A \qquad (6.80)$$

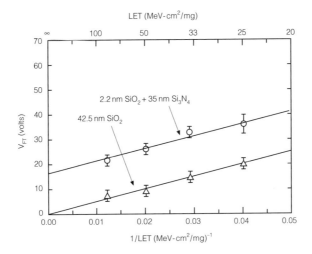

Figure 6.24. Failure threshold voltage of two dielectric gate insulators versus particle inverse LET and LET abscissas. From Refs. 40 and 41.

using the fact that n is proportional to the particle LET, which allows the interpretation of E_{tr} to be threshold energy in this context. As all the other factors in (6.80) are constant, then (LET) $V_0 \equiv$ (LET) V_{FT} = constant yields the sought-after relationship between them and is depicted in Fig. 6.24. Again, this holds for composite device gates characterized by a saturation velocity v_{sat}.

For pure single dielectric gate devices, constant mobility is the physical parameter that characterizes this unsaturated velocity system; that is, again beginning with $i = nevA$ and using $v = \mu E = \mu V_0/d_{ox}$ in the analogous expression for threshold energy E_{tr} as in (6.80) yields within a constant of proportionality

$$E_{tr} = (LET) V_0^2 \tag{6.81}$$

Using $V_0/d_{ox} = E_{FT}$, the corresponding electric field, it is seen from (6.80) and (6.81) respectively that $E_{FT} \sim 1/LET$ for composite gate dielectric devices, and that $E_{FT} \sim 1/(LET)^{1/2}$ for pure dielectric gate devices. This is shown in Fig. 6.25.

For composite gate devices, an empiricism for the threshold electric field $E_{FT(n)}$, LET and the angle of incidence is found to be [41]

$$E_{FT(n)} = \frac{1.26 \times 10^8}{(LET)|\cos(\theta + \pi/30)|} \; (V/cm) \tag{6.82}$$

whereas for pure oxide gate devices,

$$E_{FT(ox)} = \frac{4.1 \times 10^7}{(LET)^{1/2}|\cos(\theta + \pi/36)|} \; (V/cm) \tag{6.83}$$

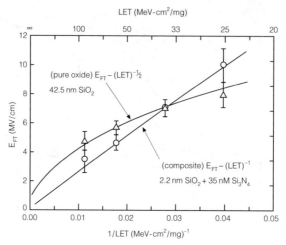

Figure 6.25. Dielectric electric field for two dielectric gate insulators versus inverse LET and LET abscissas. From Refs. 40 and 41.

The nonzero angular displacements at normal incidence ($\theta = 0$) in (6.82) and (6.83) account for the angle that the finite dimensions of the ion track subtends. This is depicted in Fig. 6.26 where the failure threshold voltage $V_{FT} = d_{ox}E_{FT}$ is plotted versus $1/\cos\theta$. The devices have composite gates ($SiO_2 + Si_3N_4$) and, here, is irradiated with 180-MeV Ge ions [41]. For normal incidence, (6.83) is usually written

$$E_{FT(ox)} \cong \frac{41}{\sqrt{LET}} \quad \left(\frac{MV}{cm}\right) \tag{6.84}$$

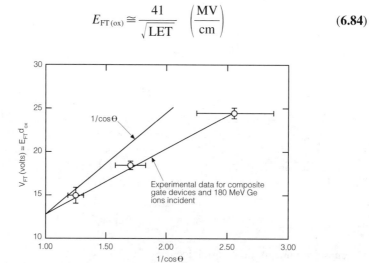

Figure 6.26. Failure threshold voltage $V_{FT} = d_{ox}E_{FT}$ versus secant of angle of incidence for composite gate devices irradiated with 180-MeV Ge ions. From Ref. 41

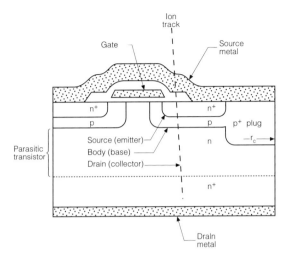

Figure 6.27. Cross section of a power MOSFET showing the parasitic bipolar transistor. From Ref. 42.

In power MOSFETs, a heavy-ion-induced burnout mechanism has been observed that occurs when the ion penetrates its gate oxide region at the instant that the device possesses, at least, a threshold value of gate-to-drain or gate-to-source bias. The subsequent gate rupture presages a permanent degradation that destroys device function. The preceding refers to n-channel MOSFETs, whereas p-channel MOSFETs are apparently immune to this phenomenon [43].

It has been shown that burnout is imminent in n-channel devices when the current density in a cell exceeds a critical current density J_{cr} [44], which is tantamount to the already mentioned threshold biases. If the transistor current density is at least J_{cr}, then the parasitic bipolar transistor, the inadvertent transistor which is part of essentially all MOSFETs and many bipolars, is turned on to support a self-sustained current-induced avalanche at the epi–substrate interface, as in Fig. 6.27. This virtual short circuit causes a local overheating of the device material (hot spot) followed by a loss of its physical integrity. The device has "burned out."

Because device burnout follows gate rupture as discussed, a circuit shown in Fig. 6.28 has been devised [43] to test the propensity of gate rupture in MOSFETs without allowing them to proceed to burnout. This then allows the easily determined device threshold to be obtained without its catastrophic destruction. The device is de-lidded, placed in a high-vacuum chamber, and irradiated with a beam of moderately heavy ions (LET ~ 40 MeV cm^2 gm^{-1}) while being connected, as shown in Fig. 6.28. A given V_{GS} and V_{DD} are applied for a given C_{ext} setting. Immediately following the instant of the heavy-ion strike, a current pulse is generated whose magnitude is governed by C_{ext}, a

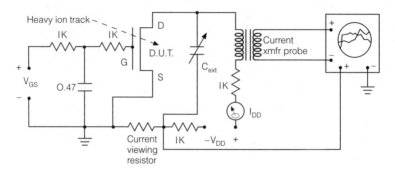

Figure 6.28. Power MOSFET test circuit to measure device burnout threshold. From Ref. 43.

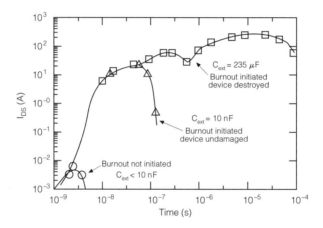

Figure 6.29. Comparison of prompt photocurrent, incipient, and burnout waveforms for an IRFF130 power MOSFET. From Ref. 43.

stiffening capacitor. This is monitored by signals produced in the current transformer probe and in the current-viewing resistor [43]. By increasing C_{ext}, the pulse amplitude can be increased to achieve an incipient burnout condition where the device is left whole, and to one where the device is destroyed, as shown in Fig. 6.29. In the figure, the three signatures correspond to (a) for low C_{ext}, a prompt primary ion-induced photocurrent insufficient to cause burnout because $J < J_{cr}$, (b) for increased C_{ext}, a prompt, secondary (parasitic transistor amplified) ion-induced photocurrent sufficient to cause burnout, $J > J_{cr}$, with the device suffering gate rupture at most, and (c) large C_{ext}, like (b) but with $J \gg J_{cr}$ sufficient to cause burnout [43].

DMOS (double diffused) is a typical high-voltage power MOSFET whose structural cross section is shown in Fig. 6.30. SEGR is a possible failure mode

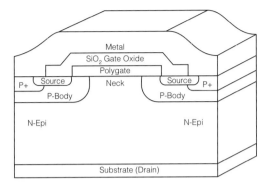

Figure 6.30. Cross section of a vertical DMOS power MOSFET. From Ref. 45.

for bias voltages that are not properly handled. DMOS devices are desirable in many applications, especially in spacecraft avionics. They are capable of fast switching times as well as low power losses [45]. As early burnout difficulties were solved allowing high-voltage operation, attention began to focus on DMOS SEGR propensities [43].

The DMOS operates as an *n*-channel transistor by varying the surface conductivity under the polysilicon gate, controlled by the gate bias V_{GS} as in Fig. 6.30. As $-V_{GS}$ increases, (a) the *n*-epi neck region approaches the state of strong population inversion while (b) the *p*-body region approaches a strong population accumulation state. These states cut off the current flow from source to drain. Only gate and drain leakage currents flow in this condition. For increasing $+V_{GS}$, (a) and (b) exchange population states. When $V_{GS} \geq V_{TH}$ (threshold voltage) in this latter state, current then flows in the device. When an incident heavy ion pierces the neck to produce an ion track from neck to the substrate drain, SEGR can occur as discussed earlier [45].

Once SEGR is manifest, current flows through the gate dielectric (SiO$_2$) to the polysilicon gate. This results in a thermal runaway situation, locally melting the SiO$_2$ in the track vicinity, causing permanent damage. The device can exhibit an early increase in gate current until the gate dielectric fails [45, 48].

After assiduous experimental effort [45], an empirical SEGR failure onset relationship among V_{GS}, V_{DS}, and the incident ion LET was obtained. Also the SOA (safe operating area) for this device type in the V_{GS}–V_{DS} plane is shown in Fig. 6.31. The empirical expression is given by

$$V_{GS} = 0.84 V_{DS} \left[1 - \exp\left(-\frac{\text{LET}}{17.8} \right) \right] - 50 \exp\left[-\ln\left(1 + \frac{\text{LET}}{53} \right) \right] \quad (6.85)$$

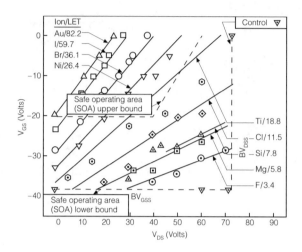

Figure 6.31. V_{GS}–V_{DS} plane showing SEGR onset LET loci for various ion types and SOA (safe operating area—upper boundary). From Ref. 45.

To accommodate the dependence of V_{GS} on gate oxide thickness, t_{ox}, and ion strike angle θ with respect to the ion beam direction, a refined empirical expression is given by [46]

$$V_{GS} = 0.87 V_{DS}\left[1 - \exp\left(-\frac{LET}{18}\right)\right] - 10^7 t_{ox}(\cos\theta)^{-0.7}\exp\left[-\ln\left(1 + \frac{LET}{53}\right)\right] \quad (6.86)$$

where $[V_{GS}]$ and $[V_{DS}]$ = V, $[LET]$ = MeV cm^2 mg^{-1}, as in (6.85), and $[t_{ox}]$ = cm. In Fig. 6.31, the gate and drain breakdown voltage line bounds BV_{GSS} (-38.5 V) and BV_{DSS} ($+73$ V), respectively, are shown for this device. The broken line bounds mark the SOA given by the manufacturer. An iso-LET line implies SEGR onset for a (V_{GS}, V_{DS}) point on that line for that LET. Decreasing V_{DS} and V_{GS} from that point provides SEGR safety. It is felt that (6.85) can be generalized to other power MOSFETs whose structures do not deviate radically from the DMOS device [45].

The temperature dependence of SEGR asserts that a temperature increase produces an increase in the gate oxide electric field due to mobility increases, which imply an increase in device SEGR tendencies [47]. It also has been shown that the greater the angle of ion incidence (from the normal) the less the SEGR possibility [47]. This can be seen to be incorporated in (6.82) and (6.83).

Problems

6.1 A funnel model discussed in Section 6.2 gives $\mathscr{L} = (1 + \mu_n \mu_p) W \sec\theta$ and $\mathscr{D} = (1 + \mu_n/\mu_p) W$ where $\mathscr{L}$ is the funnel length and $\mathscr{D}$ is the depth of its vertical component (i.e., $\mathscr{L} = \mathscr{D} \sec\theta$). The other parameters are defined in Section 6.2.1. For grazing incidence angles of the incoming ion (i.e., θ near $\pi/2$), $\mathscr{L}$ begins to approach an infinite length. How is this anomaly reconciled?

6.2 In device dimension scaling considerations, the SEU figure of merit error rate expression $E_r = 5 \times 10^{-10} abc^2/Q_c^2$ can be shown to be scale invariant as discussed in Section 6.5. If E_r were actually independent of such scaling, how would this trivialize the SEU problem?

6.3 Generally, the smaller the device cell Q_c, the greater is its SEU error rate per chip per day. However, certain devices whose Q_c is smaller than others also have a comparatively smaller SEU error rate per chip day. Why?

6.4 Why would a low-earth-orbiting spacecraft at ~ 100 miles altitude be subject to SEU from Van Allen protons as it passes over Rio de Janeiro?

6.5 Would a geosynchronous satellite orbiting the equator momentarily at a longitude passing through Rio endure enhanced galactic cosmic-ray-induced SEU?

6.6 In the initial stages of ion-track formation in the device material, its very high density implies ambipolar diffusion. Explain.

6.7 In Fig. 6.11, the accelerator ion beam cross section [49] A_{beam} is defined as the area presented to the beam by that portion of the parallelepiped containing chords of length $\ell \geq c \sec\theta_c$ (i.e., those that will produce sufficient charge to cause an SEU). Then $A_{\text{beam}} = ab\cos\theta + bc\sin\theta - 2b\ell\sin\theta\cos\theta$. The first two terms are projections onto the beam cross-sectional area of the horizontal and vertical surfaces of the parallelepiped, respectively. The third term is subtracted because the two corresponding portions have chord lengths $\ell < c\sec\theta_c$ which are of insufficient length to cause SEU. Now $L_{\text{ion}} = L_{\text{eff}}\cos\theta = L_c \cos\theta_c$, $\ell = c\sec\theta_c$ and a parallelepiped aspect ratio $k = c/a$.

(a) Using the working definition of the SEU cross section as employed in the accelerator experimental arena, (viz. $A_{\text{beam}} = \sigma_{\text{seu}} \cos\theta$), show that [49]

$$\sigma_{\text{seu}} = ab\left[1 + k\left(\tan\theta - \frac{2L_c \sin\theta}{L_{\text{ion}}}\right)\right] = ab\left\{1 + k\left[1 - \left(\frac{2L_c}{L_{\text{eff}}}\right)\tan\theta\right]\right\}$$

(b) Plot [49] σ_{seu}/ab versus L_{eff} for $L_c = 17.5$ MeV cm^2 mg^{-1} and $k = \frac{1}{4}$ for the Berkeley 88-in. cyclotron beam ions of $L_{\text{ion}} = 15, 32, 41,$ and 63, corresponding to ^{40}Ar, ^{65}Cu, ^{86}Kr, and ^{131}Xe respectively. Note that σ_{seu} vanishes for $L_{\text{eff}} < L_c$ and $\tan\theta = [(L_{\text{eff}}/L_{\text{ion}})^2 - 1]^{1/2}$.

6.8 With respect to single event gate rupture (SEGR), the threshold electric field for both pure and composite gate devices is proportional to $\cos(\theta + \theta_0)$. θ is the incident angle of the ion and $\theta_0 \ll \theta$, an initial "phase shift." For pure gate devices, $\theta_0 \cong \pi/36 = 5°$, whereas for composite gates, $\theta_0 \cong \pi/30 = 6°$. What is the reason for θ_0?

6.9 The burnout threshold electric field for a pure oxide gate at normal incidence is given by $E_{\text{FT(ox)}} \cong 41/(\text{LET})^{1/2}$ (MV/cm). What is the analogous expression for a composite gate device?

6.10 An empirical SEGR expression for failure onset as a function of V_{GS}, V_{DS}, and incident ion LET is given by (6.85).

(a) For a 50-V V_{DS} device, what is the LET that corresponds to $V_{\text{GS}} = 0$?

(b) What does $V_{\text{GS}} = 0$ imply in this case?

References

1. J.G. Rollins and J. Choma, Jr., "Single Event Upset in SOS Integrated Circuits," *IEEE Trans. Nucl. Sci.*, **NS-34** (6), 1713–1717 (1987).

2. C.M. Hsieh, P.C. Murley, and R.R. O'Brien, "A Field Funneling Effect on the Collection of Alpha Particle Generated Carriers in Silicon Devices," *IEEE Trans.*, **EDL-2** (4) (1981).

3. C.M. Hseih, P.C. Murley, and R.R. O'Brien, "Dynamics of Charge Collection from Alpha Particle Tracks in Integrated Circuits," IEEE 19th Proc. Reliability Physics Conf., 1981.

4. H.L. Grubin, J.P. Kreskovsky, and B.C. Weinberg, "Numerical Studies of Charge Collection and Funneling in Silicon Devices," *IEEE Trans. Nucl. Sci.*, **NS-31** (6), 1161–1166 (1984).

5. G.C. Messenger, "Collection of Charge on Junction Nodes from Ion Tracks," *IEEE Trans. Nucl. Sci.*, **NS-29** (6), 2024–2031 (1982).

6. F.B. McLean and T.R. Oldham, "Charge Funneling in N and P Type Silicon Substrates," *IEEE Trans. Nucl. Sci.*, **NS-29** (6), 2018–2093 (1982).

7. T.R. Oldham and F.B. McLean, "Charge Collection Measurements for Heavy Ions Incident on N and P Silicon," *IEEE Trans. Nucl. Sci.*, **NS-30** (6), 4493–4500, (1983).

8. R.M. Gilbert, G.K. Ovrebo, and J. Schifano, "Plasma Screening of Funnel Fields," *IEEE Trans. Nucl. Sci.*, **NS-32** (6), 4098–4103 (1985).

9. T.R. Oldham, F.B. McLean, and J.M. Hartman, "Revised Funnel Calculations for Heavy Particles with High dE/dx," *IEEE Trans. Nucl. Sci.*, **NS-33** (6), 1646–1650 (1986).

10. C. Hu, "Alpha Particle Induced Field and Enhanced Collection of Carriers," *IEEE Trans.* **EDL-3**, (2) (1982).

11. P.E. Dodd, F.W. Sexton, and P.S. Winokur, "Three Dimensional Simulation of Charge Collection and Multiple Bit Upsets in Si Devices", *IEEE Trans. Nucl. Sci.*, **NS-41** (6), 2005–2017 (1994).

12. L.B. Valdes, *The Physical Theory of Transistors*, McGraw-Hill Book Co., New York, 1961, Chap. 15.

13. H.C. Torrey and C.A. Whitmer, *Crystal Rectifiers*, MIT Rad. lab. series Vol. 15, McGraw-Hill Book Co., New York, 1948; Appendix C; H.C. deGraaff, "Approximate Calculations on the Spreading Resistance in Multi-element Structures," *Phillips Res. Rpts.* **24**, 34 (1969).

14. G.C. Messenger and M.S. Ash, *The Effects of Radiation on Electronic Systems*, 2nd ed., Van Nostrand Reinhold, New York, 1992, Chap. 7.

15. J. Bradford, "Non-Equilibrium Radiation Effects in VLSI," *IEEE Trans. Nucl. Sci.*, **NS-25** (5), 1144–1149 (1978).

16. J.P. McKelvey, *Solid State and Semiconductor Physics*, Harper and Row, New York, 1966, pp. 50ff.

17. J.S. Cable, N.M. Ghoneim, R.G. Martin, and Y. Song, "The Size Effect of an Ion Charge Track on Single Event Multiple Bit Upset," *IEEE Trans. Nucl. Sci.*, **NS-34** (6), 1305–1309 (1987).

18. L.W. Massengill, M.L. Alles, S.E. Kerns, and K.L. Jones, "Effects of Process Parameter Distributions and Ion Strike Locations on SEU Cross Section Data," *IEEE Trans. Nucl. Sci.*, **NS-60** (6), 1804 (1993).

19. M.A. Xapsos, T.R. Weather, and P. Shapiro, "The Shape of Heavy Ion Cross Section Curves," *IEEE Trans. Nucl. Sci.*, **NS-40** (6), 1812–1819 (1993).

20. A.M. Kellerer, "Considerations on the Random Traversal of Convex Bodies and Solutions for General Cylinders," *Radiat. Res.*, **47**, 359–376 (1971).

21. M.A. Xapsos, T.R. Weatherford, and P. Shapiro "The Shape of Heavy Ion Cross Section Curves," *IEEE Trans. Nucl. Sci.*, **NS-40** (6), 1812–1819 (1993); "Applicability of LET to Single Events in Microelectronic Structures," **NS-39** (6), 1613–1621 (1992).

22. E.A Burke, "The Impact of Component Technology Trends on the Performance and Radiation Vulnerability of Microelectronic Systems," NSRE Conf. Short Course, Sec. 4, 1985; R.W. Keyes, "Fundamental Limits in Digital Information Processing," *Proc. IEEE*, **69**, 267 (1981).

23. R.H. Dennard et al., "Design of Ion Implanted MOSFETs with Very Small Physical Dimensions," *IEEE J. Solid State Circuits*, **SC-9** (5), 256–258 (1974).

24. E.L. Petersen and P. Marshall, "Single Event Phenomena in the Space and SDI Areas," *J. Radiat. Effects, J. Radiation Effects Research and Engineering* (1989).
25. L. Petersen, P. Shapiro, J.H. Adams, Jr., and E.A. Burke, "Calculation of Cosmic Ray Induced Soft Upset and Scaling in VLSI Devices," *IEEE Trans. Nucl. Sci.*, **NS-29** (6), 2055–2063 (1982).
26. D.K. Nichols, W.E. Price, M.A. Shoga, J. Duffey, W.A. Kolasinski, and R. Koga, "Discovery of Heavy Ion Induced Latchup in CMOS-Epi Devices," *IEEE Trans. Nucl. Sci.*, **NS-33** (6), 1696 (1986).
27. D.K. Nichols, J.E. Coss, R.K. Watson, H.R. Schwartz, and R.L. Pease, "An Observation of Proton Induced Latchup," *IEEE Trans. Nucl. Sci.*, **NS-39** (6), 1654–1656 (1992).
28. J.S. Browning, J.E. Gover, T.F. Wrobel, K.J. Haas, R.D. Nasby, R.L. Simpson, L.D. Posey, R.E. Boos, and R.C. Black, "Hard Error Generation by Neutron-Induced Fission Fragments," *IEEE Trans. Nucl. Sci.*, **NS-34** (6), 1269–1274 (1987).
29. J.H. Stephen, T.K. Sanderson, D. Mapper, J. Farren, R. Harboe-Sorensen, and L. Adams, "Cosmic Ray Simulation Experiments for the Study of Single Event Upsets and Latchup in CMOS Memories," *IEEE Trans. Nucl. Sci.*, **NS-30** (6), 4464–4469 (1983).
30. J.P. Spratt and D.G. Millward, "Single Event Upset in Bipolar Technologies and Hardness Assurance Support Activities," SAIC. Inc. Rerport DNA-86-126, La Jolla, CA, 1986.
31. J.H. Stephen, T.K. Sanderson, D. Mapper, M. Hardman, J. Farren, L. Adams, and R. Harboe-Sorenson, "Investigation of Heavy Particle Induced Latchup Using a Californium-252 Source in CMOS SRAMS and PROMs," *IEEE Trans. Nucl. Sci.*, **NS-31** (6), 1207–1211 (1984).
32. W.A. Kolasinski, R. Koga, E. Schnauss, and J. Duffey, "The Effects of Elevated Temperature on Latchup and Bit Errors in CMOS Devices," *IEEE Trans. Nucl. Sci.*, **NS-33** (6), 1605–1609 (1986).
33. L.S. Smith, D.K. Nichols, J.R. Coss, W.E. Price, and D. Binder, "Temperature and Epi-Thickness Dependence of Heavy Ion Induced Latchup Using a ^{252}Cf Source on CMOS SRAMs and PROMs," *IEEE Trans. Nucl. Sci.*, **NS-34** (6), 1800–1802 (1987).
34. M. Shoga and D. Binder, "Theory of Single Event Latchup in Complementary Metal Oxide Semiconductor Integrated Circuits," *IEEE Trans. Nucl. Sci.*, **NS-33** (6), 1714–1717 (1986).
35. E.G. Stassinopolous and J.P. Raymond, "The Space Radiation Environment for Electronics," *Proc. IEEE*, **76**, 1423–1442 (1988); J. Bloxham and D. Gubbins," The Evolution of the Earth's Magnetic Field," *Sci. Am.* 68–75 (Dec. 1989).
36. K. LaBel, S. Way, E.G. Stassinopolous, C.M. Crabtree, J. Hengemihle, and M.M. Gates, "Solid State Tape Recorders: Spaceflight SEU Data for SAMPEX and TOMS/Meteor 3," 1993 Radiation Effects Data Workshop, 1993, 77–84.
37. M.S. Gussenhoven, E.G. Mullen, R.C. Filz, D.H. Brantigam, and F.A. Hanser, "New Low Altitude Dose Measurements," *IEEE Trans. Nucl. Sci.*, **NS-34** (3) 676–682 (1987).

38. L. Adams, R.H. Sorenson, and E. Daly, "Proton Induced Upsets in the Low Altitude Polar Orbit," *IEEE Trans. Nucl. Sci.*, **NS-36** (6), 2339–2343 (1989).
39. D. Falguere, S. Duzellier, and R. Ecoffiet, "SEE In-Flight Measurement on the MIR Orbital Station," *IEEE Trans. Nucl. Sci.*, **NS-41** (6), 2346–2352 (1994).
40. F.W. Sexton, "Measurement of Single Event Phenomena in Devices and ICs," IEEE NSRE Conference Short Course, Chapter 3, 1992.
41. T.F. Wrobel, "On Heavy Ion Induced Hard Errors in Dielectric Structures," *IEEE Trans. Nucl. Sci.*, **NS-34** (6), 1262–1268 (1987).
42. J.H. Hohl and K.F. Galloway, "Analytical Model for Single Event Burnout of Power MOSFETs," *IEEE Trans. Nucl. Sci.*, **NS-34** (6), 1275–1280 (1987).
43. T.A. Fischer, "Heavy Ion Induced Gate Rupture in Power MOSFETs," *IEEE Trans. Nucl. Sci.*, **NS-34** (6), 1786–1791 (1987).
44. T.F. Wrobel, F.H. Coppage, G.L. Hash, and A. Smith, "Current Induced Avalanche in Epitaxial Structures," *IEEE Trans. Nucl. Sci.*, **NS-32** (6), 3991–3995 (1985).
45. C.F. Wheatley, J.L. Titus, and D.I. Burton, "Single Event Gate Rupture in Vertical Power MOSFETs: An Original Empirical Expression," *IEEE Trans. Nucl. Sci.*, **NS-41** (6), 2152–2159 (1994).
46. J.L. Titus, C.F. Wheatley, D.I. Burton, I. Mouret, M. Allenspach, J. Brews, R. Schrimpf, K. Galloway, and R.L. Pease, "Impact of Oxide Thickness on SEGR Failure in Vertical Power MOSFETs: Development of Semi-Empirical Expression," *IEEE Trans. Nucl. Sci.*, **NS-42** (6), 1928–1934 (1995).
47. I. Mouret, M. Allenspach, R.D. Schrimpf, J.R. Brews, K.F. Galloway, and P. Calvel, "Temperature and Angular Dependence of Substrate Response in Single Event Gate Rupture (SEGR)," *IEEE Trans. Nucl. Sci.*, **NS-41** (6), 2216–2221 (1994).
48. I. Mouret, P. Calvel, M. Allenspach, J.L. Titus, C.F. Wheatley, K.A. LaBel, M.C. Calvet, R.S. Schrimpf, and K.F. Galloway, "Measurement of a Cross Section for Single Event Gate Rupture in Power MOSFETs," *IEEE Electron. Device Lett.*, **17** (4) (1996).
49. E.C. Smith, Revision of Appendix I in E.L. Petersen, J.B. Langworthy, and S.E. Diehl, "Suggested Single Event Upset Figure of Merit," *IEEE Trans. Nucl. Sci.*, **NS-30** (6), 4533–4539 (1983) Preprint, 1994.

7

Single Event Phenomena II

7.1. Introduction

This chapter continues the discussion of the other important facets of single event phenomena. It begins with multiple event effects where one incident single event upset (SEU)-inducing ion can produce more than one bit upset in the same memory array, for example. Following the above is a discussion of the effects of a device exposed to an ionizing dose of radiation prior to and/or during the occurrence of an SEU. The ionizing dose is usually construed, but not exclusively, as that from gamma rays, X-rays, protons, or electrons, corresponding to the particular hostile or benign environment in which the device finds itself.

The next two sections on redundancy, scrubbing, and error detection and correction (EDAC) are methods used to counter bit errors. These methods are principally instituted at the chip and circuit level, by using dedicated EDAC chips and specially configured memory cells built into microcircuits in their manufacture.

Section 7.6 on SEU shielding discusses the limited shielding feasibility to mitigate SEU in particular microcircuits. For example, shielding against protons from a large solar flare is a possibility in certain unique spacecraft orbital situations. The last section discusses the effects of SEU on radiation detectors and sensors.

Detectors and sensors that intercept radiation are built to be ultrasensitive to almost all signals, and so unfortunately to spurious signals such as SEU. Hence, measures must be used to differentiate between desired and extraneous signals in them. This involves associated circuitry that is far more complex than that of the detector or sensor itself. Further, this circuitry must be hardened against SEU and single event latchup (SEL) as well.

7.2. Multiple Event Upsets

Even though single event upsets are the preponderant SEU phenomenon, there are important instances in which multiple event upsets occur [1]. Multiple event upsets are those where an incident heavy ion can cause SEU in a string of memory cells, in a microcircuit, that happen to lie physically along the penetrating ion track. Multiple event upsets were detected almost as soon as the SEU phenomenon was accepted as a reality.

The role of the detailed parameters of the heavy-ion track, such as the track radius, density, and its time aspects, is a minor one with respect to the computation of SEU. However, these parameters are important in the understanding and computation of multiple event upsets. For example, one multiple event upset model used for the calculation of multiple upset cross sections depends on the ion-track radius and is discussed below [2].

To appreciate the kinematics of heavy ions within and without the ionization track, it is important to briefly investigate their transport properties. For a typical track of 4×10^{18} hole–electron pairs (hep) per cubic centimeter, of cylindrical cross section, it has been found that the initial track radius is somewhat less than 0.1 μm [3–5]. The time needed to create the initial track is less than 10 ps from the incident ion arrival, and the track possesses an initial charge density of $\sim 10^{18}\,\mathrm{cm}^{-3}$. This very high value of charge density is best described by ambipolar transport discussed earlier. Corresponding track density profiles are depicted in Fig. 7.1, showing the results of Monte Carlo

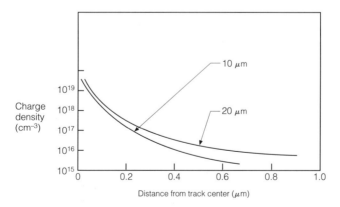

Figure 7.1. Ion-track radial charge density in silicon for various track lengths. From Ref. 2.

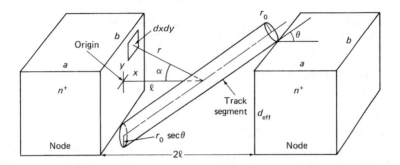

Figure 7.2. Surface layer device showing oblique cosmic ray track segment tangent to two adjacent memory nodes causing double-bit SEU. From Ref. 2.

simulations of their radial density. Implied is that track radial density gradients are very high compared to axial ones. For such ambipolar drift motion at early times, the ambipolar mobility μ^* approaches zero.

As alluded to, multiple event upsets pose a different detection and correction (of errors) problem. Even sophisticated error correction and detection (EDAC) methods must resort to elaborate and expensive schemes that can hardly cope with a very few errors per word. High-multiple-error EDAC methods must use gross redundancies such as twin or triple central processing units (CPUs) with complex voting protocols.

One scenario for the occurrence of a double-bit SEU error is that where an obliquely incident ionization track becomes tangent to two adjacent SEU-sensitive nodes [2], as for instance a RAM as depicted in Fig. 7.2. However, before calculating the corresponding double-bit cross section, it is important to appreciate some details of the general computation of the single-bit SEU cross section σ_1 of a cell node whose sensitive region is a thin parallelepiped of thickness c whose face areas are ab, as in Fig. 7.3. The single-bit error rate E_r (errors per bit second) is given in spherical coordinates by

$$E_r = \int_{L_{\min}}^{L_{\max}} dL \int_{\theta_c(L)}^{\pi/2} \Phi(L)\, ds \left(\frac{ab \cos\theta}{4\pi r^2}\right) = (2)(2)\left(\frac{ab}{4\pi}\right) \int_{L_{\min}}^{L_{\max}} dL\, \Phi(L) \int_{\theta_c(L)}^{\pi/2} d\Omega \cos\theta \tag{7.1}$$

$\Phi(L)\,ds$ is the assumed isotropic heavy-ion flux (ions cm^{-2} s^{-1}) threading the incremental area ds as in the figure. $(ab \cos\theta)/4\pi r^2$ is the flux fraction that is incident on the parallelepiped area ab. Not all of the flux incident on ab causes SEU, but only that portion between a critical angle $\theta_c = \cos^{-1}(L/L_c)$ and the grazing angle to the parallelepiped (i.e., $\pi/2$). This portion of the flux corresponds to heavy-ion track lengths sufficient to cause SEU, as discussed in

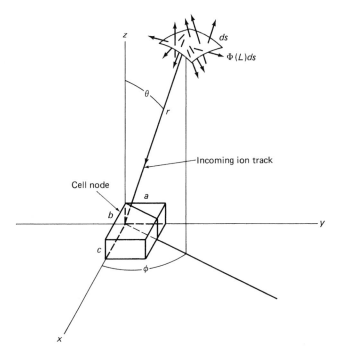

Figure 7.3. Spherical coordinate system for thin parallelepiped RAM cell node.

Section 5.2. One of the factors of 2 in (7.1) accounts for both front and back faces of the parallelepiped by doubling the front face contribution. The second factor of 2 is included to account for the parallelepiped edge effects. This is an approximation that amounts to twice the geometrical area, in the sense of an extinction cross section that has its basis in the diffraction theory of optics. Continuing the integration of the rightmost equation (7.1) over the geometry involved yields

$$E_r = \left(\frac{ab}{L_c^2}\right)\int_{L_{\min}}^{L_{\max}} \Phi(L) L^2 \, dL = (ab)\left[\int_{L_{\min}}^{L_c} \Phi(L)\left(\frac{L^2}{L_c^2}\right) dL + \int_{L_c}^{L_{\max}} \Phi(L)\, dL\right]$$

(7.2)

Because the general form for the SEU error rate $E_r = \int_{L_{\min}}^{L_{\max}} \Phi(L)\, \sigma_1(L)\, dL$, the single-bit cross section σ_1 can be identified in (7.2) as

$$\sigma_1(L) = ab \begin{cases} \left(\frac{L}{L_c}\right)^2, & L \leq L_c \\ 1, & L > L_c \end{cases}$$

(7.3)

Here, only the maximum value of the double-bit SEU cross section σ_2 is computed, as the cross section computation for all relative orientations of the track with respect to the sensitive nodes is straightforward but tedious. From Fig. 7.2, it is seen that the maximum cross-section value (worst-case upset) occurs at the track segment middle between the two node volumes. This is the case, since a solid angle Ω_2 subtended at the track segment middle by each of the two node areas bd_{eff} is a maximum, and they subtend the same solid angle due to the node area symmetry. Then, where $dA = dx\, dy$,

$$\Omega_2 = \int_{4\pi} dA \cos(\alpha)/r^2 = \int_{-b/2}^{b/2} dx \int_{-(d_{\text{eff}} + r_0 \sec \theta)/2}^{(d_{\text{eff}} + r_0 \sec \theta)/2} dy \frac{\ell}{(x^2 + y^2 + \ell^2)^{3/2}} \quad (7.4)$$

Computing the integral gives

$$\Omega_2 = 4 \tan^{-1} \left(\frac{\left(\dfrac{b}{2\ell}\right)(d_{\text{eff}} + r_0 \sec \theta)}{(b^2 + (d_{\text{eff}} + r_0 \sec \theta)^2 + 4\ell^2)^{1/2}} \right) \quad (7.5)$$

σ_2 can be written in the same general form as σ_1 [i.e., $\sigma_2 = (L/L_c)^2 \times$ "area," because of the generality implied in deriving σ_1 in (7.3). However, in this case, the "area" is reduced from the node area $b(d_{\text{eff}} + r_0 \sec \theta)$, where $r_0 \sec \theta$ is added to include the vertical projection of the track radius. The reduction comes about because each node area subtends only a fraction of the total 4π solid angle as seen from the track middle. The reduction factor is $\Omega_2/4\pi$. Thus,

$$\sigma_2 = \left(\frac{L}{L_c}\right)^2 \cdot b(d_{\text{eff}} + r_0 \sec \theta) \left(\frac{\Omega_2}{2\pi}\right) \quad (7.6)$$

The third factor is $2\Omega_2/4\pi = \Omega_2/2\pi$ because of the areas of the two adjacent nodes.

Figure 7.4 depicts the ratio σ_2/σ_1 versus 2ℓ, the node separation distance for various track radii [2]. It is seen from the figure that σ_2/σ_1 becomes significant for thick tracks and small node separations. This can be appreciated because thick tracks are due to heavy energetic ions, and small node separation implies micron and submicron feature sizes.

Multiple-bit upsets can also occur for heavy ions at normal incidence (i.e., short track segments). This is especially the case for relatively thick tracks. Such ionization tracks contain high densities of "hot" electrons (i.e., electrons in the high-energy tail of the fermi distribution). They induce a radially ambipolar diffusion as discussed earlier [6].

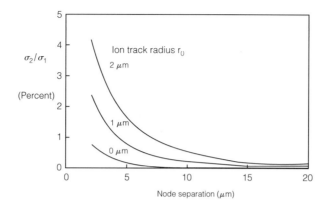

Figure 7.4. Ratio of double-bit to single-bit SEU cross section for two adjacent memory nodes for various ion track radii. From Ref. 2.

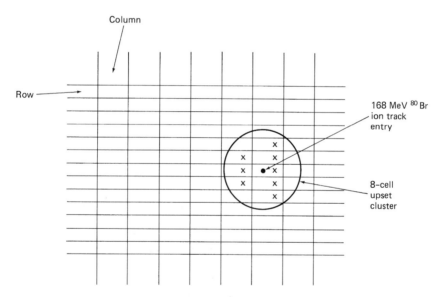

Figure 7.5. 256K DRAM multiple error memory cell bit map caused by single 168 MeV ^{80}Br ionization track. From Ref. 6.

Figure 7.5 shows a detailed multiple error bit map for a 256K-bit DRAM (dynamic RAM) generated by SEU induced by 168 MeV ^{80}Br ions [6]. It shows that eight memory cell bits clustered about the track entry puncture were upset. This is corroborated in Fig. 7.6, where the SEU cross section versus LET is plotted. This plot is obtained by using a number of heavy-ion types from a tandem Van de Graaf accelerator spread over various incident energies culmi-

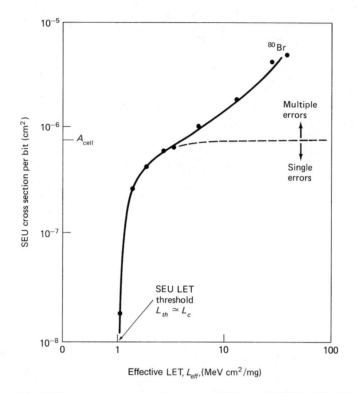

Figure 7.6. SEU cross section per bit versus LET for 256K-bit DRAM showing multiple upset from single ion track at normal incidence. From Ref. 6.

nating in the 168-MeV Br ion, which is the most energetic [6]. In Fig. 7.6, it is seen that the cross section proceeds upward with a large slope, in contrast to becoming asymptotic at the approximate area of a single cell, as is the case of ordinary SEU cross sections. Actually, the 168-MeV Br ion cross-section data end at a factor of about 8 over that of the single cell area asymptote, as seen [6]. This is in good agreement with the 8-bit charge cluster "hit" as shown in Fig. 7.5.

It is of interest to estimate the amount of charge collected that corresponds to the eight-cluster multiple upset. For this DRAM, the critical charge per cell $Q_c = C_s \Delta V$, where C_s is the cell storage capacitor available from the cell dimensions ($C_s = 0.05$ pF). $\Delta V = V_{DD} - V_{IB}$, where V_{DD} is the supply voltage and V_{IB} is the sense amplifier input bias voltage ($\Delta V = 5 - 2.5$ V). Hence, $Q_c = 0.125$ pC. In Fig. 7.6, it is seen that the threshold LET $\cong L_c \cong 1$ MeV cm^2 mg^{-1} = 0.01 pC μm^{-1} for this device. Therefore, the equivalent charge collection depth was about 0.125 pC/0.01 pC per micron = 12.5 μm; that is, the ion-track segment was at least 12.5 μm in length.

In very large scale integration (VLSIC) static RAMs, multiple-bit upsets from 2 to 5 bit upsets occur per heavy-ion incident at almost all angles. The frequency of occurrence apparently increases with the LET of the incident ion in the device material [7]. Certain 64K × 1 NMOS SRAMs, whose cell circuit pull-up resistors and V_{DD} lines are polysilicon, are extremely sensitive to SEU. This sensitivity is apparently due to their extremely high values ($\sim$ 1 gigaohm) of pull-up resistance. Such high values are used to maintain the low standby current.

However, this imposes a relatively large circuit time constant to restore a "HI" node state following an SEU event. The corresponding times are on the order of microseconds–enough to allow SEU-induced off-node diffusion currents to contribute to device upset. In the extreme, large diffusion currents flood a region of cells with charge to produce high multiple-bit upsets [8].

In terms of the determination of SEU cross sections versus LET experimentally, they can asymptote at values higher than the corresponding physical size of the SEU-sensitive regions, as in Fig. 7.6. The multiple-bit cross sections then become indicators of the range of these diffusion currents. Above a given LET, it is possible for the whole microcircuit area, including that between nodes, to become SEU sensitive. Equivalently, the critical charge for these high-packing-density NMOS SRAMs is extremely small [9].

In general, appreciable multiple-bit upsets should be anticipated for modern high-packing-density memory devices. For example, modern high-density chip recorders will contain hundreds if not thousands of megabits. Their use in spacecraft must be "designed in," with strong cautionary attention given to SEU and SEU-type multiple-bit upsets. Emphasis should be placed on device architecture as opposed to mere SEU susceptibility in the device selection process for space avionics, subject to known SEU-inducing particles at desired spacecraft orbit altitudes.

In a series of ingenious experiments to observe certain of the statistical aspects of multiple-bit upsets and the corresponding charge transport, 256K DRAMs were used as test vehicles as described below [10]. Generally, if the incident ion track is normal to the chip surface, this results in a somewhat locally confined charge track of constant charge per unit track length. If this ion track intersects a junction, the latter collects the charge promptly, due to the electric field and the funnel. In this case, the time of formation and subsequent collapse of the funnel (picoseconds to nanoseconds) is over before all the charge produced is collected. The residual charge is ultimately collected by diffusional processes. This corresponds to a much larger time scale (nanoseconds to microseconds) than the electric field–funnel-aided process above.

However, if the normal ion track does not intersect a junction, almost all of the charge is collected by diffusion over a relatively widespread region. The amount of charge collected by these two processes can differ greatly [10]. If

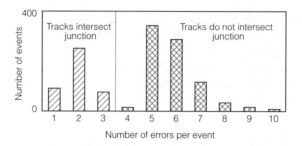

Figure 7.6'. Multiple-bit errors for two populations: those that intersect a charge collection junction and those that do not: $V_{DD} = 5$ V. From Ref. 10.

the incident ion track is far from normal (large θ), the corresponding lateral charge transport will produce multiple-bit upsets, as, of course, will the normally incident ion tracks to a lesser extent. It turns out that the multiple-bit upsets separate into two populations, depending on whether the incident ion track is close to or far from the chip normal. Another factor that distinguishes between the two populations is if a charge collection junction was intersected by a track. It is clear that the former corresponds to a more normally incident track than to a lateral track.

Figure 7.6' depicts the two populations of multiple-bit upsets corresponding to those produced by whether a junction was intersected or not [10]. As mentioned, the devices used were 256K DRAMs irradiated by low-energy bromine accelerator ions that produced the upsets. Their V_{DD} was 5 V. Fig. 7.6a shows how the multiple-bit upset population produced by normally incident ion tracks varies with V_{DD}. The accelerator ions were 295-MeV Kr (LET = 40 MeV cm^2 mg^{-1}). The population density dispersion increases as V_{DD} is decreased. This is because the critical charge decreases with lesser V_{DD}, resulting in more upsets. This can be seen from the simplistic argument that the critical charge Q_c of a particular cell in the DRAM is given by $Q_c \cong C_{cell}\Delta V$, where $\Delta V = V_{DD} - V_B \geq 0.1$ V. V_B is the sense amplifier input bias voltage and C_{cell} is the corresponding information storage capacitance of the cell node. Hence, by decreasing V_{DD}, the voltage swing ΔV decreases, and so does Q_c, resulting in a mean number increase in multiple-bit upsets, as shown in Fig. 7.6a.

7.3. Ionizing Dose Effects

Ionizing dose (total dose) is that deposited by environmental particles, usually simultaneous with incident SEU-inducing heavy ions. Ionizing dose radiation or particles cause long-term damage in devices, as opposed to SEU by which the device recovers physically in a relatively short time [11]. This can include

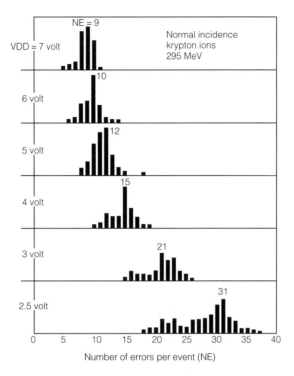

Figure 7.6a. Multiple-bit errors for normal incident ion tracks as a function of device V_{DD} as irradiated by 295-MeV Kr ions. From Ref. 10.

SEU-inducing heavy ions that can cause "hard errors." The predominant ionizing dose damage is usually construed as derived from gamma rays, X-rays, protons, and electrons. The former two are usually associated with a nuclear burst, whereas the latter two are usually associated with the Van Allen radiation belt irradiation of spacecraft electronics (avionics). Photon and other particle environments include nuclear reactor and accelerator spurious radiation.

For example, accelerator testing of devices to determine their SEU parameters, such as cross sections when they are relatively small, entails a risk of device damage due to ionizing radiation. For a relatively small SEU cross section, device exposure to the accelerator beam must be long in order to obtain "good statistics"; that is, experimentally, the SEU cross section is obtained from

$$\sigma_{seu} = \lim_{t \to \infty} \left(\frac{N_{seu}}{\Phi_{seu}} \right) (\text{events}/s \div \text{particles}/cm^2 s) \qquad (7.7)$$

where N_{seu} is the rate of observed SEU events. Φ_{seu} is the heavy-ion beam flux. For available Φ_{seu} levels, small σ_{seu} means that to obtain a reasonable N_{seu}, the exposure time must be long to provide a credible ratio of N_{seu}/Φ_{seu}. However, during that exposure time, the device can suffer permanent damage by the concomitant ionizing dose deposited in it by the heavy-ion flux. Most devices of reasonable SEU cross section can be measured with exposure times that are short compared to times needed to cause permanent ionizing dose damage in the device.

The preceding certainly holds in the operational sense, where the accumulation of ionizing dose in the spacecraft electronics of a multiyear mission can induce significant damage. This can result in the corresponding devices becoming more and more SEU susceptible even though being declared "SEU hard" prior to launch.

The ionizing dose can cause a decrease in memory device threshold LET with a concurrent increase in SEU cross section [12]. For certain CMOS SRAMs [13], it has been found that σ_{pseu}, that part of the SEU cross section attributable to the p-MOS component, increases while that due to the n-MOS component decreases. This is apparently because of the negative shift in the device gate threshold voltage, $-\Delta V_{TH}$, caused by trapped charge in the gate oxide [13]. Also occurring is σ_{seu} saturation due to the reduction of charge collection caused by displacement damage after a nuclear burst [13].

For another family of CMOS SRAMs (SA3240) not only does ionizing dose damage decrease SEU immunity, but it occurs whether the ionizing dose rate is high or low [14]. On the other hand, for high levels of ionizing dose, SEU susceptibility increases with the increase in device cell feedback resistor values. These resistors are used to mitigate SEU by imposing a "low-pass" response to the cell (nanoseconds to microseconds) antithetical to the very fast SEU phenomena (picoseconds to nanoseconds). Other devices (IDT CMOS/ NMOS SRAMs) exposed to SEU from 40-MeV protons show a similar behavior as their SEU susceptibility increases with increasing ionizing dose deposition [15].

In general, the ionizing dose absorbed by the device results in damage as evidenced by degradation of its operating characteristics. This damaging of the device causes it to become more SEU prone than its undamaged counterpart. To add quantitative substance to the preceding, investigation of the important parameters of the individual families/devices must be made experimentally. There is little in the way of theoretical or empirical predictive knowledge available in this context especially for MOSFET devices [32, 33].

As discussed above, and in Section 2.5, ionizing dose damage in spacecraft avionics is produced mainly by Van Allen belt protons and electrons.

Figure 7.7a depicts ionizing dose that spacecraft at various orbit altitudes endure versus their years of mission duration aloft [16]. It should be noted in the figure that the ionizing dose at 0.5 geosynchronous altitude ($\sim 10^4$ NM) is greater than that at 1.0 geosynchronous altitude ($\sim 2 \times 10^4$ NM). This is because 10^4 NM is close to the outer Van Allen electron belt. The global posi-

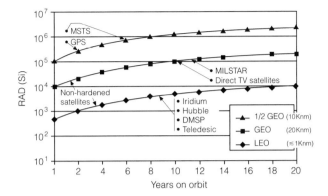

Figure 7.7a. Total ionizing dose that satellites experience in the natural environment. From Ref. 16.

tioning satellites (GPS) are in the inner Van Allen belt. After 10 years, the GPS will have experienced an ionizing dose of about 1 Mrad(Si) as seen in Fig. 7.7a. Present communications spacecraft orbits are at geosynchronous altitude, whereas the new mobile communications spacecraft are planned for low earth orbits (under the Van Allen belts). As shown in Fig. 7.7a, low-earth-orbit spacecraft will suffer a relatively benign radiation environment of about 5–10 krad(Si) for 10-year orbit missions. It is also well known that the ionizing dose can affect the sensitivity of the logic states of SRAMs in an asymmetrical manner. This is termed a memory imprint or imbalance. It is due to mobility and threshold voltage changes caused by the ionizing radiation [31].

With the cold war now definitely in the past, an unlikely exo-atmospheric nuclear weapon detonation could conceivably come from a rogue third-world developing nation. By postulating a relatively low-yield (50 kT) weapon detonated at 120 km altitude over an East Indian archipelago, the result might be devastating to spacecraft avionics that reasonably penetrate or reside in the Van Allen belts. This detonation might temporarily "pump up" the Van Allen belts by at least two orders of magnitude in their damaging particle fluxes. Spacecraft with unhardened electronics would begin to fail in a few months to 2 years following the detonation [16]. The belts might maintain their enhanced flux levels for a long interval, possibly deterring the launching of replacement spacecraft. Figure 7.7b, for comparison, shows the same three curves of dose deposited for the orbit altitudes in Fig. 7.7a. The former has two additional curves. They are the deposited doses for low earth orbits (LEOs) at altitudes of 850 km and 1200 km in an enhanced radiation environment due to the above weapon detonation. It is seen that their ionizing dose levels are considerably above those for the low earth orbit in a natural environment.

244 / *Single Event Phenomena*

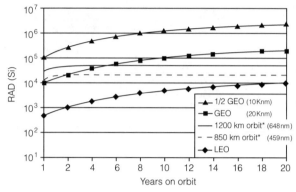

*Threat defined as 50 KT detonation at 120 km altitude.

Figure 7.7b. Total ionizing dose for natural space and threat environment. From Ref. 16.

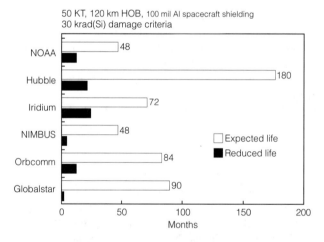

Figure 7.8. Low-earth-orbit satellite lifetime impacts for a 50-kT detonation at 120 km altitude over an East Asian peninsula. From Ref. 16.

A worst case scenario calculation is reflected in Fig. 7.8, depicting the estimated shortening of the lives of a number of low earth spacecraft due to the Van Allen radiation belt fluxes being enhanced by a low-yield exo-atmospheric nuclear detonation as discussed. It is seen from the figure that the life reduction of certain spacecraft is very great. Their mean life loss as a percentage of their pre-irradiation life varies from 66% (Iridium) up to 99% (Globalstar) [16]. Large changes from pre- to post-irradiation values of spacecraft life reduction should be construed as gross estimates.

Again, comparing Figs. 7.7a and 7.7b, it is seen that orbits at geosynchronous earth orbit (GEO), $\frac{1}{2}$GEO, and LEO altitudes are hardly perturbed due to this, assuredly unlikely, exo-atmospheric detonation. However, Fig. 7.7b shows total ionizing dose levels at 850 km and 1250 km altitudes due to the detonation, the levels of which at this magnitude are absent in the natural environment. Although not specifically indicated, the detonation-induced ionizing dose at these levels can deleteriously affect LEO spacecraft including Iridium, Hubble, DMSP, and Teledesic. This is reflected in their lifetime reduction estimations as depicted in Fig. 7.8.

"Stuck bits" are an SEU phenomenon induced by ionizing dose [17]. For certain devices, whether SEU tested using accelerator ions or instrumented aboard a spacecraft, semipermanent stored patterns of bits within the device have been observed [18]. These patterns have been detected in densely packed devices (e.g., 1-Mbit SRAMs) with small feature size ($\sim 1 \mu$m) when irradiated by heavy, high-LET (≥ 30 MeV cm^{2}mg^{-1}) ions. Some patterns anneal spontaneously within a few minutes, some anneal rapidly when irradiated with UV photons, whereas some take several months to anneal. This effect, called single hard error (SHE), is claimed to be caused by heavy-ion-induced ionizing dose damage in a device gate, causing it to remain in a semipermanent on-state. It is felt that this effect has only been observed recently because of the large feature size of chips in the past. Today's small-feature-sized transistor gates are comparable to the diameter of the incident ion track, and thus can be subject to the high ion density and therefore high ionizing dose induced by the track [18] throughout their volumes.

7.4. Redundancy, Scrubbing

One of the major techniques used to cope with electronic system reliability is redundancy. For example, it is well known that for a given noisy information channel, it is more reliable to repeat individual message symbols than to repeat groups of symbols or the whole message. Also, it is a truism that for nonmaintained (no human intervention) systems such as spacecraft, higher reliability results when redundancy is employed at successively lower levels of system design. On the other hand, it is known that for standby maintained (human intervention) systems, high-level redundancy produces high reliability. Most electronic systems considered here are in the former category, such as spacecraft, missiles, and many other military and civilian systems.

The generalized problem is that of the allocation of redundancy in electronic systems subject to the vagaries of various types of radiation constraints, at a minimal generalized cost. The latter is not necessarily financial, but it can be derived from the myriad limitations corresponding to the uniqueness of the system under consideration. Consider a computer with redundancy to be allocated within it to achieve a specified reliability at a minimum cost and within a given maintenance period. Assume that the redundancy is in the form of spare chips and an EDAC system that will switch from a SEU swamped chip to a twin if one is available. The reliability R of the computer without redundancy is given by,

$$R = \prod_{i=1}^{n} p_i, \quad p_i < 1 \tag{7.8}$$

where p_i is the reliability (survival probability) of the ith chip. The generalized cost of chip allocation is assumed to be

$$C = \sum_{i=1}^{n} k_i p_i \tag{7.9}$$

where k_i are the factors that convert chip survival probabilities to chip costs. Because $p_i < 1$, the reliability of each chip, p_i, must be greater than R as seen from (7.8). For the ith chip paralleled with a (redundant) spare to achieve an overall enhanced reliability R_c, the number of redundant chips required, on a per chip basis, is

$$R_c = 1 - (1 - p_i)^{n_i} \tag{7.10}$$

This is apparent, as $1 - p_i$ is the probability that the ith chip is temporarily nonoperable, so $(1 - p_i)^{n_i}$ is the case for n_i nonoperable chips. Then the obverse [i.e., $1 - (1 - p_i)^{n_i}$] is the enhanced reliability of the computer, using the n_i redundant chips. The number n_i of such chips for a specified R_c is

$$n_i = \frac{\ln(1 - R_c)}{\ln(1 - p_i)} \tag{7.11}$$

As an example, suppose that in between maintenance periods T_m in space, the computer reliability goal is set at

$$R_c = 1 - \exp(-\lambda_{\text{comp}} T_m) = 0.995 \tag{7.12}$$

where λ_{comp} is the mean error rate of the computer as a whole, and all chips have the identical $p_i = 0.960$. Then from (7.11),

$$n_i = \frac{\ln(1 - 0.995)}{\ln(1 - 0.960)} = 0.9634 \text{ per chip} \tag{7.13}$$

Hence, for this 1000-chip computer, about 963 redundant chips must be added to obtain the specified reliability goal R_c. This number of chips implies essentially a complete alternate memory. In an actual computer, the chips are not all identical, so that the redundancy rules would be more varied. The details for determining such protocols are well known [19].

In the sense of an optimal reliability search, i.e. how to maximize R in (7.8) consistent with (a) $p_i < 1$, and (b) the cost given by (7.9) is also well known (obtained by using Lagrange multipliers). The result is that $k_i p_i = C/n$; that is, each chip is allocated an equal portion of the cost C/n. Then the chips used must be specified such that they are constrained as $k_i p_i = C/n$.

Another illustration of how redundancy increases reliability is that of memory fault tolerance with respect to error detection and correction (EDAC) methods. Here, it is applied to mitigate neutron-induced single event upset errors, which are discussed in Section 5.4, during the flight of a memory in an airborne computer. EDAC methods for resolving *single error correction and double error detection* (SEC–DED) of bits per memory word are used frequently. This is because of the relatively high probability of single errors over multiple errors per word, and the ease with which SEC–DED can be implemented in the corresponding firmware (e.g., read-only memory) and software (computer program).

The flight time t_f can be reckoned in scrub intervals t_s, where the number of scrub intervals per word over the flight duration is simply $N_s = t_f/t_s$. A scrub or scrubbing interval is the period between scrubbings of a particular word in the memory array; thus, every word in the memory array is scrubbed once during each scrub interval. Scrubbing of a word consists of its periodic (scrub interval) inspection and, if necessary, detection and correction of a single error using a Hamming SEC–DED process. Multiple SEU errors are assumed to occur with low probability and, thus, are ignored. The actual scrubbing time is a very small fraction of the scrubbing interval. However, from the viewpoint of the memory array as a whole, its words are being scrubbed throughout each scrub interval over the flight duration. The details of the SEC–DED process in the scrubbing action are not important here, except to appreciate the competition among scrubbing, scrub intervals, and the SEU error rate.

Another fault tolerant redundancy method is parity checking of memory words, discussed in Section 7.5. Its simplicity and speed are its attractive features, requiring many fewer redundancy bits per word than the Hamming decoding procedure. Parity checking disadvantages are that it cannot detect certain periodic errors (e.g., all even-numbered errors in a word) and it does not correct errors in any case. This must be left to higher-order error recovery methods.

Consider a memory array of w words, each of which consists of a succession of one and zero bits—n bits long. The first k bits contain the information, whereas the last n-k bits are sacrificed to the EDAC process. The latter

248 / Single Event Phenomena

are called parity bits. Each word in the memory array is read as correct for the case of zero errors in it or if there is a single-bit error detected and corrected within. As mentioned, multiple-bit errors are assumed to be negligible and, thus, are ignored. Elaborate decoding schemes for handling multiple-bit errors are available, but they can be unwieldy and expensive in terms of memory space and software.

If p is the probability of zero errors per bit and $q = 1 - p$ is that of one or more errors per bit, then p^n is the probability of zero errors per word of n bits in length. For the case of one bit error in this word, $np^{n-1}q$ is the corresponding probability. This follows from the binomial probability distribution, where one of its properties is

$$1 = (p+q)^n = p^n + np^{n-1}q + \cdots q^n = \text{prob. 0 errors}$$
$$+ \text{prob. 1 error} + \cdots + \text{prob. } n \text{ errors} \quad (7.14)$$

Thus, in each scrubbing interval, the probability of obtaining each word from memory correctly, whether for a "zero or one bit error-word" to the approximation implied in the SEC–DED process is $p^n + np^{n-1}q$. Then assuming that the occurrence of errors are random, the reliability R, or the probability of reading correctly the w words in the memory, is

$$R = (p^n + np^{n-1}q)^w \quad (7.15)$$

suffering only SEU errors and with no scrubbing action.

For a mean error rate λ_s (e.g., λ_s SEUs per bit-day), the probability of a bit being correct as it stands subject only to SEU over a duration t is $p = \exp(-\lambda_s t)$. Inserting this p into (7.15) yields the corresponding no scrubbing memory reliability (but with SEC–DED) over the flight time t_f as

$$R_{ns} = \{n \exp[-(n-1)\lambda_s t_f] - (n-1)\exp(-n\lambda_s t_f)\}^w \quad (7.16)$$

It is seen in the limit of very large flight time, that R_{ns} approaches zero as the SEU error rate eventually overwhelms the memory EDAC system.

In the above case, but with scrubbing (a scrubbing interval t_s) and wN_s total scrubbing intervals, the corresponding reliability for a scrubbed memory, with SEC–DED, is, now with $p = \exp(-\lambda_s t_s)$,

$$R_s = \{n \exp[-(n-1)\lambda_s t_s] - (n-1)\exp(-n\lambda_s t_s)\}^{wt_f/t_s} \quad (7.17)$$

where the SEU error rate is random, and thus independent from one scrubbing interval to the next. Note that if $t_s = t_f$, one scrub per flight, that $R_s = R_{ns}$. This is essentially equivalent to no scrubs per flight, as $t_s \ll t_f$ must hold for

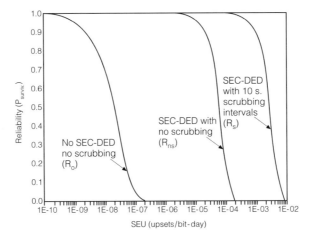

Figure 7.9. SEU error protection for a 4M × 32 memory array during a 6-h flight. From Ref. 20.

sensible scrubbing action considering all the other temporal parameters, such as memory access time, computer operating frequencies, and so forth.

If neither scrubbing nor SEC–DED is operative in the memory, then only the first term in (7.15) is applicable; therefore, the reliability becomes, using $p = \exp(-\lambda_s t_f)$ for the whole flight,

$$R_0 = p^{nw} = \exp(-n\lambda_s w t_f) \tag{7.18}$$

Figure 7.9 depicts the reliability R of a 4000K × 32-bit memory of an airborne computer versus the SEU bit error rate λ_s for a 6-h flight with 10-s scrubbing intervals t_s [20]. Each curve corresponds to one of the three cases: SEC–DED processing with scrubbing and reliability R_s, SEC–DED with no scrubbing (R_{ns}), and no SEC–DED and no scrubbing (R_0).

7.5. Error Detection and Correction

Error correcting codes first arose to handle practical problems in the reliable communication of digital information. The origins of information theory has a branch called error-detecting codes. For SEU, it is apparent that error detection and correction (EDAC) plays an important part in most successful upset mitigation schemes, not only at the system level but at the part (microcircuit) level. For example, there are a number of families of EDAC chips available for use in electronic and avionics systems. As an example of error detection, consider four 7-bit code words in Table 7.1.

Table 7.1. Four 7-Bit Code Words.

	1	2	3	4	5	6	7
	Info bits				EDAC bits		
Word No. 1	1	0	0	0	0	1	1
Word No. 2	0	1	0	0	1	0	1
Word No. 3	0	0	1	0	1	1	0
Word No. 4	1	0	1	1	0	1	0

One well-known error detection algorithm is demonstrated by the 4-bit word of information bits (viz. 1 0 1 1, which is word No. 4 in Table 7.1). To be accommodated into the 7-bit word length, 3 EDAC bits are added to make it into $\bar{x}$ = 1 0 1 1 0 1 0. Suppose that one SEU error transforms $\bar{x}$ into $\bar{x}'$ = 1 0 1 0 0 1 0. Now define three ferret vectors: **a** = 0 0 0 1 1 1 1, **b** = 0 1 1 0 0 1 1, and **c** = 1 0 1 0 1 0 1. Then form the three inner products **x**' · **a**, **x**' · **b**, and **x**' · **c**. To compute an inner product, multiply each of the corresponding components of each vector, getting a 0 or a 1 for each, as shown below. Then sum all seven of the products, subtracting out all possible two's ("modulo 2ing") from the sum to obtain remainders of 1 or 0. These results are called parity bits. Mathematically, for example, $\mathbf{x}' \cdot \mathbf{a} = (\Sigma_{i=1}^{7} x_i' a_i)$ mod 2. Now 5 mod 2 = 1, 6 mod 2 = 0, and 0 mod 2 = 0. Hence, the partity words are computed as follows:

```
     x' =  1 0 1 0 0 1 0            x' =  1 0 1 0 0 1 0
      a =  0 0 0 1 1 1 1             b =  0 1 1 0 0 1 1
   x'·a =  0 0 0 0 0 1 0          x'·b =  0 0 1 0 0 1 0
        =  1 mod 2 = 1                  =  2 mod 2 = 0

     x' =  1 0 1 0 0 1 0
      c =  1 0 1 0 1 0 1
   x'·c =  1 0 1 0 0 0 0
        =  2 mod 2 = 0
```

Now, construe these three parity bits in this order as a 3-bit binary word 1 0 0, which is the binary number for 4. The algorithm then asserts that the fourth bit in **x**' is in error; that is, **x**' = 1 0 1 0 0 1 0 should really be **x** = 1 0 1 1 0 1 0, as is already known. This procedure is known as a Hamming (7/4) decoding, as it can detect any single error in the above 7-bit word. This algorithm can be implemented almost trivially into an EDAC chip using a minimum of bytes (1 byte = 8 bits). Other more complicated EDAC algorithms use up to 11 redundancy (EDAC) bits of a 23-bit word. Those can

Table 7.2. EDAC Parameters for Various Word Lengths.

Word bit length	Info bits	Correctable errors	No. Error combos	No. bit combos	EDAC word rate
7	4	1	2^4	2^7	0.571
23	12	3	2^{12}	2^{23}	0.522
47	24	5	2^{24}	2^{47}	0.511

Source: Ref. 21.

correct up to 3 errors. However, they are so complex that besides using up a large amount of chip "real estate," it also uses a goodly amount of computer time to accomplish its task—especially for a large memory array. For a 23-bit word (i.e., 2^{23} different words from permutations of its 0's and 1's), the algorithm searches for 1–3 errors that can occur in 2^{12} different combinations. Other EDAC schemes are even more complex, as in Table 7.2.

The EDAC word bit rate is defined as the ratio of word information bits to word bit length. For example, for the 7-bit word above, the word bit rate is $4/7 = 0.571$. Note that the word bit rate in Table 7.2 approaches 0.500 as the word bit length increases. It is found from information-theoretic considerations that for word bit rates less than the system throughput (channel) capacity, if the EDAC code scheme uses sufficiently long word bit lengths, then the EDAC scheme can accomplish its task with arbitrarily small error probability. This means that if the word bit length includes a large enough number of EDAC bits, then its bit error correction algorithm can efficiently detect and correct an arbitrarily large number of bit errors. The proof is sufficiently abstract in that it offers no clues for constructing corresponding EDAC schemes as good as it predicts. A "good" EDAC scheme is one that can detect and correct many errors by its intrinsic nature. However, it is important to have a sensibly realizable EDAC scheme for practical purposes.

At the chip (microcircuit) level, SEU errors can be decreased by (a) special lithographic processing techniques during its manufacture, such as a highly doped buried layer to curtail funneling, (b) chip circuit design to include feedback resistors in the memory cell to impose a low-pass filter characteristic, and (c) SEU EDAC schemes on-chip or off-chip. As seen, a Hamming based EDAC system can detect only one error per word. Using this code, it would be prudent to physically separate the bits of a word in the chip so that multiple-bit errors are minimized, in contrast to enduring adjacent-bit-multiple-bit upsets if this practice is not employed. This is another example of a low-level reliability measure being implemented for a nonmaintained system as discussed earlier. Providing onchip EDAC reperesents savings in computing time, as well as hardware and software [22]. Consider a 256Kb × 1 DRAM that includes an on-chip Hamming error-correcting code (ECC) circuitry. Each word in the DRAM is 12 bits long, consisting of 4 EDAC bits and 8 information bits. To include the on-chip error corrector, the DRAM bit array must be expanded from 512×512 to 786×512 bits [22].

It is of interest to estimate the probability that two or more SEU-induced errors occur in the same word, say, in the ECC chip portion. Let W_2 be the number of 12-bit words that have sustained two or more upsets. Define n_T as the total number of SEU bit errors suffered by the chip during its data storage time T_s. This probability is designated as $P(W_2 > 1, n_T \geq 2)$. It is given by the binomial distribution [22]

$$P(W_2 > 1, n_T \geq 2) = \sum_{i=2}^{n_T} \binom{n_T}{i} p_s^i q_s^{n_T - i} = \sum_{i=2}^{n_T} \binom{n_T}{i} \left(\frac{1}{N}\right)^i \left(1 - \frac{1}{N}\right)^{n_T - i} \quad (7.19)$$

where, from the basic premises of the binomial distribution,

1. $N = 2^{15}$, the number of 12-bit words in the DRAM, so that $p_s = 1/N$ is the maximum likelihood probability of a particular word being subjected to one SEU, and $(1/N)^i$ is that for i SEUs
2. $q_s = 1 - p_s = (1 - 1/N)$ is the probability of that particular word escaping an SEU hit and $(1 - 1/N)^{n_T - i}$ is that for the $n_T - i$ non-SEUs
3. $\binom{n_T}{i} \equiv n_T!/(n_T - i)!i!$, the binomial coefficient, the number of combinations of n_T bit errors taken i at a time, where $2 \leq i \leq n_T$

Because multiple upset probabilities are sought, the summation index i, which counts the number of multiple error possibilities in the DRAM, starts at $i = 2$ and goes up to n_T. If $i = 0$ (no upsets) and $i = 1$ (one upset) were to be included, then the corresponding $P = 1$. Hence, (7.19) can be rewritten as

$$P(W_2 > 1, n_T \geq 2) = 1 - \sum_{i=0}^{1} \binom{n_T}{i} \left(\frac{1}{N}\right)^i \left(1 - \frac{1}{N}\right)^{n_T - i}$$

$$= 1 - \left(1 - \frac{1}{N}\right)^{n_T} \left(1 + \frac{n_T}{N - 1}\right) \cong \left(\frac{n_T}{N}\right)^2 \quad (7.20)$$

because normally $n_T \ll N$ and N is very large, so that $N - 1 \cong N$ and $(1 - 1/N)^{n_T} \cong 1 - n_T/N$. The expected number of multiple upset errors E_{av} in the DRAM is, from (7.20),

$$E_{av} = PN \cong \frac{n_T^2}{N} \quad (7.21)$$

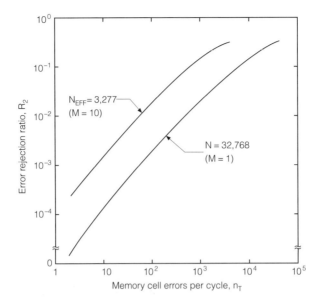

Figure 7.10. Error rejection ratio of an ECC memory chip containing $N = 32768$ code words. From Ref. 22.

The ratio of mean upset errors to the DRAM memory multiple error capacity, R_2, is given by

$$R_2 = \frac{E_{av}}{n_T} \cong \frac{n_T}{N} \tag{7.22}$$

R_2 is called the error rejection ratio. It is the ratio of two or more errors detected at the DRAM output pins to the total bit errors induced in the memory cells. It is the major parameter that characterizes the error function of the ECC chip. R_2 versus n_T is plotted in Fig. 7.10. For the case of adjacent multiple-bit upsets due to SEUs (e.g., M such upsets per SEU, the effective number of independent error correction words is reduced by the factor M so that [22] $N_{eff} = N/M$. The figure implies that the ECC error suppression capability is increased by partitioning the chip into a large number, N, of small ECC words. However, this results in a greater cost (i.e., requiring an increased percentage of EDAC bits per word), so that the chip array size must be expanded. In a single cycle time, the number of bit errors, n_T, accumulated in the array is

$$n_T = A_T T_s \dot{\Phi} \tag{7.23}$$

where A_T is the total area of all the memory cells and $\dot{\Phi}$ is the incident-ion SEU-inducing flux, where each ion causes at least one SEU in a particular experimental run. The data storage cell time T_s is given by

$$T_s = C_D \frac{2^B}{f} \tag{7.24}$$

where C_D is the number of data clock operations per bit, B is the $\log_2$ of the number of addressable bits, and f is the device clock frequency. Inserting T_s into (7.23) yields

$$n_T = \frac{A_T \dot{\Phi} C_D 2^B}{f} = \frac{0.4639}{f_{\text{MHZ}}} \tag{7.25}$$

where the rightmost equation (7.25) corresponds to parameters of the 256K ECC DRAM used in the experiment [22]. The maximum T_s in a DRAM is limited by the refresh cycle time, placing a lower limit on its operating frequency. This limitation, of course, is not a consideration for a static RAM (SRAM) because it needs no refresh cycle. Fig. 7.11 depicts R_2, the error rejection ratio, as a function of the ion flux for the case of $M = 1$ and $M = 10$, for two ion species {viz. 200-MeV ^{56}Fe ($M = 10$) and 96-MeV ^{12}C ($M = 1$) at a clock frequency of 2 MHz [22].} The preceding shows that to increase an ECC chip SEU hardness, f should be increased as well as the number of ECC code

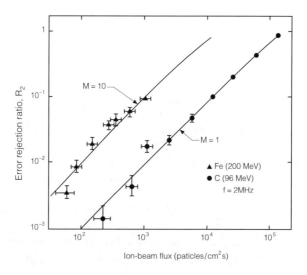

Figure 7.11. Experimental SEU data for ECC DRAM error rejection ratio versus ion beam flux at constant clock frequency. From Ref. 22.

words in the EDAC code repertoire. Of course, $\dot{\Phi}$ should be decreased if possible.

For the case of an EDAC system consisting of triple modular redundancy (TMR) and a majority (two out of three) voter, it is claimed that improvements over the preceding EDAC scheme can be realized [23]. This system is employed primarily where SRAMs are the major memory devices, although it can be matched to DRAM architecture if need be. The TMR is a decision cell where information signals are input to three modules containing three flip-flops whose outputs are fed to a majority voter. Each flip-flop in the triple modules is on a different branch of the clock circuitry tree so that an error in one branch is not propagated to the other two. As in the previous system, it is desired to determine the probability of sustaining two or more errors in one TMR cell in a given clock cycle. SEUs are scrubbed in the flip-flops during each clock cycle.

At the bit level, the SEU error statistics are assumed governed by the Poisson distribution, as SEU normally can be classed as a rare event. Thus, the probability $p(k, t; \lambda)$ of k SEUs induced by heavy ions arriving in time t is

$$p(k, t; \lambda) = \frac{(\lambda t)^k \exp(-\lambda t)}{k!} \tag{7.26}$$

where λ is the mean SEU rate. Omitting details of the TMR operation, a simple explanation of its flip-flop operation as impacted by multiple upsets follows.

The flip-flops in the TMR system, each representing one bit, exist in one of two states. State 1 (e.g., 0), if excited, can flip to state 2 (e.g., 1). In the latter state, if excited it can "flop" back to state 1. Hence, for an even number of SEUs in the scrub period, the flip-flop will ultimately reside back in state 1 after scrubbing itself; that is, (a) state 1, SEU No. 1 arrives, flip to state 2; (b) state 2, scrub, SEU No. 2 arrives, flip back to state 1, scrub; and so on for an even number of SEU arrivals. It is as if no SEUs have arrived, while the flip-flop output has been propogated through the system. Then the probability of an even number of SEU arrivals (i.e., no SEUs in this context) is,

$$p_s = \sum_{k \text{ even}}^{n_T} \frac{(\lambda t)^k \exp(-\lambda t)}{k!} \cong [\exp(-\lambda t)](\cosh \lambda t) \tag{7.27}$$

where n_T is assumed sufficiently large that terms for which the summation index k is near n_T contribute so little that n_T can be extended to infinity. This yields (7.27). This is what is meant by the "approximately equal" sign following the series in (7.27). Similarly, for the case of an odd number of SEUs, which is equivalent to one upset. Then,

$$p_f = \sum_{k \text{ odd}}^{n_T} \frac{(\lambda t)^k \exp(-\lambda t)}{k!} \cong [\exp(-\lambda t)](\sinh \lambda t) \quad (7.28)$$

with the same proviso on n_T as for k even.

At the word level, p_s and p_f can be used as parameters in the binomial distribution for the computation of the probability $P(W_2, n_T)$ of two or more SEU errors in the same word as done earlier. As before, (7.27) and (7.28) in (7.19) results in

$$P(W_2, n_T) = \sum_{k=2}^{n_T} \binom{n_T}{k} p_f^k p_s^{n_T-k} = 1 - \sum_{k=0}^{1} \binom{n_T}{k} p_f^k p_s^{n_T-k}$$

$$= 1 - [\exp(-\lambda t)][\cosh(\lambda t)]^{n_T}[1 + n_T \tanh(\lambda t)] \quad (7.29)$$

Then, in this instance, the expected number of upset errors $\bar{E}$, as earlier, is

$$\bar{E} = P(W_2, n_T) N \quad (7.30)$$

and the corresponding error rejection ratio r_2 is

$$r_2 = \frac{\bar{E}}{n_T} = P(W_2, n_T) \frac{N}{n_T} \quad (7.31)$$

7.6. SEU Shielding

Shielding for SEU usually implies shielding methods used in spacecraft. Exceptions are electronic base station systems which are normally ground based. SEU shielding methods for them are not encumbered by stringent weight considerations and constraints, as for spacecraft and, to a lesser extent, for space stations. Shielding for SEU and other corresponding environmental particles such as gamma rays have much in common. In the subdiscipline of shielding, it is common knowledge that shielding is ultimately an art and less a science, especially with respect to its practical aspects, and the same for spacecraft shielding. The fundamentals of shielding and shielding practice are covered in many references and books. The following attempts to discuss shielding aspects unique to SEU and mainly in spacecraft.

In spacecraft SEU environments, such as cosmic ray energetic heavy ions at geosynchronous altitudes, little can be done in the way of directly attenuating them with intervening material mass consistent with spacecraft weight constraints. These ions are highly charged (i.e., ionized up to "10 +" or more), with energies in the hundreds of MeV per ion nucleon, and massive, with mass numbers up to and including uranium.

Attempts to shield by positioning strategic components at the spacecraft center of mass and using the masses of the components themselves is a meager option, although it has been tried. It is well argued that all spacecraft systems are mission critical, but they cannot all be put at the spacecraft center.

For spacecraft orbits that transit substantive portions of the inner (proton) and/or outer (electron) Van Allen belts, some direct mass attenuation is possible, although not wholly practical because of decreasing shielding "returns" per increase in shielding mass.

At low earth orbits (LEOs), which usually imply that the spacecraft is essentially under the Van Allen belts at all times, shielding is not a pressing problem. This is because most particles at these altitudes do not have the required parameter values, such as sufficient mass and/or energy that can affect electronics, components and systems. Environment exceptions are the South Atlantic Anomaly (SAA) discussed in Section 6.7, in which the spacecraft spends only a few minutes for each orbit in its transit through the SAA.

For an anomalously large solar flare (Section 2.4), essentially all environmental flux levels are enhanced by the corresponding solar wind particles (mainly protons) emanating from the sun. Flares can last for hours to days. Some "consolation" can be had by the fact that the protons can be stopped by practical mass thicknesses and the spacecraft may not suffer an excessive weight penalty [3]. There is a certain modicum of natural shielding provided by the solar wind, especially during sunspot maxima when it achieves high flux levels. The solar-wind-charged particles act somewhat as a shield for the earth's larger neighborhood to divert galactic cosmic rays, including those at geosynchronous altitudes.

The interior of a spacecraft is a complex of electromagnetic and electronic systems, interconnecting cables, and metal housings, all contained in a more or less spherical shell of aluminum. Their exteriors consist of solar panels, sensors, and other vitals with their umbilicals back to the central spherical shell. All this juxtaposition of subassemblies makes approaches to shielding problems very difficult because of geometric asymmetries.

A component such as a microcircuit chip/package mounted on a printed-circuit board inside the spacecraft "sees" a varying number and density of masses in all directions. The chip package itself is made of materials of various densities, and thus affords some mass shielding. The microcircuit and its circuit board together with other boards are enclosed in a metal housing making up a subassembly which contributes to shielding the device. Usually, the shielding problem for a device is to provide it with an individual shield in the form of an enclosing capsule (RAD–PAK) [34] on the circuit board. The capsule parameters such as its material lamina, if any, density, thickness, and size depend on the type of incident SEU-inducing radiation. The computation

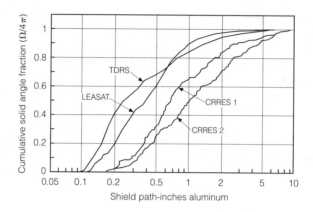

Figure 7.12. Spacecraft shield path cumulative distributions. From Ref. 25.

of the flux/fluence levels incident on the device in the spacecraft with and without the capsule is the nub of the problem. Solutions can be found using a ray tracing computer program. This program uses the idea of sectoring the volume around the device into solid angle increments that are subtended by the spacecraft masses at the device coordinates. The program "endows" the device with up to thousands of "rays," each corresponding to an incremental solid angle, impinging upon it that pierce other masses along the ray. The mass positions are obtained from construction blueprints. Then the attenuation of the incident radiation due to the masses is computed and summed to obtain the total attenuated SEU-inducing radiation incident at the device, from which the SEU error rate can be found. The preceding can also be accomplished in an approximate fashion using a modern hand-held computer [24]. Sophisticated Monte Carlo codes are also available.

The ray tracing program outputs can be extracted to yield other insights besides the results in the preceding. An interesting output from a particular ray tracing computation is depicted in Fig. 7.12. It is a plot of the cumulative distribution of solid angle fractions seen from the device reference position as a function of aluminum equivalent shield paths, strictly defined later. One device each is represented from the LEASAT and TDRS spacecraft plus two devices from the CRRESS spacecraft [25]. Table 7.3 displays the corresponding statistical outputs from the ray tracing code.

A pertinent deduction can be made from Fig. 7.12 in that, for the TDRS spacecraft, 10% or less of the ions with ranges less than or equal to 0.125 in. are transmitted to the device [25]. This path, 0.125 in., is the range in aluminum of protons, oxygen, and iron ions of 25, 54, and 99 MeV/AMU (atomic mass unit). Thus, only ions with correspondingly greater energy can be of significance to SEU.

Table 7.3. Spacecraft Shield Path Distributions.

Ser.	Spacecraft	Rays used	Incremental solid angle $\Delta\Omega$ Bins[a]	Distribution min. path (in.)	Distribution median path (in.)	Distribution mean path (in.)
1.	TDRS-1	6000	450	0.07	0.25	0.55
2.	LEASAT	1000	43	0.14	0.40	0.53
3.	CRRESS-1	240	240	0.17	0.68	1.21
4.	CRRESS-2	240	240	0.20	1.02	1.81

[a] $\sum_{\text{Bins}} \Delta\Omega/4\Pi = 1$.

Source: Ref. 25.

It is important to appreciate when, and to what degree, shielding should be taken into consideration. For example, certain shielding insights can be obtained for galactic cosmic rays in geosynchronous orbits and will be discussed. Shielding comparisons will be made with semi-infinite slab and circumscribing spherical shell shields. Expressions for the SEU error rates due to intervening shields is available from Section 5.2 involving the chord distribution function $f(s)$. Previous discussions emphasized the use of $f(s)$ for the case of parallelepiped sensitive regions. Here, $f(s)$ corresponds to that for a semi-infinite slab of thickness c. From Problem 1.3(b) for the slab, $f(s) = 2c^2/s^3$, where s is the chord length in the slab. From Section 5.2, the appropriate SEU error rate expression rewritten here is

$$E_r = \bar{A}_p \int_{s_{\min}}^{s_{\max}} \Phi(L(s)) f(s)\, ds \qquad (7.32)$$

The integration variable s is transformed to LET (L), using $c/s = L/L_c$, to obtain $f(s)\,ds = -2L\,dL/L_c^2$. From Section 1.3, $\bar{A}_p = S/4 \cong A/2$, where A is the area of one side of the slab. Inserting the preceding into (7.32) yields R_A, the SEU error rate per unit slab area, namely

$$R_A = \lim_{A\to\infty} \frac{E_r}{A} = \left(\frac{4\pi}{L_c^2}\right) \int_{L_{\min}}^{L_{\max}} \Phi(L) L\, dL = \left(\frac{4\pi}{L_c^2}\right) \int_{0.1L_c}^{L_c} \Phi(L) L\, dL \qquad (7.33)$$

where $L_{\min} \cong 0.1 L_c$ is a reasonable lower limit [21] and the 4π coefficient is the result of the additional integrations over all solid angles. The portion of the integral between L_c and $L_{\max}$ (the LET spectra cutoff) is neglected as it is very small, due to the fast decrease ($1/L^2$) of the integral flux $\Phi(L)$ with LET.

It is found from "production runs" of the ray tracing program that the above error rate $R_A(L_c)$ are logarithmic functions of the slab shield thickness c. Heuristically, this can be seen from the fact that the integral flux $\Phi(L) \cong \int \varphi(L)\,dL$,

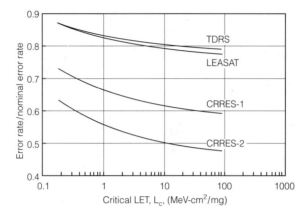

Figure 7.13. Ratio of error rate to nominal (0.1-in. Al) rate versus L_c. From Ref. 25.

and using the "figure of merit" approximation (Section 5.2), $\varphi(L) \simeq c_1/L^3$ ($c_1 = 5.8 \times 10^8$) in (7.33) yields

$$R_A \cong \left(\frac{4\pi}{L_c^2}\right) c_1 \ln\left(\frac{L_c}{L_{\min}}\right) = \left(\frac{4\pi}{L_c^2}\right) c_1 \ln\left[\left(\frac{L_c}{s_{\max}}\right) c\right] \quad (7.34)$$

where $L_{\min} \cong s_{\max}/c \simeq 0.1\, L_c$ is usually used.

A ray tracing program was used to compute an error rate for the four cases discussed in the preceding. These results were normalized using computed error rates for a 0.1-in. semi-infinite slab, the nominal error rate. The ratio of the error rates to the nominal rates (i.e., the normalization rates) versus L_c are plotted in Fig. 7.13. As seen from the figure, this ratio depends only sluggishly on L_c. This implies that computing error rates using the nominal approach with (7.33) should usually yield results that do not incur large errors, as compared to more accurate ray tracing results [25]. From the systems standpoint, SEU error rate trends are usually regarded important when they differ by an order of magnitude or more. This is especially the case for geosynchronous orbits and out to interplanetary trajectories.

Protons are the lightest and have the least charge of the ions and are the major particle type contained in solar flares and in the inner Van Allen belts. As such, they are less difficult to shield against than, say, galactic cosmic ray heavy ions. Spacecraft shielding solution methods for protons are more robust [31] than for heavy ions, which are virtually impossible to shield in practice. The SEU error rate for protons, E_{rp}, is proportional to [25]

$$E_{rp} \sim \int \sigma_{\text{seu}}(E^{(i)}) \varphi(E^{(i)}) dE^{(i)} = -\int \sigma_{\text{seu}}(E^{(i)}) d\Phi(E^{(i)}) \quad (7.35)$$

where $\sigma_{seu}(E^{(i)})$ is the SEU cross section (cm^2) and $\varphi(E^{(i)})$ is the differential proton flux energy spectrum inside the shield. $E^{(i)}$ is the proton energy variable within the shield, and $\Phi(E^{(i)}) = \int_{E^{(i)}}^{\infty} \varphi(E^{(i)}) dE^{(i)}$ is the corresponding integral flux inside. The superscript (i) refers to inside the shield material, whereas the superscript (o) refers to outside the shield.

The shield path, essentially the shield thickness x, used in the ray tracing program can be defined, where R is the range–energy function discussed in Section 3.2, by

$$R(E^{(o)}) = x + R(E^{(i)}) \tag{7.36}$$

where $R(E^{(o)})$ is the range of the particles incident at the shield whose energy variable is $E^{(o)}$ outside the shield. The particles lose energy in penetrating the shield, so their emergent energy becomes $E^{(i)} < E^{(o)}$ with corresponding residual range $R(E^{(i)})$. All the quantities in (7.36) are often expressed in units of areal density (g cm^{-2}). Also proton flux levels inside and outside the shield are approximately equal, as the major proton absorption processes are inelastic nuclear reactions. Inelastic reactions can absorb the incident particle and almost immediately emit it but at a lesser energy, however at no net loss to the flux level. Incorporating (7.36) in the SEU cross sections in (7.35) yields

$$E_{rp} \sim \int \sigma_{seu}(E^{(i)}, E^{(o)}; x) d\Phi(E^{(i)}) \tag{7.37}$$

that is, $\sigma_{seu}(E^{(i)}, E^{(o)}; x)$ now depends on x through the range–energy relation (7.36).

For a distribution of shield paths ending on the microcircuit of interest, each path can be weighted by w_i for the ith path, according to the amount of spacecraft mass the shield path threads. Then the SEU error rate is computed by an averaging process [25]. Therefore, (7.37) can be rewritten as

$$E_{rp} = \sum_k w_k \int \sigma_{seu}^k(E^{(i)}, E^{(o)}; x) d\Phi(E^{(i)}), \quad \sum_k w_k = 1 \tag{7.38}$$

The integral flux $\Phi(E)$ can be redefined as $\Phi(E) = A\Psi(E)$, where A is a dimensioned coefficient (cm^{-2}), and $\Psi(E)$ is the dimensionless integral energy spectrum. Then, (7.38) becomes

$$\bar{\sigma} = \frac{E_{rp}}{A} = \sum_k w_k \int \sigma_{seu}^k(E^{(i)}, E^{(o)}; x) d\Psi(E^{(i)}) \quad (\text{cm}^2) \tag{7.39}$$

The "shielded cross section" $\bar{\sigma}$ can be construed as an average cross section [25].

For certain solar flares, $\Psi(E)$ can be well represented by the normalized incomplete gamma function [25, 26], namely,

$$\Psi(E) = 4Ka^{11/8}\Gamma_n(11/2, (E/a)^{1/4}) \tag{7.40}$$

$$\Gamma_n(11/2, (E/a)^{1/4}) = \frac{\int_0^{(E/a)^{1/2}} x^{9/2} \exp(-x)\,dx}{\int_0^\infty x^{9/2} \exp(-x)\,dx} \tag{7.41}$$

K and a are constants obtained from solar flux spectral data [22].

Shielded average cross sections $\bar{\sigma}$ have been computed using the ray tracing code for several flare spectra and spacecraft, with the Burrell empirical range–energy relation for protons which is [25], where E is in MeV,

$$R(E) = 300\ln(1 + 3.76 \times 10^{-6}E^{1.75}) \quad \text{(cm)} \tag{7.42}$$

The preceding are normalized to a nominal 0.1-in. circumscribing spherical shell shield, and that ratio is depicted in Fig. 7.14. Numerical results are given in Table 7.4. From Fig. 7.14, it is seen that the discrepancy between the nominal and actual shielded error rates is sizable, where the latter is much less than the nominal. Table 7.4 depicts results for the two largest flares in 1989. The difference between the computed and observed SEUs may be due to the occurrence of heavy-ion upsets not included in the assumptions and other flaws in the model [25].

It is found that using the nominal 0.1-in. shield, whether slab or spherical shell, yields SEU error rates that are systematically too large compared to observations [25]. For better accuracies, thicker nominal shield models would give more accurate error rates. For example, for the spacecraft shielding distributions of mass and environments in the preceding, nominal shield thicknesses should be greater than 0.3 in. (2.1 g cm^{-2} Al). As mentioned earlier, proton-induced SEU error rates are more sensitive to shielding efforts than heavy cosmic ray ions. Hence, it is possible to provide shielding against solar

Table 7.4. TDRS Spacecraft Solar-Flare-Induced SEU.

Solar flare event	Calc. SEUs	Obs. SEUs	Obs./Calc.	Al Equiv. Sph. Thick.
Sept. 1989	75	91	1.21	0.32 in.
Oct. 1989	217	253	1.17	0.29 in.

Source: Ref. 25.

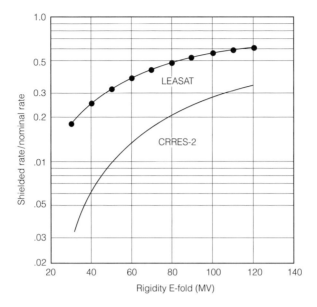

Figure 7.14. Spacecraft shielded SEU error rate to nominal shielded rate ratio at 0.1 in. versus spacecraft environment magnetic rigidity e-fold values. From Ref. 25.

flare protons in spacecraft without compomising other performance constraints like weight and space.

7.7. Radiation Sensors and Detectors

In the previous sections, emphasis has been given to SEU in various types of transistors, principally in microcircuits. SEUs also occur in different types of radiation sensors and detectors. These range from infrared (IR) to ultraviolet (UV) sensors to transistors themselves used as solid-state detectors. A simple model of a radiation sensor is that of a reversed-biased junction in a relatively large volume of material. It provides the medium in which incident radiation undergoes reactions with the sensor material and is so detected by the resulting currents flowing across the junction. The larger the volume of material for a given density, the more reactions will occur, thus increasing the detector sensitivity. Examples of detectors include *p-i-n* diodes, silicon barrier detectors, Geiger–Muller (GM) counters, and other types of mainly solid-state detectors. The large detector volume making for high sensitivity also makes it vulnerable to spurious radiation such as SEU. It is desired to distinguish between false and true signals, where, for example, a silicon detector is exposed to a cosmic ray environment among other environments aboard a spacecraft.

To establish the dynamics of these radiation detectors, an inverse range–energy operator is $E_{op}(R) = R^{-1}(E)$, such that $E_{op}R(E) = R^{-1}R = I$, the identity operator. Its operation produces zero net change of its arguments. Then, operating on (7.36) with $E_{op}(R)$ yields [27]

$$E^{(o)} = E_{op}(R(E^{(i)}) + x) \tag{7.43}$$

Note that E_{op} is not distributive, i.e., $E_{op}(R + x) \neq E_{op}R + E_{op}x$. For detectors, unlike the shield situations discussed in Section 7.6, incident energy flux spectra are assumed to be degraded. Thus, a flux degradation expression can be written as

$$\Phi(E^{(i)}) = \Phi(E^{(o)}) \frac{dE^{(o)}}{dE^{(i)}} \tag{7.44}$$

where $\Phi(E^{(i)})$ is the flux inside the detector, and $\Phi(E^{(o)})$ is that on the detector. From (7.36)

$$\frac{dE^{(o)}}{dE^{(i)}} = \frac{(dE^{(i)}/dx)^{-1} + R'(E^{(i)})}{R'(E^{(o)})} \tag{7.45}$$

The effects of detector shielding against protons and helium ions (alpha particles) are shown in Figs. 7.15a and 7.15b [27]. From these figures, it is seen that shielding mainly affects the low-energy particles and hardly perturbs their high-energy counterparts. The two shielding thicknesses in the figures are given in g cm^{-2} which are 0.25 and 36 g cm^{-2} of material. Secondary particles produced as a result of reactions in the shield with the incident ions are neglected because their cross sections are relatively small, as discussed earlier.

The assumed isotropic flux $\Phi(E^{(i)})$ that exits the spherical shield into the detector produces ionization tracks (hole–electron pairs) therein. Essentially all of the energy deposited in the detector's sensitive region creates current pulses. They are then integrated by the detector electronics to produce charge collection pulses. It is desired to obtain a distribution of ionization charge pulse heights from which SEU phenomena can be studied. Initially, define a "minimal event" which corresponds to the energy deposited E_{dep} in the assumed parallelepiped sensitive region of the detector by, for example, a proton of energy E_{pmin}; that is, the corresponding E_{pmin} ionization track is normal to the parallelepiped faces producing a minimum chord through its thickness c. E_{pmin} is the proton energy that corresponds to the minimum dE/dx, hence its subscript. Then the energy deposited E_{dep} is a maximum. This is seen by realizing that as dE/dx attains its minimum, its integral, $\bar{E}_{dep} \sim \int_0^{E_{pmin}} (dE/dx) dx$ is attaining its maxi-

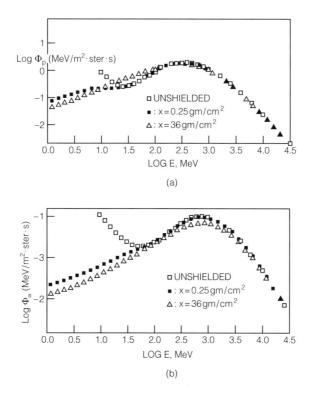

Figure 7.15 (a) The effect of shielding on the cosmic ray proton spectrum at solar minimum. (b) The effect of shielding on the cosmic ray helium spectrum at solar minimum. From Ref. 27.

mum, as in Fig. 7.16. For example, for protons at E_{pmin}, $dE/dx = 1.7$ MeV cm^2 g^{-1}. Because 1 pC μm$^{-1} \doteq 96.6$ MeV cm^2 mg^{-1}, the charge collected by the detector $Q(C)$ (coulombs) $= (dE/\rho dx)(\rho dx) = 7.56 \times 10^{-14} \cdot s$, where s is the chord length in g cm^{-2}. For the detector $s = 100$ μm (0.0233 g cm^{-2}) in silicon, the minimal event E_{dep} corresponds to a charge of 1.76 fC $= 1.76 \times 10^{-15}/1.6 \times 10^{-19} \cong 10^4$ hole–electron pairs. This magnitude is large compared to the many anticipated IR signals.

When the current pulse from the detector is fed to an operational amplifier (integrator) with feedback capacitance C_{fb}, it integrates the input current giving a voltage output $V_0 = Q_{dep}/C_{fb}$. For $C_{fb} = 0.01$ pF, $V_0 = 1.76 \times 10^{-15}$ C/10^{-14} F $= 176$ mV (millivolts), which is larger than the usual noise background [27].

For SEU error rate computation, LET $= dE/dx$ can be used if the detectors are not too thick (Section 3.8). Let the mean energy deposited in the detector be given by [27]

266 / Single Event Phenomena

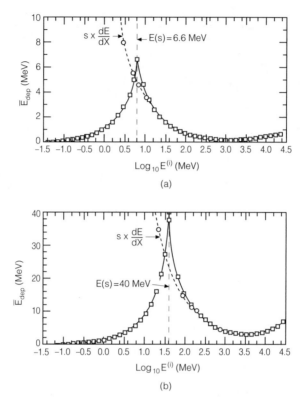

Figure 7.16. (a) Mean energy deposited in detector ($\bar{E}_{dep}$) versus proton energy ($E^{(i)}$) inside detector for chord length s = 0.1 cm² g⁻¹ (0.043 cm Si), (b) mean energy deposited in detector ($\bar{E}_{dep}$) versus proton energy ($E^{(i)}$) inside detector for chord length s = 1.75 cm² g⁻¹ (0.75 cm Si). From Ref. 27.

$$\bar{E}_{dep}(E^{(i)}) = \begin{cases} E^{(i)}, & E^{(i)} < E(s) \\ E^{(i)} - E_{op}(R(E^{(i)}) + s), & E^{(i)} \geq E(s) \end{cases} \quad (7.46)$$

Here, $E^{(i)}$ is the particle energy after penetrating the shield but before entering the detector. $E(s)$ is the particle energy deposited in the sensitive region of the detector after it has penetrated through the shield and into the detector. Equation (7.46) asserts that if $E^{(i)} < E(s)$, all the particle energy is deposited in the detector, as $\bar{E}_{dep}$ does not correspond to the range s [i.e., the required energy $E(s)$ to penetrate the detector and beyond]. However, if $E^{(i)} > E(s)$, the particle has the range to penetrate beyond the detector, with remaining energy $\bar{E}_{dep} = E^{(i)} - E_{op}(R(E^{(i)}) + s)$. The latter is $E^{(i)}$ minus the energy of the particle as it exits the detector sensitive region.

Figures 7.16a and 7.16b depict $\bar{E}_{dep}$ versus $E^{(i)}$ for chord lengths of 0.1 and 1.75 g cm⁻². Superposed are curves of the approximation for (LET) · s [i.e.,

$(dE/ds)s$. It is seen that the LET approximation is good for protons that penetrate through the sensitive region, whereas for those stopped within, $E^{(i)} < E(s)$, the LET approximation overestimates the deposited energy. Thus, the LET approximation could affect the shape of the pulse height histogram display in the maximum pulse height region for large sensitive volumes. If the maximum chord length is less than 0.75 cm (1.75 g cm^{-2}Al), then the LET approximation should be quite accurate [27].

Now $\bar{E}_{dep}(E^{(i)})$ for a given chord length s can be divided into three energy intervals corresponding to

$$0 < E^{(i)} \le E(s), \qquad E(s) < E^{(i)} \le E_{min}, \qquad E^{(i)} > E_{min} \qquad (7.47)$$

In Figure 7.16a and 7.16b, min $\bar{E}_{dep} = E_{min}$. Define a distribution function of deposited energies $N_s(E_{dep})$ for chord lengths s; that is, $N_s(E_{dep})\,dE_{dep}$ is the number of particle interaction events per cm^2 sr s that deposit energy in the detector sensitive region in the range dE_{dep} about E_{dep} by virtue of their chord track lengths. Then, the contribution to $N_s(E_{dep})$ by the above jth energy interval [$j = 1, 2, 3$ in (7.47)] is

$$N_{sj}(E_{dep})dE_{dep} = \int_{E_{jlo}^{(i)}}^{E_{jup}^{(i)}} \Phi(E^{(i)})\,dE^{(i)}, \qquad j = 1, 2, 3 \qquad (7.48)$$

where the above integral upper and lower limits, $E_{jup}^{(i)}$ and $E_{jlo}^{(i)}$, respectively, satisfy (7.46); that is,

$$E_{dep} = \bar{E}_{dep}(E_{jlo}^{(i)}, s)$$
$$E_{dep} + dE_{dep} = \bar{E}_{dep}(E_{jup}^{(i)}, s) \qquad \text{for } j = 1, 2, 3 \qquad (7.49)$$

The total energy distribution provided by the sum of the energy increments, using methods from Section (5.1), is given by

$$N_s(E_{dep}) = 4\pi\bar{A}_p \sum_{j=1}^{3} \int_0^{s_{max}} N_{sj}(E_{dep}(s))f(s)\,ds \qquad (7.50)$$

where $f(s)$ is the chord distribution function discussed in Section 5.1. The 4π factor results from integration over all solid angles, and $s_{max} = (a^2 + b^2 + c^2)^{1/2}$ is the main diagonal length of the sensitive region of the parallelepiped.

The corresponding normalized cumulative energy distribution $C_s(E_{dep}(s))$ is given by

$$C_s(E_{dep}(s)) = \frac{\int_{E_{dep}(s)}^{E_{dep}(s_{max})} N_s(E_{dep})\,dE_{dep}}{\int_0^{E_{dep}(s_{max})} N_s(E_{dep})\,dE_{dep}} \qquad (7.51)$$

where the normalizing denominator above is the total number of cosmic ray particles per cm^2 s of mean projected area that penetrate the detector sensitive

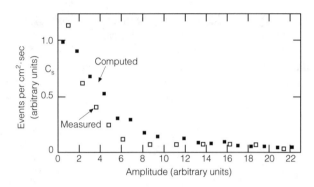

Figure 7.17. Un-normalized cumulative amplitude distribution of deposited energies in detector from cosmic rays compared with experimental results from a spacecraft at low-earth-orbit altitude. From Ref. 27.

region. This is computed [27] and yields ~ 1.64 particles cm^{-2} s^{-1}, which compares well with obtaining the same quantity by integrating Adam's LET spectra [28], thus lending an internal consistency to the preceding analysis [27].

The cumulative amplitude distribution C_s is depicted in Fig. 7.17. It is computed for an array of extrinsic detectors [27] shielded by 62 g cm^{-2} (~ 2 in. Pb) in order to detect mainly galactic cosmic rays. This array has been flight tested in a low earth orbit (LEO) where it has detected near earth cosmic rays. These results are also plotted in Fig. 7.17 and the agreement with the analytic results is quite good. Electronic means for decreasing the effects of spurious pulses were also used [27].

Previous discussions reveal that the use of a pulse height analyzing system (Section 4.3) supplies information for SEU and other single event phenomena. These systems usually begin with the particle detector/sensor to be tested, which can also be a microcircuit as well. Another simple example of a charge collection sensor is a surface barrier detector (SBD). It is shown in Fig. 7.18 [29, 30] and used as a model of the sensitive regions of a device. The SBD

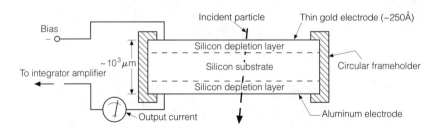

Figure 7.18. Schematic cross section of disc shaped surface barrier detector.

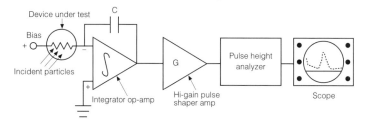

Figure 7.19. Schematic of pulse height analyzer system.

junctions are provided by the surface barrier contact potential between the silicon and the metal electrodes. Work function differences and/or surface states of the silicon produce a surface charge thereon, which results in a potential barrier with an associated depletion layer in the silicon. The potential barrier lies across the metal electrodes and the silicon bulk. The incident-particle-induced ionization currents emanate from the depletion layer and are detected as they cross the junctions. For a fully biased detector, the depletion layers essentially fill the whole silicon volume. The electrons and holes formed in the ionization processes are accelerated by the detector internal electric fields toward their respective electrode terminals. As discussed earlier, the electron–hole pairs produce currents, which are then integrated by the external electronics to yield corresponding collected charge. The latter pulses are amplified and fed to a multichannel pulse height analyzer (PHA). The PHA (Fig. 7.19) consists of a peak detector/analog-to-digital converter that transforms the charge pulse heights (amplitudes) into register address channels (bins) according to their heights. Because charge amplitude corresponds to an energy amplitude, in the sense that 22.5 MeV in ionization energy must be expended to produce 1 pC of charge in silicon ($10^{-12}/1.6 \times 10^{-19} = 6.25 \times 10^6$ electrons), the scope output of the PHA is a pulse height histogram with energy (equivalent charge) as the abscissa x. The PHA is essentially a digital Fourier transformer. The integrated input pulse amplitudes $Q(t)$ are transformed to $C(x) = \int_0^\infty \exp(ixt) Q(t) dt$ where C is the counts ordinate in Fig. 7.20, or $C(E) = \int_0^\infty \exp(-iEt) Q(t) dt$ as the abscissa where E is the energy variable.

The ordinates of the histogram, as in Fig. 7.20, yield the total number of interactions per second, each of which produces charge shown at the corresponding abscissa. Thus, the maximum (peak) ordinate must correspond to at least one of the sensitive regions of the device under study, as it represents a site at which the majority of interactions are taking place. Then, the maximum ordinate level must be proportional to the maximum interaction rate between incoming particles and the sensitive region, or maximum ordinate $\sim \sigma_{seu}\varphi_{inc}N$, where φ_{inc} is the particle flux incident on the device and N is the number of junctions affected. Therefore, the asymptotic cross section per sensitive region

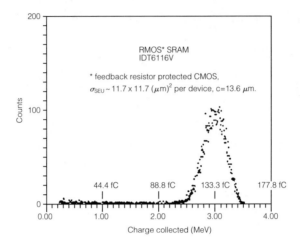

Figure 7.20. Pulse height analyzer result for the IDT 6116V RMOS SRAM. The ordinate corresponds to 2^{15} reversed-biased junctions in the device. From Ref. 30.

can be computed from $\sigma_{seu} \sim$ maximum ordinate $/\varphi_{inc} N$. This also constitutes a measurement of the cross-sectional area of the sensitive region volume.

Further, the sensitive volume thickness c can be estimated using the range–energy relation (7.26) rewritten as

$$c = R(E^{(o)}) - R(E^{(i)}) \tag{7.52}$$

where $R(E^{(i)})$ is the residual range after the particle at normal incidence has deposited energy $E(c)$ in traversing through the sensitive region volume. Hence, a sensitive region volume can be estimated as $\sigma_{seu} c$. It is seen that an accurate computation of c depends on the accurate determination of the energy lost in the sensitive volume as well as in the overlying layers of, say, a microcircuit. Also, the shape of the volume is not presently available and, importantly, whether or not it is convex (Section 1.3). Intuitively, considering the epitaxial structures of microcircuits, it could very well not be convex. For example, it could be kidney shaped with reentrant surfaces so that an exiting particle could easily reenter the chip to increase σ_{seu} over that for a convex sensitive region. However, it is felt that sensitive regions of interest are not sharply nonconvex; thus, the SEU models discussed are still assumed valid. Other tests for consistency of the preceding results are given in Ref. 30.

Problems

7.1 Multiple-bit SEU is often construed as SEUs produced in several, say, memory cells by a single SEU-inducing ion track. For very high fluxes of protons and other low-Z particles, such as of concern in Strategic Defense Initiative programs, the opposite can occur where two or more particles "in train" can produce one SEU. How is this explained?

7.2 If the SEU mitigation feedback resistor values in a memory cell increase by 50% due to prolonged exposure to high levels of spurious radiation, by how much does the cell SEU error rate change?

7.3 Assume that the SEU reliability R_{seu} (survival probability) of an n-chip memory is given by $R_{seu} = \prod_{i=1}^{n}(1 - \sigma_{iseu})$, where $1 - \sigma_{iseu}$ is the SEU survival probability of the ith chip, as σ_{iseu} is the chip SEU probability. A generalized cost of attaining this level of reliability is given by $C_{seu} = \sum_{i=1}^{n} k_i (1 - \sigma_{iseu})$, where k_i are coefficients that convert SEU cross sections to SEU hardening costs. The σ_{iseu} are nominal (e.g., saturation values) cross-section values obtained after whatever hardening measures are installed within the chip and its associated circuitry, and which modify the k_i values.

(a) Determine the coefficients k_i^* that maximize the reliability R_{seu} consistent with the cost C_{seu}. (**Hint**: use Lagrange multipliers). Show that $k_i^* = C_{seu}/n(1 - \sigma_{iseu})$.

(b) Show that the resulting reliability $R^* = (C_{seu}/n)^n / \prod_{i=1}^{n} k_i^*$.

7.4 Suppose that in a W word memory of n bits per word, each word is read correctly if (a) there are no SEU bit errors in it, or (b) a single-bit error is detected and corrected, or (c) a double-bit error is detected and corrected. What is the reliability R of this memory, where the individual bit SEU error survival probability $p = \exp(-\lambda t)$, where λ^{-1} is the mean survival time to error occurrence. Also, $q = 1 - p$.

7.5 Suppose that in an EDAC system, there is 1 error in the 7-bit word $w = 1 1 1 1 0 1 0$. Using the three vectors $\mathbf{a} = 0 0 0 1 1 1 1$, $\mathbf{b} = 0 1 1 0 0 1 1$, and $\mathbf{c} = 1 0 1 0 1 0 1$, determine which bit in $\mathbf{w}$ is in error and show the corrected word $\mathbf{w}^*$.

7.6 Why should it be relatively easier to shield spacecraft against SEU for orbits whose altitude is much less than geosynchronous?

7.7 What is the essential idea behind using the device of interest as an SEU-inducing flux detector followed by a pulse height analyzer system to determine its cross section and, ultimately, its SEU error rate?

7.8 Using a particle accelerator for the determination of the SEU cross section σ_{seu} of a device involves the measured rate of SEU events N_{seu}

and the ion beam flux Φ_{seu} in a relation given by $\sigma_{seu} = \lim_{t \to \infty} N_{seu}/\Phi_{seu}$. The limit of infinite time is not realizable, as the ion beam is incident for a finite time T. How should the above expression for σ_{seu} be changed to accommodate the finite time T?

7.9 In Problem 7.8, suppose that to satisfy the requirement that the exposure time $T \gg 1/N_{seu}$, it becomes almost absurdly long. Assuming that the accelerator can maintain the beam for this time T, what can befall the device under test to alter the test from nondestructive to destructive?

7.10 Spacecraft skin is usually thin and usually made of aluminum for a reasonable strength to weight ratio. Suppose that radiation data for an "alien" spacecraft in orbit consist of its penetration $-\Delta E/E$ of about 10% after irradiation with 40-MeV protons. Using the Burrell range–energy relation (7.42), determine the skin thickness.

7.11 In the experiment on multiple-bit upset errors discussed in Section 7.2, the devices are instrumented so that the waxing and waning of cell multiple-bit error clusters in them can be observed by sweeping V_{DD} about a nominal value determined by experiment. This varies the cell critical charge as discussed. If a cell capacitance is 0.025 pF and its critical charge is 100 fC, what must the V_{DD} be if the corresponding sense amplifier bias voltage V_B is one-third of V_{DD}?

References

1. L.D. Edmonds, "A Distribution Function for Double Bit Upsets," *IEEE Trans. Nucl. Sci.*, **NS-36** (2), 1344–1346 (1989).

2. N.M. Ghoneim, R.G. Marten, J.S. Cable, and Y. Song, "The Size Effect of an Ion Charge Track on Single Event Multiple Bit Upset," *IEEE Trans. Nucl. Sci.,* **NS-34** (6), 1305–1309 (1987).

3. G.C. Messenger, "Collection of Charge on Junction Nodes From Ion Tracks," *IEEE Trans. Nucl. Sci.*, **NS-29** (6), 2024–2031 (1982).

4. J. Bradford, "Non-Equilibrium Radiation Effects in VLSI," *IEEE Trans. Nucl. Sci.*, **NS-25** (5) (1978).

5. H.L. Grubin, J.P. Kreskovsky, and B.C. Weinberg, "Numerical Studies of Charge Collection and Funneling in Silicon Devices," *IEEE Trans. Nucl. Sci.*, **NS-31** (6), 1161–1166 (1984).

6. J.A. Zoutendyk, L.S. Smith, and G.A. Soli, "Empirical Modeling of Single Event Upset (SEU) in NMOS Depletion Mode Load SRAM Chips," *IEEE Trans. Nucl. Sci.*, **NS-33** (6), 1581–1585 (1986); J.A. Zoutendyk, H.R. Schwartz, and L.R. Nevill, *IEEE Trans. Nucl. Sci.*, **NS-35** (6), 1644–1647 (1988).

7. R. Koga, W.A. Kolasinski, J.V. Osborn, J.H. Elder, and R. Chitty, "SEU Test Techniques for 256K Static RAMs and Comparison of Upsets Induced by Heavy Ions and Protons," *IEEE Trans. Nucl. Sci.*, **NS-35** (6), 1638–1643 (1988).

8. Y. Song, K.N. Vu, J.S. Cable, A.A. Witteles, W.A. Kolasinski, R. Koga, J.H. Elder, J.V. Osborn, R.G. Martin, and N.M. Ghoneim, "Experimental and Analytical Investigation of Single Event Multiple Bit Upsets in Polysilicon Load 64K × 1 NMOS SRAMs," *IEEE Trans. Nucl. Sci.*, **NS-35** (6), 1673–1677 (1988).

9. P.M. Carter and B.R. Wilkins, "Influences on Soft Error Rates in Static RAMs," *IEEE J. Solid State Circuits*, **SC-22**, 430–436 (1987).

10. J.A. Zeutendyk, L.D. Edmonds, and L.S. Smith, "Characterization of Multiple-Bit Errors from Single Ion Tracks in Integrated Circuits," *IEEE Trans. Nucl. Sci.*, **NS-36** (6), 2267–2274 (1989).

11. G.C. Messenger and M.S. Ash, *The Effects of Radiation on Electronic Systems*, 2nd ed., Van Nostrand Reinhold, New York, 1992, Chap. 6.

12. E.G. Stassinopoulos, G.J. Brucker, O. Van Gunten, and H.S. Kim, "Variation of SEU Sensitivity of Dose Imprinted CMOS SRAMs," *IEEE Trans. Nucl. Sci.*, **NS-36** (6), 2330–2338 (1989).

13. T. Matsukawa, A. Kishide, T. Tanii, M. Koh, K. Horita, K. Hara, B. Shigeta, M. Goto, S. Matsuda, S. Kuboyama, and I. Ohdomari, "Total Dose Dependence of Soft Error Hardness in 64Kbit RAMs Evaluated by Single Ion Microprobe Technique," *IEEE Trans. Nucl. Sci.*, **NS-41** (6), 2071–2076 (1994).

14. C.L. Axness, J.R. Schwank, P.S. Winokur, J.S. Browning, R. Koga, and D.M. Fleetwood, "Single Event Upset in Irradiated 16K CM05 SRAMs," *IEEE Trans. Nucl. Sci.*, **NS-35** (6), 1602–1607 (1988).

15. A.B. Campbell and W.J. Stapor, "The Total Dose Dependence of the Single Event Upset Sensitivity of IDT Static RAMs," *IEEE Trans. Nucl. Sci.*, **NS-31** (6), 1175–1177 (1984).

16. R.C. Webb, L. Palkuti, L. Cohn, G. Kweder, and A. Constantine, "The Commercial and Military Satellite Survivability Crisis," *Defense Electronics Magazine*, Aug.1995.

17. J.A. Adolphsen, J.L. Barth, E.G. Stassinopoulos, T. Gruner, M. Wennersten, K. LaBel, and C.M. Seidleck, "SEP Data from the APEX Cosmic Ray Upset Experiment: Predicting the Performance of Commercial Devices in Space," Proc. Third European Conf. on Radiation And its Effects on Components and Systems, 1995.

18. C. Dufour, P. Garnier, T. Carriere, J. Beaucour, R. Ecoffier, and M. Labrunee, "Heavy Ion Induced Single Hard Errors in Submicron Memories," *IEEE Trans. Nucl. Sci.*, **NS-39** (6), 1693–1697 (1992).

19. G.H. Sandier, *System Reliability Engineering*, Prentice-Hall, Englewood Cliffs, NJ, 1963, pp. 162ff.

20. A. Tabor and E. Normand, "Single Event Upset in Avionics," *IEEE Trans. Nucl. Sci.*, **NS-40** (2), 120–126 (1993).

21. V. Pless, *Introduction to the Theory of Error Detecting Codes*, J. Wiley, New York, 1982, Chap 1.

22. J.A. Zoutendyk, H.R. Schwartz, R.K. Watson, Z. Hasnain, and L.R. Nevill, "Single Event Upset (SEU) in a DRAM With On-Chip Error Correction," *IEEE Trans. Nucl. Sci.*, **NS-34** (6), 1310–1315 (1987).

23. R. Katz, R. Barto, P. McKerracher, B. Karkhuff, and R. Koga, "SEU Hardening of Field Programmable Gate Arrays (FPGA) for Space Application and Device Characterization, *IEEE Trans. Nucl. Sci.*, **NS-41** (6), 2179–2186 (1994).

24. G.C. Messenger and M.S. Ash, The Effects of Radiation on Electronic Systems, 2nd ed., Van Nostrand Reinhold, New York, 1992, Section 13.2.

25. E.C. Smith, "Effects of Realistic Satellite Shielding on SEE Rates," *IEEE Trans. Nucl. Sci.*, **NS-41** (6), 2396–2399 (1994).

26. E. Normand and W.J. Stapor, "Variation in Proton Induced Upset Rates from Large Solar Flares Using an Improved SEU Model," *IEEE Trans. Nucl. Sci.*, **NS-37** (6), 1947–1952 (1990).

27. L.W. Ackerman, "Amplitude Distribution of Cosmic Ray Events in Intrinsic IR Detectors," *IEEE Trans. Nucl. Sci.*, **NS-32** (6), 4185–4188 (1985).

28. E.L. Price, P. Shapiro, J.H. Adams, Jr., and E.A. Burke, "Calculation of Cosmic Ray Induced Soft Upsets and Scaling in VLSI Devices," *IEEE Trans. Nucl. Sci.*, **NS-29** (6), 2055–2063 (1982).

29. P.J. McNulty, "Predicting Single Event Phenomena in Space," NSRE Conf. Short Course, July 1990.

30. P.J. McNulty, W.J. Beauvais, and, D.R. Roth, "Determination of SEU Parameters of NMOS and CMOS SRAMs," *IEEE Trans. Nucl. Sci.*, **NS-38** (6), 1463–1470 (1991).

31. E.G. Stassinopolous, G.J. Brucker, D.W. Nakamura, C.A. Stauffer, G.B. Gee, and J.L. Barth, "Solar Flare Proton Evaluation at Geostationary Orbits for Engineering Applications," *IEEE Trans. Nucl. Sci.*, **NS-43** (2), 369–382 (1996).

32. R.L. Pease, "Total Dose Issues for Microelectronics in Space Systems," *IEEE Trans. Nucl. Sci.*, **NS-40** (2), 442–452 (1996).

33. G.R. Hopkinson, C.J. Dale, and P.W. Marshall, "Proton Effects in Charge Coupled Devices," *IEEE Trans. Nucl. Sci.*, **NS-40** (2), 614–627 (1996).

34. L. Adams, R. Nickson, A. Kelleher, C.W. Millward, D.J. Stropel, and D. Czajkowski, "A Dosimetric Evaluation of the RADPAK^{TM} Using Mono-energetic Electrons and Protons," *IEEE Trans. Nucl. Sci.*, **NS-43** (3), 1014–1017 (1996).

8

Single Event Upset Practice

8.1. Introduction

This chapter and Chapter 5 include a reprise of the discussion in the earlier chapters pertinent to these sections, and together with Chapter 5 present guidelines for use in practical applications. Besides understanding the basic tenets of the discipline of single event phenomena (SEP), it is felt important for the reader that they be transformed into useful expressions so as to obtain practical answers to important questions. Examples include the following:

(a) how to compute SEU error rates for various radiation particle environments to satisfy electronic designer needs

(b) how to accomplish part selection with an eye to choosing those that are sufficiently SEU radiation "hard" for the particular electronics system application, within the constraints of ever fewer hardened part lines from vendors since the end of the cold war

(c) maintaining a knowledge of part-function hierarchy with respect to SEU phenomena consequences within the context of the system being built

(d) staying aware of the implication of test results as correlated with the needs of the system program

(e) understand SEP in terms of appropriate weight given to it as compared with other system engineering and radiation requirements.

Other aspects of SEP and their resolution will emerge in the course of reading the sections making up this final chapter.

This chapter contains an appreciable amount of part data covering SEU, SEL, and SEB (single event burnout) caused by cosmic rays, Van Allen protons, atmospheric neutrons, and other particles. The data herein have been selected from larger data sets and data banks, acquired over more than two decades. It behooves the readers to begin to acquire such data from sources they will encounter during their association with work in this field. In this discipline, one of the perhaps more prosaic pillars is the data acquisition task. Nevertheless, such collections are invaluable. This is especially the case when the need is to make a quick but correct decision that often cuts the Gordian knot instead of being obliged to wrestle with its knotted cordage. If one ever decides to start a consulting enterprise in this area, an immediate right-hand person should be one who has a penchant for parts, their SEP characteristics, and, importantly, their lineage.

8.2. Geosynchronous SEU Error Rate Computations

One of the earliest SEU error rate expressions is termed the "figure of merit" SEU estimation formula, as discussed in Section 5.1. It is actually an upper-bound estimator of geosynchronous error rate, although its use, in principle, is not confined to the geosynchronous environment. In practice, it usually yields a conservative overestimate. Rewritten here for convenience, it is given for the rectangular parallelepiped model of the silicon device SEU-sensitive region by [1]

$$E_r = \frac{5 \times 10^{-10} abc^2}{Q_c^2} = \frac{5 \times 10^{-10} \sigma_{\text{sat}}}{L_c^2} \text{(errors per cell-day)} \qquad (8.1)$$

where the parallelepiped dimensions a, b, and c (shortest dimension) are in microns and Q_c is in picocoulombs. For GaAs devices, replace 5 by 3.5 in the above coefficient [2]. A cell is usually construed as the equivalent parallelepiped volume-sensitive site that stores one bit of information. For example, in a SRAM, the sites (nodes) are made up of more than one transistor, and are therefore taken into account by a corresponding multiplier of, say, the SEU per bit to obtain the SEU per microcircuit device.

If the $\sigma_{\text{sat}} \equiv \sigma_\infty$ and L_c have been measured experimentally, the E_r is immediate by their substitution into the rightmost equation (8.1). However, and most frequently, the device of interest is encountered prior to its testing and, ultimately, may not be tested at all. In this case, the leftmost equation (8.1) must be employed. This means that the parallelepiped model is used; hence, the dimensions a, b, c, and Q_c must be available. Estimates of a, b, and c can be obtained from destructive physical analysis (DPA) using a scanning electron microscope (SEM), as in Table 8.1, or from the vendor when the vendor is

Table 8.1. Geosynchronous SEU Computations for Mainly TTL Microcircuits.

Ser.	Part name	Function, Vendor	Sensitive vol.[a] (abc) (µm³)	Critical chg. (pC)	$\left(\dfrac{\text{SEU}}{\text{bit-day}}\right)$	$\cdot \left(\dfrac{\text{No. of}}{\text{nodes}^c}\right)$	$= \dfrac{\text{SEU per}}{\text{device-day}}$
1	54LS90	Dec. Cntr. TI, MOT	50.8 × 50.8 × 1.2	0.44	9.6 − 6[b]	52	5.0 − 4[b]
2	54LS240	Oct. Buff. TI	50.8 × 50.8 × 1.2	0.44	9.6 − 6	40	3.8 − 4
3	5400	NAND Gte. TI	30 × 40 × 1.0	1.14	4.6 − 7	16	7.4 − 6
4	54S140	Dual Dvr. TI, NSC	33 × 17.8 × 0.4	0.17	1.6 − 6	8	1.3 − 5
5	93L16FM	Bin. Cntr. FSC	60 × 57 × 1.0	0.90	2.1 − 6	46	9.7 − 5
6	54S30	NAND Gte. TI	33 × 17.8 × 0.4	0.17	1.6 − 6	4	6.4 − 6
7	MC1692F	Line Rcr. MOT	30 × 30 × 0.5	0.84	1.8 − 7	16	2.8 − 6
8	11C91	Divider FSC	44.5 × 38.1 × 0.5	0.08	3.3 − 5	116	3.8 − 3
9	55182	Line Rcr. TI, NSC	25 × 15 × 0.7	0.17	3.2 − 6	18	5.8 − 5
10	5490	Dec. Cntr. TI	30 × 40 × 1.0	1.14	4.6 − 7	76	3.5 − 5

[a] Measured using destructive physical analysis of the part.
[b] $X.Y - Z = X.Y \times 10^{-Z}$.
[c] Sometimes multiplied by 0.5 reflecting mean node probability of on or off.

Source: Ref. 4.

willing to supply the data. Extrapolation from data already known from like parts is another possible alternative. Often, $\sigma_{sat} \cong ab$ is obtainable instead of a, b, or c. Then c must be estimated. Lore asserts that for a multitude of physical and circuital reasons that $c \cong 2\,\mu m$ for the "typical 2 micron" feature size part.

Q_c can sometimes obtained from the vendor/manufacturer using a roundabout method based on certain information provided by them. They are often reluctant to yield detailed design rules for their product lines. Often, however, they will provide values of sheet resistance for various portions of the semiconductor epitaxies (layers) that go to build the part, or at least that portion corresponding to SEU-sensitive regions. From the sheet resistance of the pertinent part fragments, their corresponding resistivity ρ is immediate from $\rho = r_s t$, where r_s and t are the sheet resistance and sensitive region thickness, respectively. Then from ρ versus N (dopant density) curves in semiconductor texts [3], N is obtained. This parameter allows the determination of the corresponding junction capacitance per junction area from $C/A_j \cong (e\epsilon\epsilon_0 N/2V_0)^{1/2}$ where V_0 is essentially the junction potential less the contact potential ϕ_0 ($\phi_0 \ll V_0$). Then Q_c is obtained using $Q_c = C\Delta V$, where ΔV is the voltage swing of (that part of) the device obtained from its cataloged electrical specifications. With the knowledge of a, b, c A_j, and Q_c from the preceding, the desired E_r can be calculated.

Another method for obtaining Q_c is through the use of SPICE-type computer simulations of the device-sensitive regions together with its electrical characteristics, plus that of its immediately surrounding circuitry as in the following two examples.

To compute the critical charge using a radiation inclusive SPICE-type code [5], the electronic and physical parameters of the device of interest are formatted as input data. The number of parameters can vary from 10 to 40 approximately, depending on device circuit complexity and model faithfulness desired. Frequently, only a typical cell of the device that represents the phenomenon to be investigated is modeled, as opposed to the complete integrated circuit. This is done in the first example, which is a typical memory cell of a CMM5114/1RZ 4K CMOS/SOS SRAM. However, in the second example, which is that of a custom multitransistor bipolar D-Flip/Flop, the whole device was modeled. Besides the preceding parameters inputted to the code, the electrical circuit connections and auxiliary current generator parameters are also entered. The latter provide the SEU current pulse excitation that suitably mirrors the proper pulse temporal characteristics. Figure 8.1 depicts these generators in the circuit diagram of the SRAM cell. In that figure, the state SET/RESET transistor and voltage source sets the desired initial state of the cell prior to the SEU pulse onset.

Once the input data are entered and the corresponding electronic configuration as represented by the above parameters is operationally checked in the code, the "production" runs begin. In these examples, trapezoidal SEU pulse

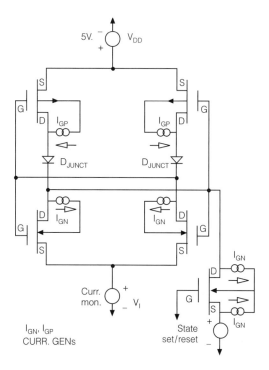

Figure 8.1. CMM5114/1RZ CMOS/SOS 4K SRAM cell circuit diagram for SEU simulation on SPICE-type code. From Ref. 5.

shapes of various current amplitudes and FWHM (full width half-maximum) pulse widths are used. For a given set of pulse parameters, pulse onset is allowed to occur, and corresponding sensitive state node voltages are tracked as a function of time following onset as shown in Figs. 8.2–8.5. For the SRAM cell, the SEU pulse current generator corresponding to the node of the "off" p-channel MOSFET is used as the pulse simulator, as in Fig. 8.1.

Various SEU pulse current amplitudes and durations are run, and results are tracked as mentioned. Initial runs for the D-Flip/Flop and the SRAM cell as depicted in Figures 8.2 and 8.4 respectively, are usually obtained. These reveal that the charge collected on the sensitive node, provided by the SEU pulse is insufficient to cause the the device to change its state. By storing and plotting the pulse current amplitudes as a function of the corresponding pulse widths for each run, the approach to the threshold pulse that just causes the device to change its state can be made. Sets of such threshold pulse current amplitude and pulse width pairs are plotted as shown in Fig. 8.6.

For very short pulse widths, of the order of 0.01–0.10 ns, which are still representative of the SEU temporal behavior, the desired critical charge Q_c is

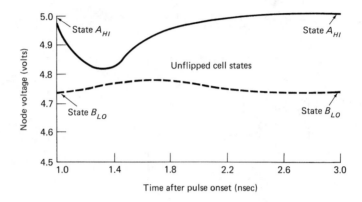

Figure 8.2. Multitransistor custom bipolar D-Flip/Flop cell excited by SEU pulse insufficient to cause cell change of state. Pulse onset at 1 ns.

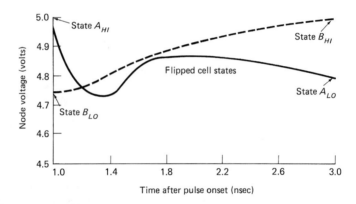

Figure 8.3. Multitransistor custom bipolar D-Flip/Flop cell excited by SEU pulse sufficient to cause cell change of state. Pulse onset at 1 ns.

available from Fig. 8.6; that is, $Q_c = \int_0^{T_p} I_p(t)\,dt \simeq I_p T_p$, where I_p and T_p are the pulse current amplitude and pulse width, respectively. The integral is well approximated by the $I_p T_p$ product for short pulse widths, as I_p is only slowly varying during the pulse duration, as seen by the device time constants implied in Figs. 8.2–8.5. In Figure 8.6, computation of the SRAM cell excited by a nominal SEU pulse whose $I_p = 20$ mA and the corresponding $T_p = 10$ ps yields $Q_c = 0.2$ pC. Similarly, for the D-Flip/Flop, $I_p = 12$ mA and $T_p = 100$ ps gives $Q_c = 1.2$ pC.

Recent work [6] has produced more accurate SEU error rate expressions, including that for a geosynchronous environment during a solar minimum, than those derived in Section 5.2. Recently modified from that given in Section 5.2, (5.8) and (8.1) become,

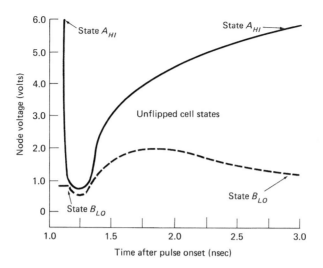

Figure 8.4. CMM5114/1RZ CMOS/SOS 4K SRAM cell excited by SEU pulse insufficient to cause cell change of state. Pulse onset at 1.1 ns.

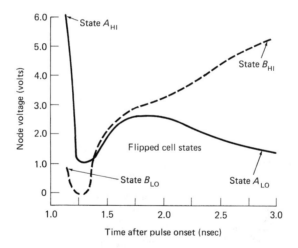

Figure 8.5. CMM5114/1RZ CMOS/SOS 4K SRAM cell excited by SEU pulse sufficient to cause cell change of state. Pulse onset at 1.1 ns.

$$E_r = \frac{2 \times 10^{-10} abc^2}{Q_c^2}, \quad [a], [b], [c] = \mu m, [Q_c] = pC\ \mu m^{-1} \quad \textbf{(8.1a)}$$

$$E_r = \frac{200 \sigma_\infty}{L_{0.25}^2}, \quad [\sigma_\infty] = cm^2, [L_{0.25}] = MeV\ cm^2\ mg^{-1} \quad \textbf{(8.1b)}$$

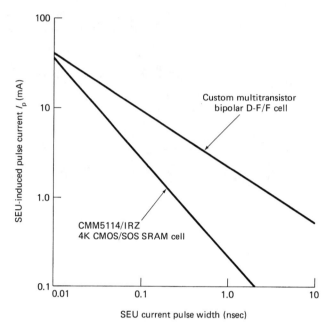

Figure 8.6. SEU-induced current pulse response for sensitive nodes of two devices versus SEU pulse width as simulated on electronic circuit-device physics SPICE-type program. From Ref. 5.

where the parameters are defined in Section 5.2 and are obtained from models, destructive physical analyses, and experimental data. $L_{0.25}$ is the value of LET on the cross section–LET curve corresponding to 25% of the "asymptotic cross section," as in Figure 4.5a.

To date, there are more or less five operational methods for computing SEU error rate from the various kinds of data available. They are given in Table 8.2.

Number 4 in Table 8.2 describes the GFOM (Generalized Figure of Merit) method. It uses (8.1a) or (8.1b) but with a multiplier coefficient $C(h, \theta)$ to account for spacecraft orbit altitude (h) and inclination angle (θ), as written at the right of Fig. 8.7. The results are plotted in Fig. 8.7. From the latter, $C(h, \theta)$ is tabulated in Table 8.3 for "hand calculations" and is fitted with a polynomial in $\theta^{1/3}$, as in (8.1c), with its coefficients tabulated in Table 8.4 for computer algorithmic purposes. Fig. 8.7, Tables 8.3 and 8.4 hold for the 93L422 only, and so must be renormalized for other parts.

One method of renormalization for the part of interest is to multiply the values in Table 8.3 by the ratio $\bar{\sigma}_{part}/\bar{\sigma}_{93L422}$ or its equivalent asymptotic value, where overbars imply mean values. Another, indirect method [6], is to compute the SEU error rate using the IRPP (Method No. 1) for a large "soft" part in the system for orbit altitude, inclination angle, solar min/max conditions, and shielding. Using this E_r,

Table 8.2. Computational SEU Error Rate Formulas[6].

1. IRPP (Integration of Rectilinear Parallelepiped)
    This is the de facto standard model. It is a numeral integration of (5.2) with the full multiterm $f(s)$. This is used in CREME code (5.10′) for refined device comparisons, and it uses the parallelepiped depth $c = 1$ μm.
2. FOM (Figure of Merit)
    Uses flux and $\int f(s)\,ds$ approximations; namely $\phi(L) = 5.8 \times 10^8/L^3$ and $C(s(L)) = \int f(s)\,ds$, respectively. Results yield (5.8) and (8.1). This is the classical approach. Defer to MFOM.
3. MFOM (Modified Figure of Merit)
    SEU error rates given in (8.1a) and (8.1b). It is used for simple computation of part SEU susceptibilities, especially during solar minimum.
4. GFOM (Generalized Figure of Merit)
    FOM, but with multiplier coefficients as functions of orbit altitude and inclination angle, from geosynchronous to low earth orbit, given in Fig. 8.7. Table 8.3 or 8.4 can also be used with your favorite interpolation method for error estimates at a given orbit altitude and inclination angle. However, the curves and tabulations must be remormalized for parts other than 93L422.
5. EFA (Effective Flux Approximation)
    Equation (5.11) and suceeding paragraphs yields SEU error rate in (5.17). Specialized use[6].

Source: Ref. 6.

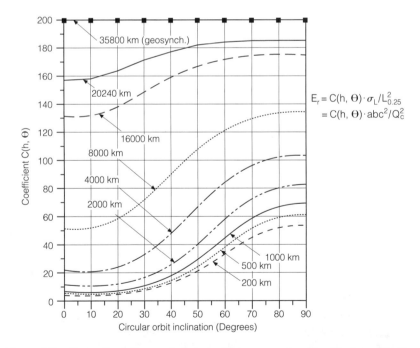

Figure 8.7. Generalized figure of merit coefficients versus orbit altitude and inclination angle for circular orbits (for 93L422). From Ref. 6.

Table 8.3. Array of Coefficients $C(h, \theta)$ from fig. 8.7.

h (km)	$\theta°$									
	0	10	20	30	40	50	60	70	80	90
35800	200	200	200	200	200	200	200	200	200	200
20240	157	158	163	172	178	182	183	184	185	185.5
16000	131	133	137	150	160	167	172	174	175	175.5
8000	51	52.5	58.5	72	92	110	123	130	133	135
4000	21	22	23.5	32	48	70	87	98	102	103
2000	12	12.5	13	15	24	40.5	60	73	80.5	81.5
1000	7	7.5	8	9.5	17	30	46	60	66.5	67
500	5	5.5	6	7	12	26	40	51	60	61
200	4	4.5	5	6	9	21.5	35	46	53	53.5

Note: Used to interpolate for orbit altitude/inclination angle between curves for "hand" calculations. Must renormalize values for devices other than 93L422.

Source: From Ref. 6.

determine the corresponding coefficient multiplier in the MFOM (Method No. 3) SEU error rate expressions. Then determine the SEU error rates of all the other parts in the system using that coefficient with the MFOM formulas.

The coefficient $C(h, \theta)$ can be expressed empirically, by,

$$C(h, \theta) = \begin{cases} C_0(h, 0) + M_h \sum_{k=1}^{9} a_k^h \theta^{k/3}, & 10° \leq \theta \leq 90° \\ C_0(h, 0), & 0° \leq \theta < 10° \end{cases} \quad (8.1c)$$

8.3. Proton-Induced SEU II

Protons are the major constituent particles of microcircuit SEU interest that make up the inner Van Allen radiation belt, as discussed in Section 2.5. Spacecraft avionics orbiting at these altitudes are targets for these particles. Some galactic cosmic rays are also trapped in the Van Allen belts. Certain recent large solar flares have apparently caused the temporary formation of a second and third maxima in the "slot" region which demarks the boundary between the inner and outer Van Allen belts [7]. Their duration is estimated to be weeks to months. The altitude of the existing proton belt peak is about 1.5 earth radii at the equator, whereas these "new" peaks are at about 2 and 2.5 earth radii, respectively [8].

Because protons principally cause SEU indirectly through complex nuclear reactions including spallation (fissionlike) reactions, a computational expression for proton-induced SEU analogous to (8.1) for the geosynchronous environment is not available. The main reasons that the geosynchronous error rate, (8.1), does not apply to protons is that it does not account for SEUs induced

Table 8.4. $C(h,\theta)$ Polynomial Fits in Fractional Powers of Inclination Angle θ as given in Table 8.3.

h 10^3 (km)	C_o	M_h	a_1^h	a_2^h	a_3^h	a_4^h	a_5^h	a_6^h	a_7^h	a_8^h	a_9^h
0.2	4.0	1505.40	1.00	-1.544	0.832	-0.139	-0.024	0.008	0.001	-(5.00 − 4)[a]	4.00 − 5[a]
0.5	5.0	304.65	1.00	-1.852	1.193	-0.267	-0.018	0.010	0.002	-(9.26 − 4)	7.55 − 5
1.0	7.0	1026.56	1.00	-1.553	0.829	-0.123	-0.036	0.011	0.001	-(4.79 − 4)	4.07 − 5
2.0	12.0	1348.65	1.00	-1.587	0.874	-0.144	-0.033	0.011	0.001	-(5.56 − 4)	4.76 − 5
4.0	21.0	416.97	1.00	-1.819	1.147	-0.221	-0.051	0.021	8.3 − 4	-(9.86 − 4)	9.43 − 5
8.0	51.0	-883.29	1.00	-1.419	0.694	-0.109	-0.010	1.3 − 3	1.2 − 3	-(2.71 − 4)	1.53 − 5
16	131.0	856.16	1.00	-1.330	0.585	-0.054	-0.027	0.005	0.001	-(3.35 − 4)	2.49 − 5
20.2	157.0	-93.06	1.00	-2.508	2.017	-0.599	-0.005	0.025	-0.001	-(7.61 − 4)	8.51 − 5
35.8[b]	200.0	0	—	—	—	—	—	—	—	—	—

[a] $X.Y - Z = X.Y \times 10^{-Z}$.
[b] Geosynchronous altitude. $C(35.8, \theta) \equiv 200$ for $0° \leq \theta \leq 90°$.

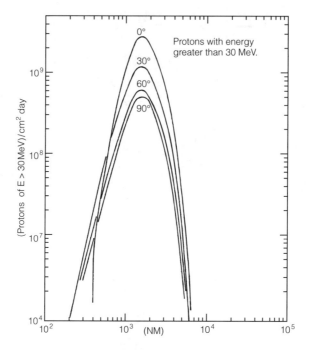

Figure 8.8. Proton integral flux averaged over circular earth orbits of various inclinations (based on AP8MIN model). From Ref. 9.

by ions produced in the sensitive region, as well as those ions that are stopped in and are slowed down in that region; that is, if the proton equivalent of a Heinrich curve-differential flux of ions were available, then an expression analogous to (8.1) might be used to compute the corresponding error rate.

The main avenue taken to compute the proton-induced SEU error rate, E_{pr}, is to use the classical approach of the integral of the product of the appropriate cross section σ_p and the corresponding proton flux Φ_p, i.e.,

$$E_{pr} = \int_A^\infty \sigma_p(A, E) \, \Phi_p(E) \, dE \tag{8.2}$$

The cross section as a function of a low-energy threshold A (MeV) can be given empirically as a one-parameter, (A), expression rewritten from Section (5.3), as

$$\sigma_p(A, E) = \left(\frac{24}{A}\right)^{14} \left\{ 1 - \exp\left[-0.18 \left(\frac{18}{A}\right)^{1/4} \right] (E-A)^{1/2} \right\}^4, \quad E > A \, (10^{-12} \, \text{cm}^2) \tag{8.3}$$

where E is the energy of the incident proton and A is the proton energy threshold parameter unique to the device of interest, and is depicted in Fig.

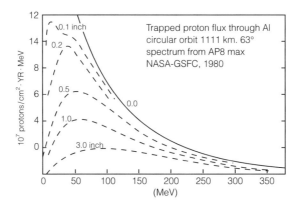

Figure 8.9. Proton flux spectrum for a 600 Nm orbit at 63° inclination. From Ref. 9.

5.3. The details of determining A and its ramifications are discussed in Section 5.3 and later in this section. This includes refinements to σ_p which result in it becoming a function of two parameters, A and B.

Figure 8.8 shows the inner Van Allen belt proton flux versus orbit altitude and inclination with its peak evidenced at about 1500 nautical miles. The figure corresponds to proton energies greater than 30 MeV, which produce the bulk of the SEU. Such protons can easily penetrate the spacecraft skin, which is usually relatively thin aluminum about 40–50 mils thick. Once inside, proton-induced SEU of the spacecraft internals depends on their mass, density, and juxtaposition of the subassemblies inside with regard to their shielding propensities. The principal result is that proton–silicon nuclear reaction products of sufficiently heavy ions cause SEU, as discussed in Section 5.3.

Figure 8.9 depicts the Van Allen proton flux spectra for a 700-mile altitude spacecraft orbit inclination at 63° with respect to the equator, after the flux passes through various spacecraft shells of aluminum thickness between 0.1 and 3.0 in. The unshielded spectrum is from AP8MAX (sunspot maximum) series from NASA–GSFC 1980 [9]. Note that the spectrum in the figure becomes "harder" after passing through increasing thicknesses of absorber; that is, the increasing mass absorbs the lesser energy flux, leaving only the high-energy component to "shine through" as evidenced by the spectrum peak energies increasing with absorber shield thickness.

Conventional thinking with regard to part SEU sensitivity lends belief to the conjecture that proton-induced SEU is the parameter that divides parts into those that are generally SEU sensitive and those that are not [10]. For example, if the part possesses a measured LET above 10 MeV cm^2 mg^{-1}, it is presumed to be immune to SEU, and so testing for proton-induced SEU is not necessary. Some workers use higher thresholds, such as 20 MeV cm^2 mg^{-1} to negate testing.

Most proton testing is done with the proton beam at normal incidence to the part. This usually corresponds to where the penetrating proton track length is a minimum (shortest chord in sensitive region), which produces a minimum SEU sensitivity. Recent analyses using established computer codes [10] have shown that for low-L_{th} devices, the beam incidence angle is not a factor in the determination of proton-induced SEU cross-section variation with proton energy. The same codes reveal that beam incidence angle is a factor for relatively high L_{th} devices. However, recent proton accelerator experiments show a marked diminution of this effect, at least for certain devices [11], and with similar results for single event latchup induced by protons.

Table 8.5 is a tabulation of proton-induced SEU cross sections for various parts for various proton energies. The energy threshold parameter A is also included. It is obtained from these data as substituted into the one-parameter proton cross-section expression (8.3); that is, using the tabulated cross section and the corresponding energy in (8.3), A is numerically extracted as its root.

As most of the data in Table 8.5 were taken using monoenergetic proton sources for each device, it is of interest to note the corresponding variation of E_{pr} as directly obtained with such energies. For a monoenergetic source, the differential proton flux can be written as $\phi(E) = \phi_0 \delta(E - E_0)$. ϕ_0 is the proton flux level and E_0 is its energy. The property of the delta function used here is that $\int_a^b f(E)\delta(E - E_0)\,dE = f(E_0)$, where $a \leq E_0 \leq b$. Inserting the above $\phi(E)$ into (8.2) with (8.3) yields the E_{pr} directly as

$$E_{pr} = \phi_0 \sigma_p(A, E_0) = \phi_0 \left(\frac{24}{A}\right)^{14} \left\{ 1 - \exp\left[-0.18 \left(\frac{18}{A}\right)^{1/4} (E_0 - A)^{1/2} \right] \right\}^4, \quad E_0 > A \tag{8.4}$$

For very large proton energies E_0, it is seen that $E_{pr} \sim \phi_0 (24/A)^{14}$, which is proportional to the asymptotic cross section $\sigma_{asy} \equiv \sigma_\infty = (24/A)^{14}$ using this model. For small E_0 near A, it is seen that (8.4) gives

$$E_{pr} = \phi_0 \left(\frac{24}{A}\right)^{14} (0.18)^4 \left(\frac{18}{A}\right) (E_0 - A)^2 \tag{8.5}$$

where the first two terms in the exponential in (8.4) suffice for E_0 near A.

Section 5.3 discusses the two-parameter proton-induced SEU cross-section expression and provides a procedure for determining the two parameters for various orbit altitudes and inclinations.

There is another method for determining A and B which involves the results of proton tests on a particular device [13]. From (8.2), with the two-parameter cross section $\sigma_p(A, B; E)$, $E \geq A$, given in Section 5.3, the proton-induced SEU error rate can be written

$$E_{pr} = \int_0^\infty \sigma_p(A, B; E)\, \phi_p(E)\, dE \tag{8.6}$$

where $\sigma_p(A, B; E) = 0$ for $E < A$. Now $\phi_p(E) = \phi_0 \psi_p(E)$, where ϕ_0 is the integrated proton flux and $\psi_p(E)$ is the corresponding normalized spectrum [i.e., $\int_0^\infty \psi_p(E) \, dE = 1$]. Then (8.6) can be written

$$E_{pr} = \phi_0 \int_0^\infty \sigma_p(A, B; E) \psi_p(E) \, dE = \phi_0 \hat{\sigma}_p(A, B) \tag{8.7}$$

where $\hat{\sigma}_p$ is the two-parameter spectrum weighted cross section,

$$\hat{\sigma}_p(A, B) = \int_0^\infty \sigma_p(A, B; E) \psi_p(E) \, dE \tag{8.8}$$

For a series of n proton-induced SEU experimental measurements of a part, at n different proton energies E_i, $i = 1, 2, \ldots, n$, the corresponding measured cross section errors $\Delta\sigma(E_i) \equiv \Delta\sigma_i$ at each measurement can be construed to be of the form

$$\Delta\sigma_i = \sigma_{imeas} - \hat{\sigma}_{pi} = \sigma_{imeas} - \int_0^\infty \sigma_p(A, B; E) \psi_{pi}(E) \, dE, \quad i = 1, 2, \ldots, n \tag{8.9}$$

where σ_{imeas} is the measured SEU cross section at each of the proton energies E_i. $\psi_{pi}(E)$ is the proton spectrum at each measurement, including its corresponding standard deviation and mean energy. Within a constant, $\psi_{pi}(E)$ is the appropriate gaussian fit to the spectra so as to explicity include the measured energy mean. $\bar{E}_i$ and standard deviation s_i at each energy; that is, $\psi_{pi}(E) \sim \exp[-(E - \bar{E}_i)^2]/2s_i^2$, and $\int_0^\infty \psi_{pi}(E) \, dE = 1$ for each $i = 1, \ldots, n$.

The parameters A and B can be obtained from the n equations (8.9) in the sense of the least squares criterion, where, from Section (5.3),

$$\sigma_p(A, B; E) = \left(\frac{B}{A}\right)^{14} \left\{1 - \exp\left[-0.18\left(\frac{18}{A}\right)^{1/4} \cdot (E - A)^{1/2}\right]\right\}^4 \quad (10^{-12} \text{ cm}^2/\text{bit}) \tag{8.10}$$

with partial derivatives,

$$\partial\sigma_p/\partial B = \frac{14\sigma_p}{B}; \quad \frac{\partial\sigma_p}{\partial A} = -\left(\frac{14\sigma_p}{A}\right)\left(1 + \frac{0.013 (18/A)^{1/4} (E + A)(E - A)^{-1}}{\exp[0.18 (18/A)^{1/4} (E - A)^{1/2}] - 1}\right) \tag{8.11}$$

The least squares approach here is to compute the optimal A and B by using them to minimize the sum of the squares of the SEU measurement errors [viz. $I(A, B) = \sum_{i=1}^n (\Delta\sigma_i)^2$]; that is,

$$\min_{A, B} I(A, B) = \min_{A, B} \sum_{i=1}^n (\Delta\sigma_i)^2 = \min_{A, B} \sum_{i=1}^n \left[\sigma_{imeas} - \int_0^\infty \sigma_p(A, B; E) \psi_{pi}(E) \, dE\right]^2 \tag{8.12}$$

Table 8.5. Proton Induced SEU Cross Sections and Energy Threshold A in Descending Order for Typical Microcircuit Parts

Ser.	Part number	Function	Vendor	Technology	Proton E (MeV)	$\sigma(10^{-12}\,\text{cm}^2/\text{bit})$	A(MeV)
1	HC6116	SRAM	HONWL	CMOS	90	3.3E−4	37.79
2	ADSP2100	µP	ADI	CMOS-EPI	200	0.033	29.47
3	5516	SRAM	TOSH	CMOS	44.8	4.1E−3	28.71
4	6147	SRAM	TOSH	CMOS	40	8.0E−3	26.89
5	6264	SRAM	HITC	CMOS	100	0.12	25.80
6	6516	SRAM	HARR	CMOS	100	0.13	25.66
7	HM1B65162-2	SRAM	MAT/HA	CMOS	100	0.25	24.51
8	KM41C400Z-8	DRAM	SAMSUN	CMOS	30	0.012	24.08
9	D431000	SRAM	NIPP	CMOS	25	5.0E−4	24.05
10	FCB61C65L	SRAM	PHILL	HCMOS	70	0.24	23.79
11	HYB514100J-1	DRAM	SIEM	CMOS	30	0.02	23.50
12	4044	RAM	TI	NMOS	130	0.80	23.09
13	2901A	4/SLC	AMD	ECL/TTL	130	1.1	22.59
14	HC62032H	SRAM	HITC	CMOS	30	0.045	22.56
15	628128R	SRAM	HITC		25	0.01	22.30
16	4416	DRAM	TI	NMOS	100	1.3	21.94
17	CXK58257	SRAM	SONY	CMOS	25	0.02	21.67
18	62383H	SRAM	HITC		25	0.03	21.40
19	HC62256H	SRAM	HITC	CMOS	30	0.20	20.85
20	6116	SRAM	HITC	CMOS/NMOS	100	2.8	20.82
21	HM4716A-4	DRAM	HITC	CMOS	4200	9.4	20.45
22	93423	RAM	FCH	TTL	130	5.0	20.35
23	2164A	D F/F	AMD	LSTTL	130	5.0	20.35
24	62256R	SRAM	HITC		25	0.12	20.16

Source: Ref. 12.

Table 8.5. Continued.

Ser.	Part. number	Function	Vendor	Technology	Proton E (MeV)	$\sigma(10^{-12}\,\text{cm}^2/\text{bit})$	A (MeV)
25	D43256	SRAM	NIPP	CMOS	25	0.20	19.68
26	71681	SRAM	IDT	CMOS/NMOS	40	1.9	19.35
27	2147	RAM			40	3.0	18.80
28	82S19	RAM	SIG	STTL	56	7.5	18.50
29	87C51FC	μP	INTL		20	0.5	17.21
30	93425	RAM		TTL	156	80.	16.94
31	82S212	RAM	SIG	STTL	130	170.	15.94
32	93419	RAM		TTL	15	1.6	14.03
33	2901C	4/SLC	AMD	ECL/TTL	15	3.0	13.80
34	EDH8832C	RAM	EDI	CMOS/NMOS	50	2E+5	9.22
35	HFE4811-014	XMTR	HONWL		60	7E+6	7.32

Source: Ref. 12.

To minimize I with respect to A and B, both $\partial I/\partial B = 0$ and $\partial I/\partial A = 0$ simultaneously; that is, for the B parameter,

$$0 = \frac{\partial I}{\partial B} = 2 \sum_{i=1}^{n} \left[\sigma_{i\text{meas}} - \int_0^\infty \sigma_p(A, B; E)\, \psi_{pi}(E)\, dE \right] \left[-\int_0^\infty \frac{\partial \sigma_p}{\partial B} \psi_{pi}(E)\, dE \right] \tag{8.13}$$

Because the second factor in the rightmost equation (8.13) is not zero, the first factor is zero for each i, $i = 1,\ldots, n$. Thus, from (8.13) and then summing gives

$$0 = \sum_{i=1}^{n} \left[\sigma_{i\text{meas}} - \int_0^\infty \sigma_p(A, B; E)\, \psi_{pi}(E)\, dE \right] = \bar{\sigma}_{\text{meas}} - \int_0^\infty \sigma_p(A, B; E)\, \bar{\psi}_p(E)\, dE \tag{8.14}$$

where the arithmetic means are given by

$$\bar{\sigma}_{\text{meas}} = \left(\frac{1}{n}\right) \sum_{i=1}^{n} \sigma_{i\text{meas}}, \qquad \bar{\psi}_p(E) = \left(\frac{1}{n}\right) \sum_{i=1}^{n} \psi_{pi}(E) \tag{8.15}$$

Equation (8.14) is the first of two equations in the two unknowns A and B. However, in this case, using $\partial I/\partial A = 0$ to obtain the second equation in A and B leads to the same equation (8.14). Hence, another condition is needed to obtain a second equation. Because B is commonly used to fit the asymptotic form of σ_p [i.e., $\sigma_{p\infty} = \lim_{E \to \infty} \sigma_p(A, B; E)$], $\sigma_{p\infty}$ will be used for $\sigma_p(A, B; E)$ in (8.14) to get the desired second equation. Then, from (8.10), $\sigma_{p\infty} = (B/A)^{14}$ and inserting it into (8.14) yields

$$0 = \bar{\sigma}_{\text{meas}} - \left(\frac{B}{A}\right)^{14} \int_0^\infty \bar{\psi}_p(E)\, dE = \bar{\sigma}_{\text{meas}} - \left(\frac{B}{A}\right)^{14} \tag{8.16}$$

because $\int_0^\infty \bar{\psi}(E)\, dE = 1$. More compactly, it is seen from (8.10) that σ_p can be written for $E > A$ as

$$\sigma_p(A, B; E) = \left(\frac{B}{A}\right)^{14} f_p(E, A), \qquad f_p(E, A) = \left\{1 - \exp\left[-0.18\left(\frac{18}{A}\right)^{1/4}\right] \cdot (E - A)^{1/2} \right\}^4 \tag{8.17}$$

Inserting (8.17) into (8.14) combined with (8.16) yields the two sought-after equations; namely A is a root of

$$\int_A^\infty f_p(E, A)\, \bar{\psi}_p(E)\, dE = 1 \quad \text{and} \quad B = A\,(\bar{\sigma}_{\text{meas}})^{1/14} \tag{8.18}$$

Table 8.6. Computed Cross-Section Parameters A and B for Three Devices.

Part No.	Function	Technology	A	B
93L422	RAM	TTL	6.79	10.34
93419	RAM	TTL	9.86	13.44
AM2901C	μp	LSTTL	10.97	15.56

Source: Ref. 13.

A and B are determined from (8.18) by numerical means. Table 8.6 depicts values of A and B for experimental measurements on three different devices using a modified version of the preceding least squares method [13].

8.4. Neutron-Induced SEU II

Neutron-induced SEU occurs mainly in two natural environments. The first is at ground level, as discussed in Section 5.6, which principally affects ground-based electronic systems. The second natural environment is at stratospheric altitudes that include the cruising altitudes of long-hop aircraft, such as the Concorde. Section 5.4 discussed neutron-induced SEU and included an avionics example.

Neutrons can cause displacement damage in the semiconductor device material. Displacement damage implies that a lattice atom is physically displaced from its position due to a neutron interaction. This is as opposed to merely ionization (electrons removed from their position around the atom) and its consequences with respect to SEU. However, neutron fluxes are so low at aircraft cruising altitude that displacement damage consequences are negligible compared to those of neutron-induced SEU.

Heavy ions, protons, and electrons cause upsets that mainly affect spacecraft, whereas neutrons principally affect aircraft and, to a lesser extent, ground systems, as already mentioned. It behooves workers in this field to provide for SEU neutron testing of parts to ascertain their SEU performance.

Adequate neutron sources that simulate well the energy spectrum of neutrons that occur in the atmospheric environment are few and far between. Besides that discussed in Section 5.6, and the comparison plutonium-beryllium (Pu-Be) source in Table 8.8, there is at least one source that is coincidentally tailored for this purpose and is beginning to be used to test parts. This neutron source is part of the Los Alamos National Laboratory Weapons Neutron Research (WNR) facility [14]. It consists of 800-MeV accelerator protons incident on a tungsten target. The resulting spallation reactions produce a continuous neutron spectrum from very low neutron energy to energies approaching 800 MeV. This spectrum is quite similar to that at high altitudes, as

294 / Single Event Phenomena

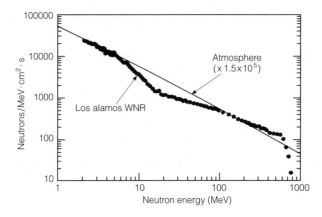

Figure 8.10. Comparison of neutron differential energy spectrum of WNR facility with that in the atmosphere at 40000 ft altitude. From Ref. 14.

seen in Fig. 8.10. Fortunately, for testing purposes, its neutron flux levels are about 2×10^5 times greater than those at stratospheric altitudes. Its matching spectral properties are due to the fact that the proton-induced neutron reactions in the above spallation process are given by $_1^1p + {}_{74}^{184}W \rightarrow {}_0^1n + {}_{75}^{184}Re$. In the context of neutron production from protons, this reaction is the same as neutrons produced in atmospheric oxygen and nitrogen from cosmic ray protons (90% of cosmic rays); that is, $_1^1p + {}_8^{16}O \rightarrow {}_0^1n + {}_9^{16}F$ and $_1^1p + {}_7^{14}N \rightarrow {}_0^1n + {}_8^{14}O$.

Table 8.7 shows the results of using the WNR neutron source to test 128K × 8 SRAMs from six vendors for SEU. It is seen that their mean SEU error rate is 6.8×10^{-10} per bit-hour, whereas their corresponding mean multiple-bit error rate is 1.13×10^{-11} per bit-hour, about 1.7% percent that of the former. The time units are in hours to lend affinity to aircraft flight hours. The neutron fluence in the table is couched in terms of equivalent (flight) hours, namely T_{eq}^i (h) for the i-th device tested. T_{eq}^i is defined as the number of equivalent flight hours of exposure to neutron flux that the ith device at altitude would require to experience the same level of neutron fluence it would endure during its test time in the WNR facility. As discussed in Section 5.2, for the ith SRAM,

$$\int_0^{T_{test}^i} dt \int_{10\text{ MeV}}^{\infty} \phi_{WNR}^i (E, t) \, dE = T_{eq}^i \left\langle \int_{10\text{ MeV}}^{\infty} \phi_{40} (E, t) \, dE \right\rangle_{da} \quad (8.19)$$

being exposed to WNR neutrons, where

$\phi_{WNR}^i (E, t)$ = neutron flux in the WNR facility incident on the ith device
T_{test}^i = measured test time in the WNR facility for the ith device (h)
$\phi_{40} (E, t)$ = neutron flux at an altitude of 40,000 ft (and 45° latitude)
$\langle \int_{10\text{ MeV}}^{\infty} \phi_{40} (E, t) \, dE \rangle_{da} = (1/24) \int_0^{24\text{ h}} dt \int_{10\text{ MeV}}^{\infty} \phi_{40} (E, t) \, dE$; the daily flux at 40,000 ft is ~ 1.6n cm^{-2} s^{-1} [14]

Table 8.7. Experimental Values of Neutron-Induced SEU and Multiple Event Error Rates For 128k × 8 SRAMS

Ser.	Part number	Function	Vendor	Neutron fluence equiv. hrs (T_{eq})	SEU error rate per 10^{-10} bit-hr	Multiple event rate per 10^{-12} bit-hr
1	HM628128LP	SRAM	HITC	7.45×10^5	6.75	7.0
2	TC551001P	SRAM	TOSH	7.45×10^5	11.5	20.0
3	MT5C1008	SRAM	MICR	7.45×10^5	10.4	30.0
4	EDI188128CS	SRAM	EDI	7.40×10^5	4.0	4.5
5	CXK581000P	SRAM	SONY	7.40×10^5	2.2	2.5
6	SRM20100L	SRAM	S-MOS	7.40×10^5	6.15	4.0
			Mean		6.8×10^{-10}	1.13×10^{-11}
			std. dev.		3.6×10^{-10}	1.11×10^{-11}

Source: Ref. 14.

Table 8.8. Experimental Values of SEU Error Rate and σ_{SEU} From Neutrons Using a Pu-Be Source.

Ser.	Part number	Function	Technology	Neutron flux (cm^{-2} s^{-1})	Rate (bits s^{-1})	σ_{seu} (cm^2)
Using a Pu-Be neutron source[a].						
1	MM54C929	SRAM	CMOS	1.95×10^5	1.5×10^{-6}	7.7×10^{-12}
2	AM29705APC	SRAM	BIPOLAR	2.03×10^5	3.65×10^{-6}	1.8×10^{-11}
3	AM2901BPC	4/SLICE	BIPOLAR	1.98×10^5	1.65×10^{-6}	8.3×10^{-12}
4	93L422	RAM	BIPOLAR	7.05×10^5	1.30×10^{-4}	6.4×10^{-10}
Experimental Values of SEU Error Rate and σ_{SEU} from Neutrons at 33,000 ft.						
1	D43256A6U	SRAM[b]	1.3 μm CMOS[c]	2.0	1.05×10^{-11}	5.25×10^{-12}
2	HM628128LR12	SRAM[b]	0.8 μm CMOS[c]	2.0	2.99×10^{-12}	1.50×10^{-12}

[a] Data from Ref. 15.
[b] Data from Ref. 21.
[c] n-channel with polysilicon feedback resistors.

The lower limit of integration over energy of the fluxes is 10 MeV, as it is well known that lesser energy neutrons produce negligible SEU and SEL. Corresponding neutron-induced SEU cross sections in the form of burst generation rates (BGR), discussed in Section 5.4, are also measured, but agreement with their analytical counterparts presently leaves much to be desired [14].

8.5. Single Event Latchup II

The latchup phenomenon in integrated circuits (ICs) of all types is an unfortunate concomitant of their manufacture and operation. As is well known, the epitaxial manufacturing process used for ICs also produces inadvertent or parasitic combinations of *np* and *pn* junctions, as well as parasitic bipolar transistors, that can be excited in train to produce latchup "paralysis" in transistors of interest. For example, *npnp* or *pnpn* four-layers can form an SCR device that is turned on to preempt the normal function of the IC. The resulting low-impedance, high-current path thus formed can cause the IC to burn out unless its power input is quickly shut down. Prior to the advent of SEU and SEL (single event latchup) phenomena, latchup was well known and occurred throughout IC technology. It still does, although modern mitigation techniques have been introduced in their manufacture.

Latchup produced by the dose rate is different from SEL. Dose rate latchup is usually meant as that caused by current transients due to the gamma ray pulse or X-ray efflux from a nuclear detonation, or similar phenomena. The incident gamma rays cause photocurrents to flow in the ICs that excite the latchup therein. These spurious current transients can simultaneously affect portions or all of an integrated circuit [14].

Single event latchup is much more localized than dose rate latchup and can be characterized as causing SEL in a single *pnpn* or *npnp* path, caused by protons, heavy ions, and other particles. Another major difference between dose rate latchup and SEL are the time epochs involved. The dose rate-inducing transients typically last about 10^{-7}–10^{-9} s, whereas corresponding SEL durations are about 10^{-10}–10^{-12} s long. It is also well known that purely electrical transients, such as caused by aberrant circuit behavior, can produce latchup. Latchup triggering from single-particle-induced charge is similar to electrically introduced latchup because the corresponding transient currents are highly localized [16].

Single event latchup does not correlate well with the idea of a latchup "cross section" (i.e., as related to a device-sensitive area, as is the case for SEU). This lack of correlation also holds for parameters like the associated L_{th} or σ_{seu} of a transistor, as seen in Table 8.9. It is noted there that SEL occurs in devices whose LET threshold varies from low to high, and higher values that do

Table 8.9. Devices Susceptible to Latchup from Heavy Ions

Ser.	Part number	Vendor	Technology	Function	SEU L_{th} (MeV cm^2 mg^{-1})	σ_{seu} (cm^2/device)
1	28C256	XIC	CMOS	EEPROM	$\ll 37$	1.0E-3
2	D4464D	NEC	CMOS	SRAM	~ 2	1.5E-1
3	CXK58255P	SONY	CMOS	SRAM	25	8.0E-4
4	5504	TOSH	CMOS	SRAM	< 37	1.4E-2
5	56001	MOT	CMOS	µP	9–16	—
6	7204	AMD	CMOS	FIFO	26	—
7	80386	INTL	CHMOS	µP	28	7.0E-5
8	80C86	HAR	CMOS	µP	7	4.0E-4
9	ADSP2100A	ADI	CMOS-EPI	µP	3.4	1.6E-5
10	TC5516	TOSH	CMOS	SRAM	27	1.0E-4
11	CY7C401	CYP	CMOS	FIFO	10	6.0E-4
12	EP1800	ALT	CMOS	PAL	15	1.5E-3
13	HD6434	HAR	CMOS	DRVR	14	8.0E-5
14	HM6504	HAR	CMOS	SRAM	13	1.0E-4
15	HM6514	HAR	CMOS	RAM	< 15	1.2E-4
16	HM65162S	HAR	CMOS	SRAM	< 2.8	1.5E-2
17	HM6614	HAR	CMOS	PROM	21	8.0E-4
18	IDT6116	IDT	NMOS/CMOS	SRAM	12	3.5E-4
19	IDT6167	IDT	NMOS/CMOS	SRAM	< 15	4.0E-4
20	IMS1601	INM	NMOS/CMOS	SRAM	5	4.0E-4
21	L64730	LSI	CMOS	µP	8	2.0E-3
22	L64814	LSI	CMOS-EPI	µP	10	2.0E-3
23	LTC485CN8	LTC	CMOS	XCVR	3	8.0E-5
24	MM54C929	NSC	CMOS	RAM	14	2.5E-4
25	MT5C2568C	MCN	CMOS	SRAM	23	9.0E-3
26	NSC810	NSC	CMOS	RAM	14	—
27	87C51FC	INTL	CHMOS	µP	15	5.0E-4
28	P4C164L	PFS	CMOS	SRAM	8	1.0E-2
29	PACE422	PFS	CMOS	SRAM	10	2.0E-3
30	R3000	IDT	CMOS-EPI	µP	3.3	—
31	HS82C59A	HAR	CMOS	µP	16	2.0E-3
32	68881	MOT	HCMOS	µP	6	4.0E-3
33	TMS8338	TMS	HCMOS	ADC	12	2.0E-3
34	X28C256	XIC	CMOS-EPI	EEPROM	18	1.0E-3
35	AD9048TQ	ADI	ECL	CNVTR	6	1.0E-5
36	MHS90C601	MTA	CMOS-EPI	µP	12.7	1.8E-3
37	ZR34161	ZOR	CMOS-EPI	µP	22	6.0E-3

Source: Ref. 12.

not appear in the table [12]. All of this means that it is more difficult to attempt to characterize SEL, although some inroads in this direction have been made [16].

It is enlightening to compare the relative amount of ion-induced charge deposited in the case of SEL with that for dose-rate-induced latchup in CMOS technology. The latter is the principal type of IC in which SEL occurs. To produce SEL, the required charge, Q_{SEL}, is given by the product of the LET delivered by the incident ion and an effective charge generation length d_{eff} [16]. Because 96.6 MeV cm² mg^{-1} of LET produces 1 pC μm^{-1} of charge deposited in silicon,

$$Q_{SEL} = \left(\frac{1}{96.6}\right)(\text{LET})\, d_{eff} \quad (\text{pC}) \tag{8.20}$$

where d_{eff} is in microns. It is estimated [13] that $d_{eff} \sim 5\text{–}10\,\mu$m. In a CMOS transistor, one MOS transistor is built into a well of dopant material that, in turn, is embedded in a somewhat larger structure of oppositely doped material also occupied by the complementary MOS transistor. Usually, this assembly contains two parasitic *npn* or *pnp* bipolar transistors that span the well. One does so laterally and the other vertically (Fig. 6.19). [14] Either parasitic transistor can comprise part of a *npnp* or *pnpn* chain to provide the spurious SCR that is the latchup-producing agent.

To compute the deposited charge for dose rate latchup, a photocurrent computation can be made as [3]

$$I_{photo} = 4.05 \times 10^{18}\ \text{hep cm}^{-3}\,\text{rad}^{-1}\,(\text{Si})\,(1.6 \times 10^{-19}\ \text{C hep}^{-1})\,\dot{\gamma}_D A_{well} w_{eff} F \quad (\text{amps}) \tag{8.21}$$

where $\dot{\gamma}_D$ is the incident dose rate [rads (Si) s^{-1}] of the prompt pulse and hep $\equiv$ hole–electron pairs. A_{well} is the well area which comprises the parasitic transistor junction depletion region volume $A_{well} w_{eff}$. w_{eff} is the corresponding charge collection depth and F is the fraction of the current that flows toward the parasitic junction area. The charge deposited in this case, Q_D, is obtained from (8.21), where the dose rate is assumed to be a rectangular pulse of width t_{pulse} s; that is, from (8.21),

$$Q_D = I_{photo} t_{pulse} = 6.5 \times 10^{11}\,\gamma_D A_{well} w_{eff} F \quad (\text{pC}) \tag{8.22}$$

where the dose of the pulse $\gamma_D = \dot{\gamma}_D t_{pulse}$. To ascertain the equivalent LET (viz. LET$_{eq}$ of dose rate latchup), equating (8.20) and (8.22) results in

$$\text{LET}_{eq} = 6.28 \times 10^{13}\,\gamma_D F A_{well} \left(\frac{w_{eff}}{d_{eff}}\right) \tag{8.23}$$

For $F \cong 0.5$ and $w_{eff}/d_{eff} \cong 5$, (8.23) gives [13]

$$\text{LET}_{eq} = 1.57 \times 10^{13} A_{well} \gamma_D \quad (\text{MeV cm}^2\,\text{mg}^{-1}) \tag{8.24}$$

An estimate for well dimensions yields about a factor of 10 greater than the device feature size, so that for a feature size of $\sim 2\,\mu\mathrm{m}$, $A_{\mathrm{well}} \sim 400\,\mu\mathrm{m}^2$. $\gamma_D = 100\text{–}1000$ rads(Si) for $\dot{\gamma}_D \simeq 10^8\text{–}10^{10}$ rads(Si) s^{-1} and $t_{\mathrm{pulse}} \simeq 10^{-6}\text{–}10^{-7}$ s. Hence, it is seen that L_{eq} is much larger than 1000 MeV cm^2 mg^{-1}, orders of magnitude greater than the usual critical LETs, implying that a much greater amount of charge is deposited due to dose rate latchup than for SEL. This is borne out by SRAM experiments [16].

SEL can be caused by turning on either the vertical or lateral parasitic transistor. Their excitation also depends on the diffused charge component as well as that produced by the prompt pulse (latchup-inducing) charge from the incident ion track. It is important to normalize experimentally obtained SEL cross sections to a device region area to lend substance to their value. Also, in general, SEL involves deeper regions in the device as contrasted with SEU [13]. Table 8.10 provides a tabulation of SEL-prone devices due to protons [12].

When the SEL cross section is available, the BGR method used in Section 5.4 for neutron-induced SEU error rate can be also used to determine the latchup rate [17]. The BGR functions are modified volumetrically with respect to sensitive volume, and energetically with respect to the recoil particle range–energy relationships for latchup computations. This compares favorably with experimentally determined latchup rates, at least for certain gate arrays [17] using the Weapons Neutron Research (WNR) experimental facility discussed in Section 8.4.

8.6. Single Event Burnout II

Both power MOSFETs and power bipolar transistors are subject to SEU-induced burnout. The former devices are also discussed in Section 6.8; the latter are discussed below. It will be noted that close similarities exist between the burnout mechanisms for both types of transistors.

In both transistors, burnout can be initiated by the SEU ion track that produces a plasma filament in the device upon penetration. Recalling Fig. 6.27, in the MOSFET, electrons generated by track ionization flow down through the plasma filament from the n^+ source region to the n^+ substrate. The corresponding holes flow up and out of the filament through the body p region to the body contact. For the track filament located, as in Fig. 6.27, near the channel, the hole current provides a voltage drop laterally along the p region. This forward biases the junction between the p region and the source, which turns on the parasitic transistor, as in the figure. Especially near the parasitic transistor collector depletion region, the applied voltages provide sufficiently strong electric fields to produce avalanche multiplication. The latter can provide regenerative feedback, which will cause a rapid increase in the collector

Table 8.10. Devices Susceptible to SEU Latchup from Protons

Ser.	Part number	Vendor	Technology	Function	Proton E (MeV)	Latchup fluence (cm^{-2})	σ_{seu} (cm^2/device)
1	ADSP2100A	ADI	CMOS-EPI	µP	200	—	—
2	D431000	NEC	—	SRAM	200	2.0E+6	—
3	D4464G-15L	NEC	CMOS/BULK	RAM	33	—	1.5E-9
4	L64811	LSI	CMOS/BULK	µP(SPARC)	50	9.8E+10	1.0E-11
5	L64814	LSI	CMOS-EPI	µP(SPARC)	200	5.0E+8	9.1E-11
6	MHS90C601	MTA	CMOS-EPI	µP(SPARC)	50	1.9E+10	5.3E-11
7	MHS90C602	MTA	CMOS-EPI	µP(SPARC)	70	1.2E+10	8.3E-11
8	MT4C400L	MIC	—	DRAM	30	1.0E+10	—
9	MT5C400L	MIC	—	SRAM	200	2.0E+6	—
10	NSC32C016	NSC	CMOS-EPI	µP	50	—	2.0E-8
11	R3000	IDT	CHMOS IV	µP	200	1.9E+9	—
12	R3000A	IDT	CHMOS	µP	200	7.3E+10	—
13	R3000A	SIEM	CHMOS	µP	200	1.6E+11	—
14	R3000A	PFS	PACE II	µP	200	5.6E+10	—
15	R3000A	LSI	HCMOS	µP	200	9.7E+10	—

Source: Ref. 12.

current, a precursor for causing second breakdown. Second breakdown is usually defined as the sudden collapse of the collector–emitter impedance with accompanying high voltage and current which will produce burnout failure [18]. The structurally inherent feedback mechanism of the device, MOSFET or bipolar, determines whether the particular voltage and current levels increase regeneratively or not, propelling the transistor toward burnout failure.

A feedback model [19] is derived from electron injection from emitter to collector (spanning the base). This is followed by the avalanche of hole current from collector to base, including a lateral hole current through the base to its contacts, resulting in a base–emitter voltage drop. This model is applicable to both MOSFET and bipolar transistors. In the former, the base, emitter, and collector refer to the parasitic bipolar transistor elements, whereas, in the latter, they are the actual elements.

The corresponding geometric and physical model used is a right circular cylinder, with a central axis containing the ionization track plasma as in Fig. 6.27. The principal one-dimensional coordinate is the radius r about the central axis. The current density injected by the emitter, $j_{EC}(r)$ is well known to be given by

$$j_{EC}(r) = \left(\frac{en_i^2 D_n}{N_{AB} W_B}\right) \exp\left(\frac{V_{BE}(r)}{V_T}\right), \quad V_T = \frac{kT}{e} \cong 26 \text{ mV} \quad (8.25)$$

where

D_n = the diffusion coefficient in the transistor n^+ emitter

N_{AB} = the dopant density of the corresponding p base

W_B = the base width

and the remaining symbols are universally known in the solid-state engineering field. V_{BE} is the base–emitter potential sufficient to turn on the parasitic transistor following the ion strike. The avalanche hole current density, $j_{HC}(r)$ is given by

$$j_{HC}(r) = M j_{EC}(r) \quad (8.26)$$

where the avalanche multiplication factor [14], M, is the ratio of hole to electron current during avalanche. V_{BE} provides a forward bias to produce the "reverse-injected" hole current density, $j_{HE}(r)$, namely

$$j_{HE}(r) = \left(\frac{en_i^2 D_p}{N_{DE} L_p}\right) \exp\left(\frac{V_{BE}}{V_T}\right) \quad (8.27)$$

where

D_p = is the diffusion coefficient of holes
N_{DE} = the dopant density in the emitter
L_p = the diffusion length of holes

As mentioned, there is a lateral flowing hole current density, $j_{HB}(r)$, through the base to ground, thus developing a voltage drop that keeps the base–emitter junction forward biased [19]. With (8.26), it is simply

$$j_{HB} = j_{HC} - j_{HE} = M j_{EC} - j_{HE} \tag{8.28}$$

The corresponding incremental lateral base current, i'_B is [19],

$$i'_B(r) = j_{HB}(r) \tag{8.29}$$

where the "prime" denotes the derivative with respect to r. Thus, the emitter voltage drop is

$$V'_{BE}(r) = -R_B(r) i_B(r) \tag{8.30}$$

Eliminating $i_B(r)$ between (8.29) and (8.30) yields

$$V''_{BE}(r) = -(R_B j_{HB} + R'_B i_B) \cong -R_B j_{HB} \tag{8.31}$$

where R_B is the sheet resistance of the p base, and the second term is neglected; that is, R_B is slowly varying with r, so that R'_B and, hence, $R'_B i_B$ is negligible with respect to the first term in (8.31). Using (8.26)–(8.30) yields a differential equation in $V_{BE}(r)$ from (8.31),

$$V''_{BE} = -K \exp\left(\frac{V_{BE}}{V_T}\right) \tag{8.32}$$

with boundary conditions [18], $V'_{BE}(0) = 0$, $V_{BE}(0) = V$, and $V_{BE}(r_c) = 0$, where r_c is the distance to the body contact (Fig. 6.27).

K is given by

$$K = e n_i^2 R_B \left[\left(\frac{M D_n}{N_{AB} W_B}\right) - \left(\frac{D_p}{N_{DE} L_p}\right)\right] \cong \frac{e n_i^2 R_B M D_n}{N_{AB} W_B} \tag{8.33}$$

The second term is very small compared to the first because of the large difference between the two dopant densities, therefore it is neglected. For both

R_B and M assumed slowly varying, K is considered essentially constant. Then the integral of (8.32) is immediate,

$$V_{BE} = V_0 - 2V_T \ln[\cosh(mr)], \qquad m = \left[\left(\frac{K}{2V_T}\right)\exp\left(\frac{V_0}{V_T}\right)\right]^{1/2}, \qquad (8.34)$$

$V_{BE}(r_c) = 0$ implies a relation between V_0, m and r_c, given from (8.34) by

$$V_0 = 2V_T \ln[\cosh(mr_c)] \qquad (8.35)$$

The lateral base current i_B is then, with (8.34) and (8.30),

$$i_B = -\frac{V'_{BE}}{R_B} = \left(\frac{2mV_T}{R_B}\right)\tanh(mr) \qquad (8.36)$$

and, likewise, the corresponding base current density, j_{HB}, is, with (8.31),

$$j_{HB} = -\frac{V''_{BE}}{R_B} = \left(\frac{2m^2V_T}{R_B}\right)\text{sech}^2(mr) \qquad (8.37)$$

It is seen that the functional variation of the above approximate V_{BE}, i_B, and j_{HB} are clearly closely similar to those given in Fig. 8.11. The latter are obtained

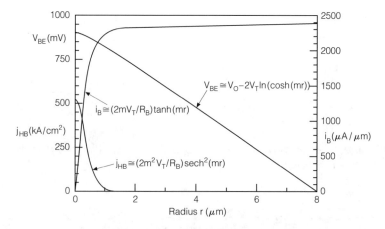

Figure 8.11. Approximate solutions superimposed on numerical solutions of burnout model for hole current density, j_{HB}, base current i_B, and potential V_{BE}. Approximate and numerical solution curves are almost coincident in that differences are irresoluble. From. Refs. 18 and 19.

by numerically integrating the preceding equations, where K is not constant but is refined to include nonconstant R_B, and M which is obtained numerically [18]. However, the parameters in the "constant K" case can be adjusted to make the resulting approximate curves virtually coincident with the refined curves in Fig. 8.11. One confirming reason is that for electron concentrations in the transistor depletion region that are much greater than the dopant density therein, M approaches the constant value of unity [18].

The curves in Fig. 8.11 also represent boundaries between burnout critical and noncritical regions. Exceeding those voltages or currents presages avalanching and consequent second breakdown leading to device burnout. Approximate and numerical solutions are so close that they virtually coincide.

Extensive experimental testing was done to determine burnout susceptibilities in samples of six different power bipolars, two of which were power darlingtons. Of those tested, four exhibited persistent burnout, revealing that the darlingtons were the most susceptible, as evidenced by their lower V_{CE} failure voltages (Table 8.12). Four heavy-ion types (^{58}Ni, ^{79}Br, ^{127}I, ^{197}Au) were used in the Brookhaven Van de Graaf accelerator to induce burnout in these devices [19]. Tables 8.11 and 8.12 provide burnout data induced by heavy ions and protons for selected power MOSFET and power bipolar devices.

8.7. Guidelines

In the field of single event phenomena, there are a number of useful guidelines from the viewpoint of the practitioner obtained from associations with and experience in the building of spacecraft and aircraft systems that are enumerated below. An immediate guideline springs from the often mistaken urge to be concerned with every active device in the system with regard to its susceptibility to SEU and SEL, but without regard to where and in what part of the particular electronic system it is. For example, it is realized that subsystems like power supplies are almost immune to SEU. This is because a possible very momentary SEU-induced transient leaves it virtually unaffected with respect to its purpose. There are a large number of subsystems in this category. Examples include solar cells, pulse-width-modulated power conditioners, voltage regulators, comparators, GaAs RF amplifiers, and so forth.

Because almost all passive components are inert to SEU, beginning at the active component level, an SEU-susceptible hierarchy of part technologies from most to least is given by [22]

(a) DRAMs, SRAMs
(b) Flip-flops, latches
(c) Sequential gates
(d) Combinational gates
(e) Other

Table 8.11. Heavy–Ion and Proton–Induced SEU Burnout in Power MOSFETs

Ser.	Part number	Vendor	HEAVY IONS σ_{seu} (cm^2/device)	Ion failure energy (MeV)	LET thresh. (MeV cm^2 mg^{-1})
1	25N20 (chip)	RCA	—	130	30
2	2N6666-1	—	—	48	37
3	2N6660	SIL	2.2E-6	36	12
4	2N6756	IRF	1.0E-6	—	30
5	2N6758	IRF230	3.0E-3	130	32
6	2N6762	IRF430	4.0E-2	—	30
7	2N6764	IRF150	1.3E-1	—	30
8	2N6766	RCA250	—	175	—
9	2N6768	IRF350	3.0E-3	130	30
10	2N6782	IRF110	1.3E-2	—	30
11	2N6784	INR	9.4E-7	133	30
12	2N6788	IRF120	4.2E-2	—	30
13	2N6792	UNITRO	—	210	30
14	2N6796	IRF130	4.9E-2	—	30
15	2N6798	IRF9130	—	100	30
16	2N6849	IRFF9130	—	100	30
17	2N6897	RCA	—	100	30
18	2SK725	FUJI	—	—	10
19	B350	SGS	1.0E-4	325	30
20	B450	SGS	3E-4 / 6E-2	310	—
21	B530	SGS	1.0E-4	35	—
			PROTONS V_{DS} failure (V)		Failure fluence (cm^{-2})
22	25N20 (chip)	RCA	176–200	50,150	9.0E+10
23	2N6660	SIL	40–54	50,150	4.5E+8

Source: Ref. 12.

Table 8.12. Heavy Ion Induced Burnout in Power Bipolar Transistors

Ser.	Part number	Vendor	V_{CE} failure (V)	Ion E (MeV)	Ion species	L_{th} (MeV cm^2 mg^{-1})
1	2N5629-10	—	195	265	^{58}Ni	27
2	2N6056-07[a]	—	50	285	^{79}Br	37
3	2N6059-11[a]	—	48	320	^{127}I	60
4	2N6284-12	—	100	320	^{127}I	60
5	MJ15003-11	—	210	350	^{197}Au	80

[a] Power darlingtons.

Source: Ref. 19.

Refined Susceptibilities [22]

High-risk devices	Lower-risk devices
(1) Bipolar RAMs	(1) Certain CMOS bulk
(2) Low-power logic (54Lxxx)	(2) Certain CMOS/SOS
(3) LS, ALS low-power schottky	(3) Standard power logic
(4) μ processors, bit slices	(4) PROMs
(5) NMOS, PMOS technology	(5) Low-speed devices
(6) DRAMs	(6) $\geq 10\mu$m feature size

Associated with the above chip-level hierarchy are methods that can be used to improve SEU hardness by microcircuit manufacturers. They include the following:

1. The use of cross-coupled resistors in the feedback loops of memory cells for partial decoupling. The resistors can impose a low-pass characteristic into the cell for immunity against the extremely fast SEU transient (10^{-10}–10^{-12} s).
2. Maximize memory cell physical size to increase cell capacitance creating a high-Q_c cell and, therefore, a minimum error rate.
3. Use dielectric chip coatings to stop alpha particles originating in the package from entering the chip proper.
4. Minimize memory bit line floating time, and use metal bit lines.
5. Use a folded bit line structure to counter collected charge.
6. Use junction isolated structures to sweep out alpha particle ionization charge.
7. Use "pure" packages that contain minimum amounts of trace actinides.
8. Lower substrate resistivity by using heavily doped buried layers to truncate funneling effects.
9. Fractionate memory words by their physical dispersion over the memory array.
10. Use SRAMs where feasible, as they contain no floating bit lines; anything "floating" is anathema regarding improving SEU hardness.

At the circuit and systems level, a number of SEU-hardening methods are potentially useful. They include the following:

1. Judicious part selection at the circuit design level eases the burden of achieving a substantive measure of SEU hardness using the tabu-

lations in Fig. 8.12 and Tables 8.13–8.17. Attempt to choose parts whose $L_c \geq 40$ MeV cm^2/mg, which is the LET spectrum approximate upper cutoff point.

2. Use error detection and correction (EDAC) chips and/or part of each memory word for this purpose.
3. Lower the circuit/subsystem operating frequency, which is another "low-pass filter" simulation measure.
4. Maximize supply voltages by choosing appropriate parts and maintain a constant ambient temperature, as some SEU deleterious phenomena increase their effectiveness with changing temperatures.
5. Use *p*-channel power MOSFETs where possible instead of *n*-channel power FETs. If obliged to use *n*-channel power FETs, derate their V_{DD} and/or their maximum operating voltages.
6. Minimize impedances of bit line circuitry that connect directly to sense amplifiers.
7. Maintain "clean" board layout with respect to decoupling measures.
8. Employ circuit design that provides heavy I/O decoupling, again to slow SEU response.
9. Utilize relatively high currents in SEU-sensitive circuits in order to maintain low SEU transient current to operating current ratios.

At the system and mission level, there are a few rules of thumb to aid in mitigating SEU and its effects:

1. If possible, employ a low-altitude earth equatorial orbit to stay under the Van Allen belts and away from both magnetic poles where cosmic rays more easily leak through into the earth neighborhood.
2. For short missions of ≤ 5 years, attempt to launch during a sunspot maximum. This exploits the corresponding maximum solar wind–earth magnetic field coupling that diverts incident SEU-inducing particles.
3. Try to coexist with SEU-inducing environments by examining the overall system from the subsystem aspect, to delineate only those subsystems whose function is seriously impaired by single event phenomena. An SEU susceptible system may still function adequately even when enduring appreciable error rates.
4. If possible, construct systems using older (larger) part technologies if, fortuitously, there are no conflicting constraints. Older parts have larger SEU geometries mitigating against large SEU error rates.

5. Ensure use of SEU-hard memory buffering circuits immediately connecting to memory I/O, so as to preserve the memory SEU hardness.
6. Consider circumvention methods for solar flares and for the South Atlantic Anomaly.

Aside from the preceding SEU-hardening methodology, SEU hardening uses somewhat the same thought processes as hardening against other radiation environments. Again, a good share of the SEU-hardening burden rests with judicious part selection at the substantive level, as opposed to merely using simple "go–no go" approaches. Toward that end, Fig. 8.12 depicts a bar chart compilation of SEU data [20] representative of technology over the last one and a half decades. It depicts SEU error thresholds for a number of modern part technologies, divided into mainly MOSFETs and bipolar digital families. It shows that generally MOSFET devices are harder than bipolar devices. Further part data are given in Tables 8.13–8.17.

Vendor claims, both oral and written, of SEU hardness of devices they manufacture and/or sell should be received with caution. Their technical/business representatives are often woefully unaware of many important practical aspects of SEU part susceptibility. This is even more so than for other radiation environments, perhaps because SEU is the most recent damage mechanism to become part of the family of such environments. Also, because a number of vendors are "start up" companies within the past decade or so, pressure because of demanded short-term return on investment from their backers often enjoins sales representatives to produce short-term results (i.e., quick sales). Unfortunately, this engenders exaggerated radiation hardness claims for all environments including SEU. Virtually raw, but well corroborated, SEU data from established sources, as opposed to magazine-slick company advertisements touting hardness claims, must be sought to evaluate vendor parts. On the other hand, experiental evidence reveals that often vendors will strive to obtain radiation data on their parts if they are directed to that database but, usually short of any financial outlays for sources with which to obtain their own data on their product. In actuality, this is beginning to change for the better in recent years. However, certain vendors refuse to deal with these difficulties by now confining their business to commercial users with no radiation requirements almost exclusively, satisfying radiation-hard needs minimally or not, especially since the close of the cold war. Generally, vendors adjust their manufacturing policies accordingly with respect to (a) the relatively large but per-unit inexpensive commercial part market in contrast to (b) the much smaller radiation-hard market with very much more stringent radiation and other product control requirements, resulting in fewer but more expensive military components.

Besides part selection, certain SEU-hardening measures have been put forward. One such, still under contention, pertains to hardening flip-flops, latches,

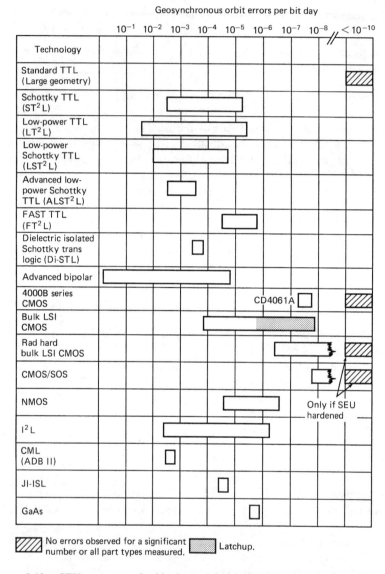

Figure 8.12. SEU error rates for bipolar and MOSFET digital integrated-circuit families. From. Ref. 20.

and other like two-state or multistate integrated circuits. Mentioned previously, it is that of inserting resistors or diodes or diode-like devices monolithically into the feedback loops of, for example, memory cells, to slow their response, thus making them insensitive to the picosecond time scale of the SEU-inducing cosmic ray heavy-ion track, and corresponding charge collection ramifications. There a number of reasons why this method can fall short, even though it has been put into production. The first is that slowing the device response to act as a low-pass filter is the antithesis of the modern direction of component manufacture and operation, which is to make such devices achieve higher speeds to handle the anticipated massive throughput data loads of the present and near future. The second is that some military requirements include very high neutron fluence levels. Neutrons cause displacement damage, resulting in resistivity increases in high-resistivity silicon and gallium arsenide. For high neutron fluences, this resistivity increase will increase the value of the feedback resistors to place their ultimate values outside the useful boundaries for proper time constants in the trade-off between SEU hardness and the dimunition in the operating speed of the device. The third is a more practical one of manufacturing parts with these resistances emplaced monolithically within the integrated circuit cell that results in a maintenance of the stability of their value in the hardness assurance sense. The fourth is that the SEU-hardness, speed-decrease trade-off is not yet robust as evidenced by investigating various feedback resistance configurations [3].

8.7.1. How to Use Part Index Tables 8.13 Through 8.16

Tables 8.13–8.17 comprise indices for SEU tests of devices that appear in the *IEEE Transactions of Nuclear Science* for 1985–1993 of more than 600 parts of about 50 part types. The majority of tests used the synchrocyclotrons of the University of California, Berkeley as the heavy-ion source. The tests were conducted by a consortium of personnel from the Jet Propulsion Labs (JPL), Pasadena, CA, and the Aerospace Corp., El Segundo, CA under the contractual auspices of the U.S. Air Force.

Table 8.13 provides part counts for about 30 part types with respect to their reference sources in the IEEE Trans. Nucl. Sci., NS – XX (6), Dec. 1985-1991. Table 8.14 is a similar table, but for parts whose individual part count is unity. From 1993 on, these compilations appear in the *NSRE Conference Workshop Record*. Its part index appears in Tables 8.15–8.17. Each Workshop Record is a collection of articles that mainly provide results of part tests and their vagaries. It appears concurrently with the annual December issues of the IEEE Transactions. Its articles are presented simultaneously with those at the IEEE NSRE annual summer conferences.

More recent SEU data on parts that do not appear in the tables can be obtained from the NASA/Jet Propulsion Labs. "RADATA" database. Their world wide web URL is http://www.radnet.jpl.nasa.gov.

Detailed part SEU results, such as their critical LET, SEU cross section, SEU-induced latchup propensities, and so forth can be found in the above references by first finding the part type in the above tables. From the

Table 8.13. Index to JPL Part Data From IEEE Trans. Nucl. Sci., Dec. 1985, 1987, 1989, 1991

Ser.	Page	ADC	Analog SW	Array (gate)	Array (logic)	Bit slice	Buffer	Countr	DAC	Decod/encod	DRAM	DSP	EAROM	F/F-D	F/F-JK	FIFO reg
NS-38 (6) Dec. (1991)																
1	1532															
2	1533											3				
3	1534										3					
4	1535	1		9												
5	1536	4				1			1					2		
6	1537											9				
7	1538	3														4
8	1539															
NS-36 (6) Dec. (1989)																
9	2391															
10	2392															
11	2393		5				2							1		3
12	2394															
13	2395															
14	2396	1		6	2						1					
15	2397	2							2					3	2	

Table 8.13. continued.

Ser.	Page	ADC	Analog SW	Array (gate)	Array (logic)	Bit slice	Buffer	Countr	DAC	Decod/ encod	DRAM	DSP	EAROM	F/F-D	F/F-JK	FIFO reg
NS-34 (6) Dec. (1987)																
16	1334			1							4					
17	1335															
18	1336	2						2	2					5	2	
19	1337							2		2	1				3	
NS-32 (6) Dec. (1985)																
20	4190					2		4						4	4	
21	4191										1					
22	4192					7		2					7		3	
23	4193							7						13	7	
	Totals	13	5	16	2	10	2	17	5	2	10	12	7	28	21	7

Table 8.13 continued.

Ser.	Page	Gate	Latch	Line drvr	Logic	Micro proc.	Multi-plier	MUX	PROM	PROM-EE	PROM-UVE	RAM	Regist/data	Regist/shift	SRAM	VP int cont
NS-38 (6) Dec. (1991)															
1	1532					12										
2	1533					3									25	
3	1534								2	6	2				12	
4	1535				13											
5	1536		2			4			2						3	
6	1537						2									
7	1538					10				3	1				10	
8	1539				26											
NS-36 (6) Dec. (1989)															
9	2391					15										
10	2392				38											
11	2393		5		5			3								
12	2394														20	
13	2395														20	
14	2396								3	1						
15	2397		2			2			1			3				

Table 8.13 continued.

Ser.	Page	Gate	Latch	Line drvr	Logic	Micro proc.	Multip-lier	MUX	PROM	PROM-EE	PROM-UVE	RAM	Regist/data	Regist/shift	SRAM	VP int cont
NS-34 (6) Dec. (1987)																
16	1334								3			8				
17	1335					7	1									
18	1336					2				1		1				
19	1337	4	5	2								11		5		
NS-32 (6) Dec. (1985)																
20	4190		6			14						4	2	4		
21	4191									1		50				
22	4192		4			8							2	5		2
23	4193	1							1			13	2	5		
	Totals	5	24	2	82	77	3	3	12	12	3	90	6	19	90	2

Table 8.14. Index to JPL SEU Part Data for Single Part Count From IEEE Trans. Nucl. Sci., 1985–1991

Ser.	Page	ASIC	Bus cntl	Bus Xcvr	Clk cntl	Coder	Comp artr	DMA cntl	EA PROM	EE ROM	FET Drvr	FFT	Inv rttr	Manc encr	OP amp	PAL	Par chkr	Rcvr	Rcvr ROM	Tim cntl	Xcvr
NS-38 (6) Dec. (1991)																					
4	1535														120 104	EP 910					PCT 245
5	1536																		2910		
6	1537	ES 1005	DDC 1553																		
8	1539					SOR 5053					TSC 430	TMC 2310						MC10 115F			
NS-36 (6) Dec. (1989)																					
9	2391			SA 3297				82C3 7ARH													
14	2396				CO 422									H515 530							

Table 8.14. continued

Ser.	Page	ASIC	Bus cntl	Bus Xcvr	Clk cntl	Coder	Comp artr	DMA cntl	EA PROM	EE ROM	FET Drvr	FFT	Inv rtr	Manc encr	OP amp	PAL	Par chkr	Rcvr	ROM	Tim cntl	Xcvr
NS-34 (6) Dec. (1987)																					
16	1334								2864												
17	1335																			MS32 201	
19	1337						54HC T688						54HC T04				54HC 280				
NS-32 (6) Dec. (1985)																					
21	4191									X281 6AM											

Table 8.15. Index to JPL et al. SEU Part Data from 1993 IEEE Workshop Record

Ser.	Page	ADC	Analog SW	Array (gate)	ASIC	Bus contr	Co-Proc.	Counter	DAC	DRAM	DSP	EDAC	F/F-D	FPGA
2	7	17					7		4		4			
3	8													
4	9				2	2				12				
5	10			2	1			2				4	3	5
6	11	9				1	5				10			
7	12	3	3			1		1	3					1
	Totals	29	3	2	3	4	12	3	7	12	14	4	3	6

Ser.	Page	FIFO rep.	Latch	Line drvr	Logic	Micro-proc.	PROM	PROM-EE	PAL	RAM	Rcvr	SRAM	Timer
1	6					31							
3	8	3						7		2		33	
4	9						4		7			2	2
5	10		2										
6	11					12							
7	12	4		2	2		1	3			2	11	
	Totals	7	2	2	2	43	5	10	7	2	2	46	2

Table 8.17. Index to JPL et al. SEU Part Data from 1995 IEEE Radiation Effects Data Workshop Record

Ser.	Page	ADC	Analog switch	Array (gate)	Comprtr	Cntrlr	DAC	Decod/ encodr	DRAM	DSP	EAROM	Filter (lopass)	FPGA	FIFO reg
5	12					9								
6	12									10				
8	12													5
10	13								13					
10	14								4					
14	14												10	
14	14			8										
15	14										4			
16-17	14								3					
18	15				6									
18	15		3											
19	15						8							
19	15	3												
19	16	19												
21	16					5								
21	16											2		
21	16							1						
21	16						2							
Totals		22	3	8	6	14	10	4	17	10	4	2	10	5

319

Table 8.17 continued.

Ser.	Page	Latch	Line DVR/RCR	Micr proc.	Op AMP	PLA	PROM	PROM (EE)	PROM (UVE)	Pwr conv	REG (data)	REG (shft)	SRAM	SCR	Volt reg.	XMTR/RCVR
2-4	12			18												
7	12															
9	12												6	5			
9	13													50			
12	14						2									
12	14								1							
12	14							18								
15	14					6										
18	14															4
18	15															12
18	15														5	
18	15		10													
18	15				16											
18	15													2		
20	16									9						
21	16										3					
21	16	1														
21	16															4
Totals		1	10	18	16	6	2	18	1	9	3	6	55	2	5	20

320

Table 8.16. Continuation of SEU Single Part Count Data. (*From:* 1993 IEEE Workshop Record)

Ser.	Page				
1	6	UART 82510			
2	7	DMA 82380			
5	10	JK F/F 54LS112			
6	11	Xputer T800			
7	12	GA 5000	MUX IH6208	SAR 25HCT04	Xcvr LTC485

corresponding reference page number, the detailed information on that part can be obtained by consulting the actual reference in that IEEE Transactions issue.

The glossary below is provided as an aid to interpret the part function to find it readily in the references. Some of the part type names in the glossary do not always coincide with the conventional descriptors. The glossary references, though not exhaustive, consist of the majority of devices tested worldwide during the aforementioned era. Other SEU testing is ongoing at the above, as well at other accelerator sites. References for these results are given in most annual December issues of the *IEEE Transactions for Nuclear Science* as well as the corresponding Workshop Records.

Glossary for Tables 8.13–8.17

1. ADC—analog-to-digital converter
2. Analog SW—analog switch
3. Array (gate)—also field programmable (FPGA)
4. Array (logic)—logic array
5. Bit slice—4-bit slice
6. Buff—octal buffer
7. Countr—counter
8. DAC—digital-to-analog converter
9. Decod—decoder
10. DRAM—dynamic random access memory
11. DSP—digital signal processor
12. EAROM—electrically addressable ROM
13. F/F-D—flip-flop D type
14. F/F-JK—flip-flop JK type
15. FIFO—first-in first-out register
16. Gate—gate
17. Latch—latch
18. Line drvr—line driver
19. Logic logic
20. Microproc—microprocessor
21. Mult.—multiplier
22. MUX—multiplexer
23. PROM—programmable ROM (fusible)
24. PROM-EE—electrically erasable PROM
25. PROM-UVE—ultraviolet-light erasable PROM
26. RAM—random access memory
27. Register, data—data register
28. Register, shift—shift register
29. SRAM—static RAM
30. VP Int Cont.—voltage/power integral controller
31. ASIC—application-specific microcircuit
32. Bus Cntl—bus controller
33. Bus Xcvr—bus transceiver
34. Clk cntl—clock controller
35. Coder—coder
36. Comprtr—comparator
37. DMA Cntl—direct memory access controller
38. EA PROM—electrically addressable PROM
39. EE ROM—electrically erasable ROM
40. FET drvr—FET transistor driver
41. FFT—fast fourier transform chip
42. Invrtr—inverter
43. Manc Encr—manchester encoder
44. Op Amp—operational amplifier
45. PAL-PLA—programmable logic array
46. Par. Chkr—parity checker
47. Rcvr—receiver
48. ROM—read-only memory
49. Tim Cntl—timing controller
50. Xcvr—transceiver
51. Xputer—transputer

Appendix 8.1. Manufacturer Abbreviations (Ref. 21)

ACT	Actel Corp.	LSI	LSI Logic Corp.
ADA	Advanced Analog	MCN	Micron Technologies
ADI	Analog Devices Inc.	MDI	Modular Devices, Inc.
AFX	Aeroflex Labs	MED	Marconi Electronic Devices
ALS	Allied Signal	MIC	Micrel Semiconductors
ALT	Altera Corp.	MIR	Micro-Rel Corp.
AMD	Advanced Microdevices Corp.	MIT	Mitsubishi
APX	Apex	MMI	Monolithic Memories Inc.
ATM	Atmel	MOC	Mosaic Semiconductor
ATT	American Tel & Tel	MOT	Motorola Semiconductor Products Inc.
BAL	Bal Efratom		
BUB	Burr-Brown Research	MPS	Micro Power System
CPT	Crosspoint	MTA	Matra Harris Semiconductor
CRY	Crystal Semiconductor Inc.	MXM	MAXIM
CYP	Cypress Corp.	NCR	National Cash Register
DAT	Datel	NEC	Nippon Electric Corp.
DDC	Data Device Corp.	NSC	National Semiconductor Corp.
DEC	Digital Equipment Corp.		
EDI	EDI Corp.	NTL	Natel Engineering
ELN	Elantec	OKI	Oki Semiconductor, Inc.
FAS	Fairchild Semiconductor Div.	OWI	Omni-Wave, Inc.
FER	Ferranti	PFS	Performance Semiconductor Inc.
FUJ	Fujitsu Ltd.		
GAZ	Gazelle	PLS	Plessey Semiconductors
GEC	General Electric	PMI	Precision Monolithics, Inc.
HAR	Harris Corp., Semiconductor Div.	RAY	Raytheon Co., Semiconductor Division
HTC	Hitachi Ltd.	RCA	Radio Corporation of America
HON	Honeywell Inc.		
HPA	Hewlett-Packard Corp.	RMT	Ramtron
HYB	Hybrid Systems	SAM	Samsung
IBM	IBM	SEI	Seiko
IDT	Integrated Device Technologies, Inc.	SEQ	SEEQ Technology Inc.
		SGN	Signetics Corp.
INM	INMOS Corporation	SGP	Signal Processing
INT	Intel Corp.	SIE	Siemens Components, Inc.
ISM	INMOS Corp.	SIL	Siliconix
ITP	Interpoint	SIP	Sipex
LDI	Logic Devices Inc.	SLG	Silicon General
LTN	Linear Technology	SNL	Sandia National Laboratories

SNY   Sony Corp.
SOR   Soreq NRC, Israel
SSD   Solid State Devices
STC   Silicon Transistor Corp.
TEL   Teledyne Crystalonics
TIX   Texas Instruments Inc.
TMS   Thomson Military & Space, France
TOS   Toshiba
TRW   TRW Inc.

UTM   United Technologies Microelectronics Center
WAF   Waferscale
WDC   Western Digital Corp.
WEO   Westinghouse Electric Corp.
XIC   Xicor Inc.
XIL   Xilinx Corp.
ZOR   Zoran
ZYR   Zyrel

Appendix 8.2.   Test Organizations (Ref. 21)

A   The Aerospace Corporation; El Segundo, CA
ADI   Analog Devices Semiconductor, Wilmington, MA
ALC   Alcatel Espace, Toulouse, France
Ball   Ball Aerospace Systems Division, Boulder, CO
BDS   Boeing Defense & Space Group, Seattle
BPS   Boeing Physical Sciences Research Center, Seattle
CERT   2, Avenue Edouard Belin, Toulouse, France
CLM   Clemson University; Clemson, SC
CNES   Centre National d'Etudes Spatiales; Toulouse, France
DASA   Deutsche Aerospace AG, Munich, Germany
ESA   European Space Agency—several facilities
GD   General Dynamics
GDD   NASA Goddard Space Flight Center; Greenbelt, MD
GE   GETSCO, Philadelphia
HAR   Harris Semiconductor

HAC   Hughes Aircraft
HON   Honeywell
IBM   IBM
J   Jet Propulsion Laboratory (JPL); Pasadena, CA
JH   John Hopkins Applied Physics Laboratory; Laurel, MD
LIN   Lincoln Laboratories, M.I.T.; Cambridge, MA
Loral   Loral, Manassas, VA
LLNL   Lawrence Livermore National Laboratory; Livermore, CA
MM   Martin Marietta Astrospace, Valley Forge, PA
MMS   Matra Marconi Space; Vélizy, France
MOT   Motorola GSIG
NASA   National Aeronautics & Space Administration
NRL   Naval Research Laboratories, Washington D.C.
NWSC   Naval Weapon Support Center, Crane, IN
PHY   Physitron, Inc., San Diego
R   Rockwell International, Anaheim, CA

SNL	Sandia National Laboratory, Albuquerque	TRW	TRW Space and Defense Sector; Los Angeles
Soreq	Soreq Nuclear Res. Labs., Israel	UTM	United Technologies Microelectronics Center, Colorado Springs, CO
SSS	S-Cubed, San Diego, CA		

Appendix 8.3. Test Facilities (Ref. 21)

88-in.	88-in. cyclotron, Lawrence Berkeley Laboratory, Berkeley, CA	UW	Tandem Van de Graaff, University of Washington, Seattle
BNL	Tandem Van de Graaff, Brookhaven National Laboratory, Long Island, NY		
Cf-252	A Cf-252 fission source. The data from this type of source are rarely given because of inaccuracies inherent in the limited range of the fission ions.		
ESA	European Space Agency—several sites		
GANIL	Cyclotron for High Energy Heavy Ions. Caen, France		
GSI	Cyclotron for High Energy Heavy Ions, Darmstadt, Germany		
HAR	Van de Graaff, Harwell England.		
IPN	Tandem Van de Graaff, Institut de Physique Nucleaire, Orsay, France		

RADATA (Ref. 18) may be accessed *via WWW* using URL: *http://radnet.jpl.nasa.gov*. or *via MODEM* at BBS# (818) 393-1725 up to 14.4K baud—8N1 or *via INTERNET* with FTP: *radnet.jpl.nasa.gov* (137.79.152.10), USERID: anonymous, PASSWORD: your-email-address. RADATA BBS also supports Remote Imaging Protocol (RIP) and is available for users who use a communication software that supports RIP.

For support, call Sam Farmanesh:

Voice: (818) 354-1968

FAX: (818) 393-4559

e-mail: farmanesh @ jpl.nasa.gov

Problems

8.1 Destructive physical analysis (DPA) of a part of interest from a geosynchronous spacecraft avionics reveals a cell size of $10 \times 10 \times 2 \,\mu m^3$. Its critical charge is obtained by a SPICE circuit simulation and turns out to be 10 fC. Using the "figure of merit" SEU error rate expression, compute its corresponding value in errors per bit-day.

8.2 Suppose that in Problem 8.1, the part cross section is obtained experimentally as 10^{-6} cm^2 and its critical LET is $L_c = 0.5$ MeV cm^2

mg^{-1}. What is the error rate in errors per bit-day using the "figure of merit" formula?

8.3 For σ_L in cm^2 and LET in $MeV\,cm^2\,mg^{-1}$, how should the coefficient in the above SEU error rate expression be modified?

8.4 For a device in a spacecraft at 500 miles altitude and 51.6° inclination angle orbit, Table 8.3 yields a coefficient $C \cong 28$ for a 93L422 RAM to be used in the generalized "figure of merit" SEU error rate expression ($E_r = C\sigma_L/L_c^2$). Table 8.1A was computed for the notoriously high SEU error rate device, the 93L422 RAM. The cross section for the device of interest here is $\sigma_L = 5 \times 10^{-3}\,cm^2$, whereas for the 93L422, it is assumed here as $0.1\,cm^2$. How should the coefficient C obtained from Table 8.1A be changed for the part of interest?

8.5 In their eagerness, electronic system designers find that the computed SEU error rates are apprehensively high for certain voltage regulators, pulse width modulators, comparators, and GaAs FET RF amplifiers in their system. How should the designers handle this problem?

8.6 During a solar minimum, a spacecraft orbit (800 km, 35° inclination angle) whose major exposure to SEU-inducing flux occurs during its travel in the inner Van Allen belts (mainly protons) contains SRAMs whose characterictics are close to the Harris CMOS 6500 series family. What is their anticipated SEU error rate due to these protons?

8.7 In problem 8.6, the typical LET for these parts is $L_{0.1} = 11\,MeV\,cm^2\,mg^{-1}$. $L_{0.1}$ is that corresponding to the σ_L value that is 10% of its asymptotic value from an experimental σ_L–LET curve. What is the corresponding E_{pr}?

8.8 Using Table 8.4, and that the Weapons Neutron Research (WNR) facility flux is 2×10^5 times the daily average neutron flux at 40,000 ft., show with (8.18) that the corresponding WNR device exposure time T_{exp} is about 4 h.

8.9 In Problem 8.9, what is the mean WNR flux if the neutron flux at 40,000 ft is 1.6 neutrons $cm^{-2}\,s^{-1}$?

8.10 Figure 8.8 depicts proton flux spectra in the Van Allen belts at an altitude of 600 NM shining through various thicknesses of aluminum. Why do the spectra levels fall and their peaks shift to higher energies with increasing material thickness?

References

1. E.L. Petersen, J.B. Longworthy, and S.E. Diehl, "Suggested Single Event Upset Figure of Merit," *IEEE Trans. Nucl. Sci.*, **NS-30** (6), 4533–4539 (1983).

2. P.J. McNulty, "Predicting Single Event Phenomena in Natural Space Environments," IEEE NSRE Conference Short Course, Chap. 3, 1990.
3. G.C. Messenger and M.S. Ash, *The Effects of Radiation on Electronic Systems*, 2nd ed., Van Nostrand Reinhold, New York, 1992.
4. M.S. Ash, TRW interoffice correspondence, 10 Dec. 1985.
5. RADSPICE owned be SAIC San Diego CA and META Software, Sunnyvale CA.
6. E.L. Petersen, "SEE Rate Calculations Using the Effective Flux Approach and a Generalized Figure of Merit Approximation," *IEEE Trans. Nucl. Sci.*, **NS-42** (6), 1995-2003 (1995).
7. E.G. Mullen, M. Gussenhoven, K. Ray, and M. Violet, "A Double Peaked Inner Radiation Belt: Cause and Effect as Seen on CRRESS," *IEEE Trans. Nucl. Sci.*, **NS-38** (6), 1713-1717 (1991).
8. M.S. Gussenhoven, E.G. Mullen, M. Sperry, and K.J. Kerns, "The Effect of the March 1991 Storm on Accumulated Dose For Selected Satellite Orbits: CRRESS Dose Models," *IEEE Trans. Nucl. Sci.*, **NS-39** (6), 1765-1772 (1992)
9. E.L. Petersen, "Soft Errors Due to Protons in the Radiation Belt," *IEEE Trans. Nucl. Sci.*, **NS-28** (6), 3981-3986 (1981).
10. R.A. Reed, P.J. McNulty, and W.G. Abdel-Kader, "Implications of Angle of Incidence in SEU Testing of Modern Circuits," *IEEE Trans. Nucl. Sci.*, **NS-41** (6), 2049-2054 (1994).
11. J. Levinson, J. Barak, A. Zentner, A. Akkerman, and Y. Lifshitz, "On the Angular Dependence of Proton Induced Events and Charge Collection," *IEEE Trans. Nucl. Sci.*, **NS-41** (6), 2098-2102 (1994).
12. "SEE-Quest" Data Base, Exciton Science Corp., Aliso Viejo, CA.
13. Y. Shimano, T. Goka, S. Kuboyama, K. Kawachi, T. Kanai, and Y. Takami, "The Measurement and Prediction of Proton Upsets," *IEEE Trans. Nucl. Sci.*, **NS-36** (6), 2344-2348 (1989).
14. C.A. Gossett, B.W. Hughlock, M. Katoozi, G.S. La Rue, and S.A. Wender, "Single Event Phenomena in Atmospheric Neutron Environments," *IEEE Trans. Nucl. Sci.*, **NS-40** (6), 1845-1852 (1993).
15. E. Normand, J.L. Wert, W.R. Doherty, D.L. Oberg, P.R. Measel, and T.L. Criswell, "Use of Pu–Be Source to Simulate Neutron-Induced Single Event Upsets in Static RAMs," *IEEE Trans. Nucl. Sci.*, **NS-35** (6), 1523-1529 (1988).
16. A.H. Johnston and B.W. Hughlock, "Latchup in CMOS from Single Particles," *IEEE Trans. Nucl. Sci.*, **NS-37** (6), 1886-1893 (1990).
17. E. Normand, J.L. Wert, P.P. Majewski, D.L. Oberg, W.G. Bertholet, S.K. Davis, M. Shoga, S.A. Wender, and A. Gavron, "Single Event Upset and Latchup Measurements in Avionics Devices Using the WNR Neutron Beam and a New Neutron-Induced Latchup Model," 1995 IEEE Radiation Effects Data Workshop Record, 33-38.
18. J.H. Hohl and G.H. Johnson, "Feature of the Triggering Mechanism for Single Event Burnout of Power MOSFETs," *IEEE Trans. Nucl. Sci.*, **NS-36** (6), 2260-2266 (1989).

19. J.L. Titus, G.H. Johnson, R.D. Schrimpf, and K.F. Galloway, "Single Event Burnout of Power Bipolar Junction Transistors," *IEEE Trans. Nucl. Sci.*, **NS-38** (6), 1315–1327 (1991).
20. M.A. Rose, *Updated Bar Chart, Device SEU Thresholds*, Physitron Corp., San Diego, 1990.
21. D.K. Nichols, J.R. Coss, K.P. McCarty, H.R. Schwartz, L.S. Smith, G.M. Swift, R.K. Watson, R. Koga, W.R. Crain, K.B. Crawford, and S.J. Hansel, "Overview of Device Susceptibility from Heavy Ions," IEEE 1993 Radiation Effects Workshop Record, 1993 NSRE Conf., p. 4.
22. ASTM F1192-90, "Standard Guide for the Measurement of Single Event Phenomena (SEP) Induced by Heavy Ion Irradiation of Semiconductor Devices," Section 7.1.2, April 1990.

Appendix A: Answers to Problems

Chapter 1

1.1 (a) Flux is the *total* number of particles threading a sphere of unit cross sectional area. In this case, the total is $10^5 + 0.5 \times 10^5 = 1.5 \times 10^5$ particles cm^{-2} s^{-1}. (b) The current is the *net* number of particles threading the same sphere. Thus, the current is $10^5 - 0.5 \times 10^5 = 0.5 \times 10^5$ particles cm^{-2} s^{-1}. (c) The fluence is the time integral of the flux. In this case, the flux is constant, the fluence is 1.5×10^5 multiplied by the number of seconds in 5 years, or $(1.5 \times 10^5)(1.58 \times 10^8) = 2.4 \times 10^{13}$ particles cm^{-2}.

1.2 Let dA be an incremental area on a spherical surface of radius r surrounding the source. Then $S_0 = \int \Phi \, dA = \int \Phi \, r^2 \, d\Omega = \Phi r^2 \int_0^{2\pi} d\phi \int_0^\pi \sin\theta \, d\theta = 4\pi r^2 \Phi$, so that $\Phi = S_0/4\pi r^2$.

1.3 The particle survival probability per lamina is $1 - p = 1 - \Delta x/\lambda$. For N lamina, it is $(1 - \Delta x/\lambda)^N = \left[1 - (x/\lambda)\left(\dfrac{\Delta x}{x}\right)\right]^{\frac{x}{\Delta x}}$. Then the limit as $u = x/\Delta x$ and N approach infinity is $\lim_{u \to \infty} [1 - (x/\lambda) u^{-1}]^u = \exp(-x/\lambda)$.

1.4 (a) $\lambda = (N\sigma)^{-1}$, $N_{si} = \rho_{si} N_0/A_{si} = (2.33)(6 \times 10^{23})/28 = 5 \times 10^{22}$ atoms cm^{-3}. Then $\lambda = [5 \times 10^{22} (10)(10^{-24})]^{-1} = 2$ cm. (b) $A_{air} = 0.2 A_{O_2} + 0.8 A_{N_2} = (0.2)(32) + (0.8)(28) \cong 29$. Thus, $N_{air} = (1.24 \times 10^{-3})(6 \times 10^{23})/29 = 2.57 \times 10^{19}$ atoms cm^{-3}. Then $\lambda = [2.57 \times 10^{19} (10)(10^{-24})]^{-1} = 3891$ cm. N_0 is Avogadro's number.

1.5 A *cascade* of 1000 Si recoil atoms per 10^{-16} cm^3 per incident neutron implies that about 10^{16} neutrons is needed per cubic centimeter of silicon. Hence, the required fluence $\Psi_n = (10^{16}$ neutrons cm$^{-3})/(2.33$ g

$cm^{-3}) = 4.3 \times 10^{15}$ neutrons g^{-1}. It is seen that the total number of silicon secondaries produced is $4.3(10^{15} \times 10^3) = 4.3 \times 10^{18}$ per gram of Si.

1.6 Rewrite equation in Section 1.3 as $I'/I = -\sigma N(x)$ and integrate to get $I = I_0 \exp[-\sigma \int_0^x N(x)\,dx] = I_0 \exp\{-(\rho_0 \sigma N_0/A) \int_0^x \rho(x)\,dx/\rho_0\}$ to obtain $I = I_0 \exp\{-[\int_0^x \rho(x)\,dx/\rho_0 \lambda_0]\}$, where $\lambda_0 = (\rho_0 N_0 \sigma/A)^{-1}$.

1.7 The meteorological rule of thumb implies $\rho(R+10) = 0.1\rho(R)$, where R is the altitude in miles. Trying $\rho = \rho_0 \exp(-mr)$ in this difference equation yields $m = 1.43 \times 10^{-6}$ cm^{-1}, where r is in cm, and ρ_0 is in g L^{-1} (1.1 g L^{-1} at sea level). So that $\exp[-\int_0^{R_0} \rho(r)\,dr/(2.38 \times 10^4)] = \exp - [\rho_0/(2.38 \times 10^4 m)][1 - \exp(-mR_0)]$ $= \exp(-32.27) = 9.63 \times 10^{-15}$. Then $\Phi_n = (5 \times 10^{22})(200)/[(1.5 \times 10^5) 30.5]^2 (9.63 \times 10^{-15}) = 4.6 \times 10^{-3}$ neutrons $cm^{-2} s^{-1}$ at ground level. This is in good agreement with row 7 of Table 5.8.

1.8 (a) Using (1.42), $\bar{\ell} = (1/\bar{R} + 1/\bar{r})^{-1} = (1/10\bar{r} + 1/\bar{r})^{-1} = (10/11)\bar{r} = 0.91\bar{r}$. (b) For a sphere, $\bar{r} = 4V/S = 4r/3$, so that $\bar{\ell} = (10/11)(4r/3) = (40/33)r$. (c) $\bar{\ell} = 0.91\bar{r}$ implies a reduction of less than 10% in the number of incomplete chords, or that more than 90% of tracks in the sensitive region are chords. To this accuracy, all tracks can be considered as chords. Insofar as this result is concerned, linear energy transfer (LET) is a lone valid parameter by which deposited energy in the sensitive volume can be computed; or, because the mean proton range $\bar{R} \gg \bar{r}$, then any change in the particle LET during penetration in the material can be neglected, as discussed in Section 3.9. As will be seen, all the expressions involving LET in the computation of SEU imply that all ion tracks penetrating the sensitive volume are chords. Other criteria for the suitability of employing LET are derived in Section 3.9.

1.9 In two dimensions, Φ is defined as the mean number of traversals in a circle of unit diameter. Thus, for a figure of mean diameter $\bar{d} = C/\pi$, the number of traversals is $\Phi C/\pi$. Then $\bar{r}_2 =$ (total track length: $S_2 \Phi$)/(number of chords penetrating the figure: $\Phi C/\pi$) $= S_2 \pi/C$.

1.10 (a) Common sense asserts that the probability of any angle of incidence θ on the surface element is independent of the orientation of the latter. Then, for an isotropic flux of particles, a surface element on any convex body will do, such as a sphere as in the leftmost part of Fig. 1.7. The disc subtending a sector of that sphere, of area $\pi(r \sin \theta)^2$, represents a cross section for flux particles whose angles of incidence lie between 0 and θ (i.e., $0 \le \theta_1 \le \theta$). The total cross section of the sphere is πr^2. A corresponding cumulative probability distribution is $P(\theta \le \theta_2 \le \pi/2)$ or, equivalently, the fraction of flux particles whose angles of incidence θ_2 are greater than θ (i.e., $\theta \le \theta_2 \le \pi/2$). So that intuitively, $P(\theta \le \theta_2 \le \pi/2) \equiv \int_\theta^{\pi/2} p(\theta') d\theta' = [\pi r^2 - \pi(r\sin\theta)^2]/\pi r^2 = \cos^2\theta$. Differentiating P, as represented by the integral, with respect to θ gives the probability density as $|p(\theta) d\theta| = 2\cos\theta\sin\theta d\theta$. (b) A chord length ℓ in the one-dimensional infinite slab of thickness c is given by $\ell = c/\cos\theta$, where θ is the angle of incidence as in the rightmost part of Fig. 1.7. Because the orientation is arbitrary, as discussed in (a), substituting for $\cos\theta = c/\ell$ in P in (a) gives the corresponding cumulative probability distribution for the chord length ℓ or $P(\ell \le \infty) = c^2/\ell^2 = \int_\ell^\infty p(\ell') d\ell'$. Differentiating P with ressect to ℓ yields the probability density distribution $|p(\ell)| = 2c^2/\ell^3$. (c) Using $p(\ell) d\ell$ from (b), the mean chord length $\bar\ell$ in the slab is given by $\bar\ell = [\int_c^\infty \ell p(\ell) d\ell]/[\int_c^\infty p(\ell) d\ell] = 2c$. The serious reader will verify this integration. (d) The mean cosine of the incident angle in (a) is given by $\overline{\cos\theta} = [\int_0^{\pi/2} \cos\theta p(\theta) d\theta]/[\int_0^{\pi/2} p(\theta) d\theta] = 2/3$. Needs serious reader verification. (e) The mean angle of incidence $\bar\theta = [\int_0^{\pi/2} \theta p(\theta) d\theta]/[\int_0^{\pi/2} p(\theta) d\theta] = \pi/4$. (f) The value of the cosine of the mean angle $\bar\theta$ in (e) is $\cos\bar\theta = 1/\sqrt{2}$. Note that the cosine of the mean angle of incidence in (f), is not equal to the mean cosine of the angle of incidence in (d).

1.11 (a) Using the property of the delta function that $\int g(x) \delta(x - x_0) dx = g(x_0)$, write $f(r_s) = [\int d\Omega \int \cos\theta dA \, \delta(r - r_s)]/[\int d\Omega \int \cos\theta dA]$; thus, $\bar r = \int rf(r) dr/\pi S = \int dr \, r \int d\Omega \int \cos\theta dA \, \delta(r - r_s)/\pi S$; or $\bar r = (1/\pi S)\int dr \, \delta(r - r_s) \int d\Omega \int r \cos\theta dA = (4/S) \int dr \, \delta(r - r_s) V(r_s) = 4V/S$. (b) $\int f(r) dr = \int dr \int d\Omega \int \cos\theta dA \, \delta(r - r_s)/\pi S$. As $\int dr \, \delta(r - r_s) = 1$, $\int f(r) dr = \int d\Omega \int \cos\theta dA/\pi S = \pi S/\pi S = 1$.

1.12 The total track length in the body is ΦV, whereas the mean number of chords threading it is $\Phi S/4$, so $\bar s = \Phi V/(\Phi S/4) = 4V/S$.

Chapter 2

2.1 The ion under the influence of the magnetic field in an orbit as described above satisfies $Mv^2/r = H(Ze) v$, or Hr (rigidity) $= Mv/Ze = (M/Ze)(2E_{max}/M)^{1/2}$ where $E_{max} = \frac{1}{2}Mv^2$ is the ion kinetic energy. Thus $E_{max} = (HrZe)^2/2M$. For neon whose $Z = 10$, its $A = 20$, so that its mass is $1.66 \times 10^{-24} A \cdot g = 3.32 \times 10^{-26}$ kg. For (a), a fully ionized ion, $Z = 10$, then $E_{max}^{10} = [10^{-7}(1.6 \times 10^{12}) \cdot (10$

$(1.6 \times 10^{-19})]^2/(2)(3.32 \times 10^{-26}) = 0.988$ J $= 0.988(6.25 \times 10^{12}) = 6.17 \times 10^{12}$ MeV. **(b)** For a singly ionized ion "$Z = 1$", $E_{max}^1 = 10^{-2} E_{max}^{10} = 6.2 \times 10^{10}$ MeV. 1 J $= 6.25 \times 10^{12}$ MeV.

2.2 $\Omega/4\pi = \frac{1}{2}[1 - (1 + 2R_e/h)^{1/2}/(1 + R_e/h)]$. **(a)** $2\pi/4\pi = \frac{1}{2} = \frac{1}{2}[1 - (1 + 2R_e/h)^{1/2}/(1 + R_e/h)]$, giving $h = 0$. Hence, at launch, the earth subtends half the "sky." **(b)** $0 = \frac{1}{2}[1 - (1 + 2R_e/h)^{1/2}/(1 + R_e/h)]$, expanding to give $h = \infty$; thus, the earth still subtends some portion of the heavens from the spacecraft except at "infinite" altitude. **(c)** $\Omega/4\pi = \frac{1}{2}[1 - (1 + (2)(4 \times 10^3)/2.5 \times 10^5)^{1/2}/(1 + (4 \times 10^3/2.5 \times 10^5)] = 6.2 \times 10^{-5}$. For small solid angles, $\Omega_{EM} \cong \theta_{EM}^2$, so $\theta_{EM} = ((4\pi)(6.25 \times 10^{-5}))^{1/2} = 2.79 \times 10^{-2}$ radians, or 1.6°. **(d)** $\Omega_{ME} \cong \theta_{ME}^2 = (2\pi)(3.15 \times 10^{-5})$, so that $\theta_{ME} = 1.41 \times 10^{-2}$ radians or 0.81°.

2.3 Muons produced in the upper atmosphere by cosmic rays are traveling at very high speed so that those traveling toward the earth, with respect to a ground observer, enjoy a time dilation due to relativistic effects. Time slows down for them by a factor of about 15, giving them ample time to reach the surface of the earth. This is as compared to their natural decay time of 1 μs in this relativistically reduced track length through the atmosphere. This is an example of special relativity.

2.4 Gamma rays cannot directly cause SEU, therefore any SEU from them must be induced indirectly. Gamma rays can produce (γ, α), (γ, n), and (γ, p) reactions in device materials (i.e., production of alpha particles, neutrons, and protons, respectively). Each of these particle's propensity for producing SEU has been discussed in corresponding sections. Generally, their SEU production is less than that of galactic heavy ions, mainly because the ions are "primary" particles, whereas the others are mainly secondaries, which often affects their energetics negatively and, thus, their SEU production. However, a spacecraft environment mainly characterized by one SEU-producing particle type, such as protons in the Van Allen belts, can be susceptible to that particle even though it is a secondary because of earlier nuclear reactions by which it was created.

2.5 The concentration of ^{233}U atoms cm^{-3}, $N_{233}(t)$, decays according to $\dot{N}_{233}(t) = -\lambda N_{233}(t)$, so that $N_{233}(t) = N^0 \exp(-\lambda t)$. Its half-life $T_{1/2}$ is obtained from $N_{233}(T_{1/2}) = \frac{1}{2}N^0 = N^0 \exp(-\lambda T_{1/2})$, where λ is the decay constant. Then, from the preceding: **(a)** $\lambda = \ln(2)/T_{1/2} = 0.6932/1.6 \times 10^5 = 4.33 \times 10^{-6}$ yr$^{-1} = 1.374 \times 10^{-13}$ s^{-1}. **(b)** $N_{233} = \rho_{233}N_0/A_{233} = \rho_{233}N_0/233$ is the concentration of ^{233}U, in atoms cm^{-3}; thus, in atoms g^{-1}, $N_{233}/\rho_{233} = N_0/233 = 6 \times 10^{23}/233$

$= 2.58 \times 10^{21}$ atoms g^{-1} $= 2.58 \times 10^{18}$ mg^{-1}. The disintegration rate is $\lambda N_{233}/\rho_{233} = (1.374 \times 10^{-13})(2.58 \times 10^{18}) = 3.55 \times 10^{5}$ atoms s^{-1} mg^{-1}, decaying into as many alpha particles. **(c)** From Figure 2.9, for $T_{1/2} = 1.6 \times 10^{5}$ years. $E_\alpha = 4.9$ MeV. **(d)** For $E_\alpha = 4.9$ MeV, Fig. 2.11 shows the range to be $(20)(1.16) = 23.2$ μm $= 0.9$ mils. Not very penetrating. Alphas will not even go through newsprint.

2.6 $\lambda_f = 24/7 = 3.43$, $\sigma_f = (3.43)^{1/2} = 1.85$. So, the "flare count" per year is taken as 3.43 ± 1.85.

2.7 Figure 2.19 depicts the comparison between galactic cosmic ray ion SEU rate curves with those due to the Van Allen belts. Their intersection about the "minimum" altitude occurs at about 3000-NM. Spacecraft with only cosmic ray SEU measures (e.g., part selection, special EDAC, etc.) should not orbit continually below about 6000 NM. Figures 2.20 and 2.21 should be also be perused in this regard. This is, of course, no concern for highly eccentric orbits, where the spacecraft darts in and out of the belts.

2.8 The dip is not an intrinsic attribute of the ion, as it comes about because of the sum of the components present in essentially all such ion spectra. The major component is composed of multiply-ionized ions, whereas the minor (anomolous) component is believed to be made up of singly-ionized counterparts.

2.9 **(a)** Eliminating v from $E = mc^2 = m_0c^2(1 - v^2/c^2)^{-1/2}$ and $p = mv = m_0v(1 - v^2/c^2)^{-1/2}$ yields, after some algebraic manipulation, $E^2 = p^2c^2 + m_0^2c^4$ or $E^2 = p^2c^2 + E_0^2$. **(b)** After the ion enters the part and is stopped at the end of its track, its momentum $p = 0$; thus, from (a), $E = E_0 = m_0c^2$.

2.10 Io, one of Jupiter's close moons is volcanically active ejecting sulphur, oxygen, and sodium and other ion types which are ultimately captured by the Jupiter radiation belts.

2.11 Because $[A/N_0]$ = (mass/"mole")(particles/"mole")$^{-1}$ = mass/particle.

Chapter 3

3.1 **(a)** $N = \rho N_0/A$ = number of material atoms per cubic centimeter, as seen by dimensional analysis. As Z is the number of protons in the atomic nucleus, it is also the number of electrons per atom. Hence, NZ is the electron concentration in electrons per cubic centimeter. **(b)** The energy loss per unit mass (g cm^{-2}) in the material (i.e., $-dE/d\xi = NZ/\rho$), which is proportional to the number of electrons per gram. Penetration is measured in units of $\xi = \rho x$ (g cm^{-2}) instead of distance x. For most materials of interest, their mass number A is in the region where $A/Z \cong \frac{1}{2}$. In this region, it is seen that $-dE/d\xi = N_0 Z/A$, which is then seen to be constant to "first order." For high values of A, such as for the trans-uranium elements, Z/A decreases to less than $\frac{1}{2}$ as A increases. Also, the refined expression for dE/dx is much more complicated than the preceding, [e.g., as in (3.5), (3.6)].

3.2 From Fig. 3.6 it is seen that as shielding thickness increases, the incident electron flux is, of course, attenuated. However, beyond a certain material shield thickness depending on material and electron parameters, the bremsstrahlung electron production concentration begins to surpass that of the incident electrons, so that further shield thickness is at most a marginal effect. This is called bremsstrahlung limited.

3.3 LET implies a maximum energy transfer from the incident particle to the material it is penetrating. The Bragg–Gray cavity theory asserts that this maximum energy transferred corresponds to the energy of the incident particle, whose range is equal to the mean chord length in the material body volume. Stopping power implies no such energy transfer upper limit.

3.4 The charge of an electron is 1.6×10^{-19} C. For Si, 3.6 eV/1.6×10^{-19} C = 22.5 MeV pC^{-1}. Thus, E_{si} (MeV)/22.5 = Q_{si} (pC). For GaAs, $4.8/1.6 \times 10^{-19}$ = 30 MeV pC^{-1}, or E_{ga} (MeV)/30 = Q_{ga} (pC).

3.5 For Si, 1 pC μm^{-1} = (1 pC μm^{-1})(22.5 MeV pC^{-1})(10^4 μm cm^{-1}) (2330 mg cm^{-3}) = 96.57 MeV cm^2 mg^{-1}. For GaAs, (1 pC μm^{-1}) (30 MeV pC^{-1})(10^4 μm cm^{-1})(5320 mg cm^{-3}) = 56.39 MeV cm^2 mg^{-1}.

3.6 Consider two particles in one dimension (head-on collisions only), with the heavy ion mass m_h and speed v_h and the electrons of mass m_e and speed v_e. Using subscripts i implies "before the collision" and subscript f means "after the collision," for the following:

(a) Conservation of momentum before and after the collision qives

$$m_h v_{hi} + m_e v_{ei} = m_h v_{hf} + m_e v_{ef} \quad \text{or} \quad m_h (v_{hi} - v_{hf}) = m_e (v_{ef} - v_{ei})$$

All collisions are assumed head-on elastic (billiard balls), so that kinetic energy is conserved, giving

(b) Conservation of energy before and after the collision gives

$$\tfrac{1}{2} m_h v_{hi}^2 + \tfrac{1}{2} m_e v_{ei}^2 = \tfrac{1}{2} m_h v_{hf}^2 + \tfrac{1}{2} m_e v_{ef}^2 \quad \text{or} \quad m_h (v_{hi}^2 - v_{hf}^2) = m_e (v_{ef}^2 - v_{ei}^2)$$

Dividing (a) by (b) and rearranging gives

(c)
$$v_{hf} = \left(\frac{m_h - m_e}{m_h + m_e}\right) v_{hi} + \left(\frac{2m_e}{m_h + m_e}\right) v_{ei}$$

and similarly for v_{ef},

(d)
$$v_{ef} = \left(\frac{2m_h}{m_h + m_e}\right) v_{hi} + \left(\frac{m_e - m_h}{m_h + m_e}\right) v_{ei}$$

For an incident heavy-ion of mass m_h ionizing (colliding with) an electron of mass m_e, $m_e \ll m_h$, from (c) and (d), respectively,

(e)
$$v_{hf} \cong v_{hi}$$

(f)
$$v_{ef} \cong 2v_{hi}$$

This means that (e) the speed of the incident heavy ion is virtually unchanged by the collision with the electron, but that (f) the electron can "rebound" with up to twice the speed of the ion. This implies that straggling of the ion is a mimimum. This is seen from the straggling angular deviation parameter; that is, the root mean square collision deflection (scattering) angle, $\theta_{rms} = \frac{1}{2}[21 \text{ MeV}/E \text{(MeV)}] (x/2\lambda_R)^{1/2}$. It is seen to vary inversely with the incident particle energy E. For the collision of two electrons, this is not the case, where for two equal masses, (c) and (e) yield $v_f = v_i$ (subscripts e and h dropped) and 21 MeV/E(MeV) is usually much larger than for heavy ions, so θ_{rms} is also larger.

As an aside, if the atomic electron is stationary (valence electron), then $v_{ei} = 0$ in collision with an incident electron ($m_h = m_e$). Then, from (c) and (d), $v_{hf} = 0$ and $v_{ef} = v_{hi}$ (i.e., the incident electron is stopped "cold"), whereas the stationary electron now bounds away with all of the momentum the incident electron originally had.

When oblique collisions (in two dimensions) are accounted for, the results are moderated by the angles involved and the preceding yields minimum and maximum results for the corresponding angles. For example, the fractional decrease in kinetic energy, $-\Delta K$, of the incident particle in colliding with a stationary electron is $-\Delta K = (\frac{1}{2}m_h v_{hi}^2 - \frac{1}{2}m_h v_{hf}^2)/(\frac{1}{2}m_h v_{hf}^2) = 1 - v_{hf}^2/v_{hi}^2$. From (c) with $v_{ei} = 0$, $v_{hf} = [(m_h - m_e)/(m_h + m_e)] v_{hi}$ and inserting above gives $|\Delta K| = 4m_h m_e/(m_e + m_h)^2$. $|\Delta K|$ is actually the maximum fractional change in kinetic energy when the angles are taken into account. It also results in (3.25) in Section 3.9.

3.7 The higher the energy of the incoming particle, the less its direction is changed by an interaction (collision) in the medium of interest. The scattering is more forward for higher energies in the parlance of particle collision theory and experiment, so that its random walk from collision to collision approaches its crow-flight distance from the point of incidence in the material to its end point (i.e., its range). A low-energy particle can have collisions which can radically change its direction, causing it to careen randomly about and resulting in a much longer random-walk distance than its range.

3.8 The abscissa in the usual particle spectrum is the energy E, whereas the abscissa in the LET spectrum is $|dE/dx|$ (i.e., the LET).

3.9 Most segments of the track distance are very small, over which the particle LET is reasonably constant and normally not near the the Bragg peak (Fig. 3.1). Thus, for an incident particle of given energy, there corresponds a nominal LET value.

3.10 Identical ions with identical speeds do not all have the same range. The distribution of ranges about the mean range is called straggling. A normal distribution fits these distributions quite well. Another type of straggling is energy straggling with an energy distribution about a mean residual energy, after the incident particle has penetrated the medium. Straggling theory solves problems like the fraction of emerging particles from a material with energies greater than an energy $E_0 \leq E_i$, where E_i is the particle incident energy.

Chapter 4

4.1 The spiraling xenon ions in the Dees have a centrifugal force balanced by the magnetic field force [i.e., $H(Z'e)v = mv^2/r = mr\omega^2$; $\omega = 2\pi f$, where f is the oscillator frequency in Hz]. In MKS units, 1 weber m^{-2} = 10,000 G, 1 AMU = 1.6×10^{-27} kg.

(a) From $HZ'ev = mv^2/r$, $Z' = 2\pi fm/He = 2\pi(1.5 \times 10^7)(131)(1.6 \times 10^{-27})/(2.2)(1.6 \times 10^{-19}) \cong 56$. Yes.

(b) $E_{xe} = \frac{1}{2}mv^2 = (HrZ'e)^2/2m = [(2.2)(1.118)(56)(1.6 \times 10^{-19})]^2/[(2)(131)(1.6 \times 10^{-27})] = 1.16 \times 10^{-9}$ J; or $(1.16 \times 10^{-9}$ J$)(6.25 \times 10^{12}$ MeV J$^{-1})(131) = 55.2$ MeV per nucleon.

(c) $(1 - v^2/c^2)^{1/2} = (1 - (2\pi fr/c)^2)^{1/2} = \{1 - [(2\pi)(1.5 \times 10^7)(1.1176)/(3 \times 10^8)]^2\}^{1/2} = 0.9363$. No.

4.2 $E_p = \int_0^V Q(V') dV'$, where V is the voltage atop the accelerator dome with respect to ground. The dome acts as a capacitor also with respect to the system electrical ground. Because $Q = CV$, $E_p = \int_0^V CV' \, dV' = \frac{1}{2} CV^2 = \frac{1}{2} QV$. The ratio of the energy of the modified to the unmodified accelerator is $E/E_p = 2QV/\frac{1}{2}QV = 4$.

4.3 From $E = 2QV$, $(400 \text{ MeV})((1.6 \times 10^{-13} \text{ J/MeV}) = 2Q(15 \times 10^6)$, giving $Q = (400/30)(1.6 \times 10^{-19}) \text{ C}$. Because the value of a unit of $\pm$ charge is 1.6×10^{-19} C, the uranium ion charge state is $400/30$ or $+13$ (i.e., U^{13+}).

4.4 An "SEU-hard" part is characterized by a high critical LET, L_c. The larger the L_c of the part, the greater is the required charge per unit track length collected to cause SEU. For example, the galactic cosmic ray LET spectrum, discussed in Sections 3.7 and 3.8, cuts off to negligible levels beyond LETs of 40–50 MeV cm² mg⁻¹. This upper limit means that virtually no galactic cosmic ray particle in the earth's neighborhood has a LET greater than 50 MeV cm² mg⁻¹. Hence, if the critical LET of a part is greater than 50 MeV cm² mg⁻¹, then this part is immune to SEU caused by galactic cosmic rays.

4.5 $A_L = C_1/(1+r) = 248.2/2.303 = 107.8$ AMU

$A_H = rC_1/(1+r) = (1.303)(248.2)/2.303 = 140.4$ AMU

$V_L = (rC_3/C_1)^{1/2} = [(1.303)(3.35 \times 10^{20})/248.2]^{1/2}$
$= 1.33 \times 10^9$ cm s⁻¹.

$V_H = (C_3/rC_1)^{1/2} = [(3.35 \times 10^{20})/(1.303)(248.2)]^{1/2}$
$= 1.02 \times 10^9$ cm s⁻¹.

4.6 (a) The LET for the light group $(\text{LET})_L = -dE_L/d\xi = 63.69 - 2(7.04)\xi$, whereas that for the heavy group $(\text{LET})_H = -dE_H/d\xi = 72.89 - 2(20.35)\xi$. (b) Because the density of air in the chamber is 0.247×10^{-3} g cm⁻³, $x_{\max} = \xi_{\max}/\rho_{\text{air}} = 1.235/0.247 \cong 5.0$ cm. (c) The LET values of an ion usually implies its value at incidence to the medium (i.e., at $x = \xi = 0$). Then from (b), $-E'_L(0) = 63.69$ MeV cm² mg⁻¹, $-E'_H(0) = 72.89$ MeV cm² mg⁻¹.

4.7 The expression $L = L_{\text{eff}} \cos \theta = L_c \cos \theta_c$ is actually three relations viz. (a) $L = L_{\text{eff}} \cos \theta$, (b) $L = L_c \cos \theta_c$, and (c) $L_{\text{eff}} \cos \theta = L_c \cos \theta_c$. From (a), $L \leq L_{\text{eff}}$, as $\cos \theta$ is always less than or equal to 1. From (c), $L_{\text{eff}} = L_c (\cos \theta_c / \cos \theta)$, and $\cos \theta_c / \cos \theta$ is also always less than or equal to 1, because the experimental procedure for device exposure to the heavy-ion beam begins at normal incidence ($\theta = 0$), then rotates the device to $\theta_c > 0$. Thus $\theta_c > \theta$ always; then $\cos \theta_c / \cos \theta < 1$, hence $L_{\text{eff}} \leq L_c$. In the case where L_c is apparently achieved at normal incidence (i.e., $\theta = \theta_c = 0$, so that $L_{\text{eff}} = L_c$), it is considered questionable because there is virtually no experimental latitude to "home-in" on a $\theta_c = 0$ result. Then the procedure is to substitute an ion of higher L, so that the device under test must be rotated to reach L_c, as discussed. This is described in detail in ASTM SEU Measurement Guide, F1192-90, Sections 7.3.1.4(2) and 7.3.1.5(2), April 1990. Hence, with $L \leq L_{\text{eff}}$ and $L_{\text{eff}} \leq L_c$, the answer is (1), $L \leq L_{\text{eff}} \leq L_c$.

4.8 (a) Using (4.33) i.e., $1/\bar{r}_{\text{pr}} = 1/\bar{R} + 1/\bar{r}$ or $1/20 = 1/40 + S/4V$ (Section 1.3). $S = V/10 = \ell^2 w/10 = 2\ell^2 + 4\ell w$. This yields for $w = \ell/3 = 100/3$, $\ell = 100$, not quite a "domino" ($w < \ell/3$) parallelepiped. (b) From (4.33), $S/V = 4(1/\bar{r}_{\text{pr}} - 1/\bar{R}) = (2\ell w + 2\ell h + 2wh)/\ell wh$, so that $1/h + 1/w + 1/\ell = 2(1/\bar{r}_{\text{pr}} - 1/\bar{R})$.

4.9 $|(dE/dx)|/h\nu$ gives the number of laser photons per micron that produce charge from ionization. $h\nu = hc/\lambda$ is the energy quantum per photon of wavelength $\lambda = c/\nu$. The charge e of the electron is 1.6×10^{-7} pC, so $e \cdot |(dE/dx)|/h\nu$ yields LET in terms of pC μm^{-1} within a constant of proportionality.

4.10 From $\ln(L_{\text{eff}}) = \ln(L) - \ln(\cos\theta)$, $dL_{\text{eff}}/L_{\text{eff}} = \tan\theta \, d\theta \cong (d\theta)^2 \cong (\Delta\theta)^2$ for small θ, and L is the LET (constant) of the incident ion.

4.11 If $\eta \sim \cos\theta$, then $L \cong \Delta E \cos\theta/c \sec\theta = L_{\text{eff}} \cos\theta/\sec\theta$, yielding $L_{\text{eff}} = L \sec^2\theta$.

Chapter 5

5.1 The cell critical charge $Q_c = C\Delta V = 0.05 \times 10^{-12}(5.5 - 2.5) = 0.15$ pC. (a) Using the "figure of merit" expression $E_r = 5 \times 10^{-10} abc^2/Q_c^2 = 5 \times 10^{-10}(10)(10)(3^2)/(0.15)^2 = 2 \times 10^{-5}$ errors per cell-day. (b) For this SRAM, "1K" = 1024, so that $(1.024 \times 10^3)(2 \times 10^{-5})\frac{1}{2} = 1.02 \times 10^{-2}$ errors per device-day. The factor of $\frac{1}{2}$ implies that "on the average" only 50% of the cells are biased so as to be SEU susceptible at any given time.

5.2 (a) From Section 1.3, the mean chord length $\bar{s} = 4V/S = 4abc/2(ab + ac + bc)$. A thin parallelepiped is characterized by $c \ll a,b$. Neglecting c in the denominator yields $\bar{s} \cong 2c$. (b) $Q_c = 10^6(1.6 \times 10^{-19}) = 0.16$ pC; $L_c = Q_c/c\rho_{\text{si}}$. Because the dimensions of $c\rho_{\text{si}}$ is g/cm^2, $L_c = (0.16 \text{ pC})(22.5 \text{ MeV pC}^{-1})/(1 \times 10^{-4}\text{cm} \, \mu\text{m}^{-1})(2330 \text{ mg cm}^{-3}) = 15.5 \text{ MeV cm}^2 \text{mg}^{-1}$.

5.3 (1) An ICBM, especially on an over-the-pole trajectory, because the polar magnetic lines of force allow even galactic cosmic rays to intrude into the earth's neighborhood to become incident on the ICBM. (2) High-flying aircraft exposed to cosmic ray atmospheric reaction products, mainly neutrons. (3) Ground-based systems especially on an atmosphereless satellite like the moon. This is also the case with an atmosphere at ground level, but at much reduced SEU-inducing flux.

5.4 For a monoenergetic flux, the classic integral over energy of the product of SEU cross section with the corresponding flux reduces to a simple product of "cross section times flux," assuming that the cross section is essentially constant. Thus, (a) the SEU error rate $E_r = \sigma_{SEU}\phi_{accel}$ where σ_{SEU} is the device SEU cross section and ϕ_{accel} is the accelerator beam flux. Then $\sigma_{SEU} = \#SEUs/$ fluence $= 10/(1.6 \times 10^8)(10) = 6.25 \times 10^{-9}$ cm^2 per device.
(b) The corresponding SEU error rate $E_r = \sigma_{SEU}\phi = (6.25 \times 10^{-9})(10^5) = 6.25 \times 10^{-4}$ SEUs per device-day. (c) These results strictly hold for this device only, as the device cross section is unique. However, engineering extrapolations can be made to encompass other ions. Extrapolation across other part types is much more difficult to make with confidence.

5.5 Because the discrepancy between the known charge produced by a heavy-ion track in a device, without considering funneling, falls far short of that known to be collected at an SEU-sensitive information node.

5.6 The sensitive region volume depletion-layer equivalent flat-plate capacitance $C = \epsilon\epsilon_o ab/c$. Because $Q_c = C\Delta V$, where ΔV is the voltage swing across the depletion layer/junction, $E_p = 5 \times 10^{-10}abc^2/(C\Delta V)^2 = 5 \times 10^{-10}c^4/ab(\epsilon\epsilon_o\Delta V)^2$.

5.7 No; for example, one contradictory factor, as shown in Fig. 6.14, Section 6.5, is that the critical charge Q_c scales as $Q_c = 0.023\ell^2$, where ℓ is the feature size. So that from Problem 5.6, $E_r = 5 \times 10^{-10}abc^2/(0.023)^2\ell^4$. As the device-sensitive region thickness c is comparable to ℓ, then $E_r \sim 1/\ell^2$. Hence, the smaller the device (feature size), the greater will be the SEU error rate.

5.8 Because protons, as yet, can only cause SEU indirectly because their LET is too small to cause it directly. Indirect causes include proton-induced nuclear reactions in the device material that produce ions with sufficient LET to, in turn, cause SEU. In the future, submicron (≤ 0.5-μm) devices will possess small enough Q_c that can be provided by protons directly. The preceding manifests a complication that cannot be overcome by an analogous "figure of merit" expression. (b) Resort is then made to the aforementioned classic integral over energy of the SEU cross section–flux product. Here, the proton cross section with energy is obtained from empirical fits to experimental cross-section data, and using a threshold energy parameter as a lower limit to the above integral. Future submicron device direct proton-induced SEU cross sections are orders of magnitude larger than the present levels.

5.9 The 150-kft curve cuts off at $Q_c \cong 0.2$ pC. This means that at this altitude, no incoming cosmic ray particle has sufficient LET to provide more than 0.2 pC of critical charge. However, at the lower altitudes, the Q_c at 75 kft and 55 kft are respectively 1.5 pC and 0.4 pC. At these altitudes, in addition to any incoming cosmic ray flux, neutrons also can cause SEU which corresponds to Q_c increases over that at higher altitudes.

5.10 (a) In silicon, 1 pC μm^{-1} = 100 MeV cm^2 mg^{-1}; then Q_c/c (pC μm^{-1} = L_c (MeV cm^2 mg^{-1})/100. Thus, $Q_c = cL_c/100 = (2 \times 5)/100 = 0.1$ pC. From Fig. 5.17 (plot #3), the corresponding NIE = 3×10^{-15} cm^2 μm^{-3}. The SEU-sensitive volume per cell $V = abc (\mu m^3) = \sigma_d c/16K = (0.2 \times 10^8)(2/16,384) = 2.44 \times 10^3 (\mu m)^3$. Then from $E_r = \epsilon V \phi_n$ (NIE) = (0.3)(2.44 $\times 10^3$)(0.5)(3 $\times 10^{-15}$) = 1.1×10^{-2} errors per bit per second. (b) $(1.1 \times 10^{-12})(8.64 \times 10^4)$s day$^{-1} \times 10^7$ bits = 0.95 errors per memory day or $(1/0.95)(24$ h day$^{-1}) = 25.2$ h between SEU errors in the memory.

5.11 The flux from an 8-μCi monoenergetic alpha source at 1 mm, where E_0 is the alpha particle energy, is $\phi_\alpha(E_0) = S_0 \exp(-\mu r)/4\pi r^2 \cong S_0/4\pi r^2$, because the exponential factor is essentially unity at these short distances from the part in air. Thus, $\phi_\alpha(E_0) = (8 \times 10^{-6})(3.7 \times 10^{10})/4\pi(0.1)^2 = 2.36 \times 10^6$ alphas cm^{-2} s^{-1}. Because the flux is monoenergetic and the cross section is essentially constant, $\sigma_\alpha = E_{r\alpha}/\phi_\alpha(E_0) = (1/60 \times 20 \times 13)(2.36 \times 10^6) = 2.72 \times 10^{-11}$ cm^2 bit^{-1}. This assumes that all of the 4π flux source is incident on the part, which it is not. The computation is approximate in this sense at least. The above cross section is very small compared to the sensitive region area of a single transistor (bit), which is 3×10^{-6} cm^2. Hence, the corresponding Q_c is probably very much larger than what could be delivered by alpha particles of reasonable energy. Therefore, the part is very probably hard to alpha-induced SEU.

5.12 The integral of the product $\Phi'(L)\sigma_u(L) dL$ is proportional to the SEU cross section per bit per unit time, as $\sigma_u(L)$ is the SEU cross section per memory cell, each of which stores one bit. For N equal to the number of, say, memory bits per device, multiplication of the above error rate by N yields the SEU error rate per device per unit time; that is, "targets per cm^3" becomes "bits per device".

Chapter 6

6.1 θ is bounded by the dimensions of the device. The funnel length cannot protrude beyond the SEU-sensitive region depletion layer dimensions. Thus, $0 \leq \theta \leq \theta_{max}$, where $\theta_{max} = (\tan^{-1} X_j)/W$ and X_j is the sensitive region distance to the junction edge.

6.2 If E_r were independent of scale as construed in Section 6.5, the device size would make no difference. Submicron devices would enjoy the same SEU error rate (at least for geosynchronous cosmic rays for which E_r is derived) as larger devices. Hence, there would be no bar to tiny devices, which could be made as small as feasible from the SEU standpoint. In actuality, this scaling breaks down for large scaling ratios (≥ 10). Hence, the SEU problem is unfortunately a nontrivial one.

6.3 Because there are many more cells per chip in the small cell device; that is, the chip SEU error rate is the product of the cell SEU error rate and the number of cells per chip, within about a factor of 2. This assumes that the cell Q_{cs} are independent and equal which is usually the case within engineering approximations.

6.4 The geographical coordinates of Rio place it near the center of the South Atlantic Anomaly, where the inner Van Allen belt (protons) descends to very low altitudes, thus producing mainly proton-induced SEU.

6.5 Not especially because galactic cosmic rays incident at geosynchronous altitudes, even though affected, are remote from the South Atlantic Anomaly with respect to altitude and somwhat with respect to latitude. The former can be appreciated qualitatively from Fig. 5.5.

6.6 Initially, the ion-track particle density is so high that the electrons and holes are forced to diffuse in train. Also, very early, the ambipolar mobility vanishes, because the particles are momentarily frozen with zero net field motion.

6.7 As $A_{beam} = ab(\cos\theta)[1 + k\tan\theta - (2\ell/c)\sin\theta] = \sigma_{seu}\cos\theta$, then inserting for $\ell = c\sec\theta_c = cL_c/L_{ion}$ yields $\sigma_{seu} = ab\{1 + k[\tan\theta - (2L_c/L_{ion})\sin\theta]\}$ or, inserting for $\ell = c\sec\theta_c = c(L_c/L_{eff})\sec\theta$ gives $\sigma_{seu} = ab\{1 + k[1 - (2L_c/L_{eff})\tan\theta]\}$.

6.8 The bias shift angle θ_0 accounts for the finite dimensions of the track. For example, for pure gates, $\theta_0 = \tan^{-1}[d/(t_{ox} + 15)]$ where $d(\mu m)$ is the track diameter, $t_{ox}(\mu m)$ is the gate thickness, and $15\,\mu m$ corresponds to the device layers exclusive of the track.

6.9 From (6.82) at normal incidence (cosine term assumed to be unity), then $E_{FT(c)} \cong 126/\text{LET}$ (MV/cm).

6.10 (a) LET = 27.37 is the root of (6.85) for $V_{GS} = 0$. (b) As shown in Fig. 6.31, $V_{GS} = 0$ places the device outside its safe operating area (SOA). This implies that the device will probably suffer SEGR under these bias conditions.

Chapter 7

7.1 For dense beams of low-Z particles, although not necessarily limited to them, a multiple-particle single SEU can occur due to the coherent action of two or more incident particles which enhances the corresponding SEU coherent cross section (Chap. 5, [16]).

7.2 Because the cell susceptibility to SEU varies inversely as the resistor values, increasing them by 50% (i.e., increasing R to $1.5R$) for R not too high, then the error rate is three-halves of what it was prior to the spurious radiation.

7.3 (a) Using the Lagrange multiplier method, write the Lagrangian $L = \prod_{i=1}^{n}(1 - \sigma_{iseu}) + \lambda \sum_{i=1}^{n} k_i (1 - \sigma_{iseu})$, where the Lagrange multiplier λ is to be determined. Now calculate $\partial L / \partial \sigma_{jseu} = 0 = -\prod_{i=1}^{n}(1 - \sigma_{iseu})/(1 - \sigma_{jseu}) - \lambda k_j$, or

(1) $\prod_{i=1}^{n}(1 - \sigma_{iseu}) + \lambda k_j (1 - \sigma_{jseu}) = 0$, $j = 1,\ldots,n$. Summing these equations on j yields,

(2) $n\prod_{i=1}^{n}(1 - \sigma_{iseu}) + \lambda C_{seu} = 0$ using the cost contraint. Inserting λ from (2) into (1) yields the k_i^* [viz. $k_i^* = C_{seu}/n(1 - \sigma_{iseu})$].

(b) $R_{seu}^* = \prod_{i=1}^{n}(1 - \sigma_{iseu}) = (C_{seu}/n)^n / \prod_{i=1}^{n} k_i^*$. Note that efforts made to reduce the "overall SEU cross section" of a part increases the reliability, but at an increased cost C. Usually $\sigma_{iseu} \ll 1$ so that $k_i^* \cong C_{seu}/n$, implying that the cost be allocated about evenly over the n chips. To realize a stronger relation between k_i^* and σ_{iseu}, the reliability function above is modified by replacing σ_{iseu} in it with $c_i \sigma_{iseu}$. c_i provides an enhanced cross section represented by an additional cost K given by $K = \sum_{i=1}^{n} c_i \sigma_{iseu}$. This results in a modified cost conversion coefficient $k_i = k_i^* [1 + (a/C_{seu})(K - nc_i \sigma_{iseu})]$, where a is a constant.

7.4 From the binomial probability distribution of an n-bit word, $1 = (p + q)^n = p^n + np^{n-1}q + \frac{1}{2}n(n-1)p^{n-2}q^2 + \cdots =$ prob. of 0 errors in n bits + prob. of 1 error in n bits + prob. of 2 errors in n

bits $+\cdots$. The reliability for the W word memory is $R = [p^n + np^{n-1}q + \frac{1}{2}n(n-1)p^{n-1}q^2]^W$. Inserting for p from the preceding gives $R \cong \{(n^2/2)[\exp(\lambda t) - 1]^2\}^W \cdot \exp(-nW\lambda t)$, $n \gg 1$, which is the usual case; or after a long time into the mission $[\exp(\lambda t) \gg 1]$, then $R \sim (n^2/2)^W \exp(-nW\lambda t)$.

7.5 Form the inner products $\mathbf{w} \cdot \mathbf{a}$, $\mathbf{w} \cdot \mathbf{b}$, and $\mathbf{w} \cdot \mathbf{c}$ by multiplying corresponding component columns; then sum each result horizontally, subtracting all possible 2's leaving a remainder.

```
                 1 1 1 1 0 1 0
                 0 0 0 1 1 1 1
Σ mod 2:         0 0 0 1 0 1 0   = 2 mod 2 = 0,

                 1 1 1 1 0 1 0
                 0 1 1 0 0 1 0
Σ mod 2:         0 1 1 0 0 1 0   = 3 mod 2 = 1,

                 1 1 1 1 0 1 0
                 1 0 1 0 1 0 1
Σ mod 2:         1 0 1 0 0 0 0   = 2 mod 2 = 0
```

The binary number from the remainders is $010 = 2$; therefore, the second bit is in error. The corrected word is then $\mathbf{w}^* = 1\,0\,1\,1\,0\,1\,0$.

7.6 Mainly, because at altitudes much less than geosynchronous, the major SEU-inducing particles are Van Allen belt protons. They are much easier to stop in shielding materials than the high-energy heavy cosmic ray ions at geosynchronous altitude.

7.7 The pulse height analyzer (PHA) systems displays essentially the number distribution of interaction-producing charges in the device versus their energy and number of these charges. The peak in the distribution corresponds to the SEU cross-sectional area of the sensitive node volume, by definition of sensitive node (i.e., its cross section). This is in the sense that the maximum ordinate is equivalent to the (maximum) number of sensitive volume interactions $\sigma_{seu}\phi$, where ϕ is the SEU-inducing flux incident on the device. The maximum number of interactions is obtained from the PHA display and ϕ is obtained from the detector part of the PHA system, so that the quotient yields $\sigma_{seu} = \max \text{ ordinate}/\phi$.

7.8 For a finite beam time T, the SEU cross section is simply given by $\sigma_{seu} = n_{seu}(T)/\varphi_{seu}(T)$, where n_{seu} is the total number of SEUs meaasured during the time T $(n_{seu} = \int_0^T N_{seu}\,dt)$, and the corresponding

344 / Single Event Phenomena

fluence $\varphi_{seu}(T) = \int_0^T \Phi_{seu} dt$. It is assumed that T is long with respect to $1/N_{seu}$, the time between SEUs, to obtain "good measurement statistics."

7.9 For large T, the beam ions can begin to cripple some devices from ionizing dose to cause them to suffer permanent damage.

7.10 From (7.36), the skin thickness satisfies $x = R(E^{(o)}) - R(E^{(i)})$. The Burrell range–energy relation for protons is $R(E) = 300 \ln(1 + 3.76 \times 10^{-6} E^{7/4})$ in cm and E is in MeV. $E^{(o)}$ is the proton energy on the skin and $E^{(i)}$ is the proton energy after penetration. Inserting $R(E)$ into the skin thickness relation yields $x = 300 \cdot \ln[(1 + 3.76 \times 10^{-6} E^{(o)7/4})/(1 + 3.76 \times 10^{-6} E^{(i)7/4})]$. For a thin skin, $3.76 \times 10^{-6} E^{7/4}$ is small compared to unity. Then, using the fact that $\ln(1+u) \cong u$ for small u and that $E^{(i)} \cong E^{(o)} - \Delta E$ results in the simpler relation $x \cong 300 (3.76 \times 10^{-6})(7/4)(\Delta E/E) E^{(o)7/4} = 300 (3.76 \times 10^{-6})(7/4)(0.1)(40)^{7/4} = 0.125$ cm or about 50 mils.

7.11 $Q_c = C_{cell}(V_{DD} - V_B) = C_{cell}(V_{DD} - V_{DD}/3)$ or $V_{DD} = (3/2)(Q_c/C_{cell}) = (3/2)(100/25) = 6$ V.

Chapter 8

8.1 With cell dimensions in microns and the critical charge in picocoulombs, the error rate is $E_r = 5 \times 10^{-10} abc^2/Q_c^2 = 5 \times 10^{-10}(10)(10)(2)^2/(10^{-2})^2 = 2 \times 10^{-3}$ errors per bit-day.

8.2 Here, $E_r = 5 \times 10^{-10} \sigma_L/L_c^2$. However, σ_L must be expressed in μm^2 and L_c in pC μm^{-1}. So $\sigma_L = (10^{-6})(10^8)(\mu m\ cm^{-1})^2 = 100\ \mu m^2$. Recall that 1 pC μm^{-1} is equivalent to approximately 100 MeV cm^2 mg^{-1}, then $L_c = 0.5/100 = 5 \times 10^{-3}$ pC μm^{-1}. Thus, $E_r = 5 \times 10^{-10}(10^2)/(5 \times 10^{-3})^2 = 2 \times 10^{-3}$ errors per bit-day, the same as in problem 8.1.

8.3 $E_r = 5 \times 10^{-10} \sigma_L((\mu m)^2)/L_c^2 (pC\ \mu m^{-1})^2 = 5 \times 10^{-10}(10^8)\sigma_L(cm^2)/10^{-4}(MeV\ cm^2\ mg^{-1})^2 = 500\sigma_L/L_c^2$. Actually, the most recent coefficient value should be 200 instead of 500.

8.4 One way to renormalize all the entries in Table 8.3 for the part of interest is to multiply them by $\sigma_{Lpart}/\sigma_{L93L422}$. Here, this ratio is $5 \times 10^{-3}/0.1 = 0.05$. This method implies a cross-section correction in the "canonical" expression $E_r \sim \int \sigma_L(L)\phi(L) dL$. Hence, the modified coefficient should be $C = (28)(0.05) = 1.4$ for this part.

8.5 They should do nothing with regard to SEU considerations, because these parts following SEU onset remain essentially unaffected and certainly will not affect system operation. They are far down in the hierarchy of SEU-vulnerable part types.

8.6 Table 8.5 yields a nominal $A = 26$. To obtain the SEU error rate, this A is used to compute the empirical cross section $\sigma_p(A, E)$ in (8.3). Substituting $\sigma_p(A, E)$, including an appropriate proton flux at this altitude and inclination angle in a solar minimum into (8.2), yields the classical SEU error rate integral. For $A = 18$, this integral has been computed and tabulated for a number of altitudes and inclination angles in Tables 5.5 and 5.6. For $A = 26$, a correction must be made in order to use the $A = 18$ tables. This is accomplished using

$$E_{pr}(A) = \frac{E_{pr}(18)\,(18/A)^{14}\,(18 + B)}{A + B}$$

where B is a correction factor obtained from the rightmost columns in Tables 5.5 and 5.6. For a solar minimum, the tables give $E_{pr}(18) = 4.69 \times 10^{-4}$ upsets per bit-day. The corresponding $B = 23.5$ from the tables. Then the desired SEU error rate due to these protons is $E_{pr}(26) = 4.69 \times 10^{-4}\,(18/26)^{14}\,(18 + 23.5)\,(26 + 23.5)^{-1} = 2.29 \times 10^{-6}$ errors per bit-day. Is this error rate high or low? See Fig. 8A.1 for the Galileo Orbiter spacecraft.

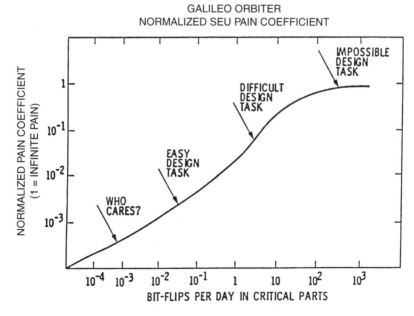

Figure 8A.1. Normalized pain chart for the Galileo Orbiter (circa 1983). Courtesy of JPL Pasadena, CA.

8.7 From Section 5.3.1, the expression $A = 15 + L_{0.1} = 15 + 11 = 26$. This is the same A as in problem 8.6, so the E_{pr} here is the same.

8.8 Inserting $(2 \times 10^5) < \int_{10 \text{ MeV}}^{\infty} \phi_{40}(E,t)\,dE >_{da}$ for ϕ_{WNR}^i on the left-hand side of (8.19) yields $(2 \times 10^5) \int_0^{T_{exp}} dt < \int_{10 \text{ MeV}}^{\infty} \phi_{40}(E,t)\,dE >_{da} = T_{eq} < \int_{10 \text{ MeV}}^{\infty} \phi_{40}(E,t)\,dE >_{da}$. Canceling the daily average flux at 40,000 ft leaves $2 \times 10^5 T_{exp} = T_{eq} = 7.45 \times 10^5$ from Table 8.8, or $T_{exp} = 7.45 \times 10^5 / 2 \times 10^5 = 3.725$ h.

8.9 $\phi_{WNR} = 2 \times 10^5 (1.6) = 3.2 \times 10^5$ neutrons per cm^2 s.

8.10 The spectra levels fall with increasing material thickness because more incident particles can be absorbed in the thicker, and thus more massive, material. The peaks shift toward higher energies with increasing material thickness because the less energetic (soft) incident particles are more preferentially absorbed (higher-absorption cross section at lower energies), leaving the higher-energy particles (lower-absorption cross section) to penetrate the material with their higher-energy peaks. In the parlance, the penetrated spectrum is becoming "harder" as more and more of the softer component is being absorbed "out" of the incident flux energy spectrum.

Index

A energy parameter, 144, 148
   LET correlation, 153ff
accelerators, 77, 271
actinide nuclei, 162
Adam's
   10 percent flux, 130
   90 percent flux, 130
alpha particle, 23, 38, 57, 62, 104
   cross section, 39
   decay chain, 39
   induced SEU, 39ff, 162
   range, 42, 163
   reactions, 39, 164
ambipolar
   diffusion/transport, 195, 227
   mobility, 196, 197
anomolously large solar flare, 43, 157
Argus test series, 52
asymptotic (saturation) cross section, 95, 129, 147
atomic
   mass unit (AMU), 80
   number, 62, 145
   weight, 62
avalanche multiplication factor, 71ff, 302

B L coordinates, 30
beam
   area, 227
   incidence angle, 134, 136
   intensity, 90
   ion, 90, 227, 272
binaries, stellar, 36
binomial distribution, 252
bit data lines, 166
Bragg curves, 62, 66
bremsstrahlung, 61, 68
burnout, 223, 228, 300, 302ff
   current density, 322
burst generation rate (BGR), 158, 300

californium-252, 100ff, 123
charge collection, 193
   by drift, 182
   diffusion, 141, 181, 239
   funnel, 193ff
   internal bipolar action, 181
   deposited, 121
charge coupled devices (CCDs), 57, 67, 174
chord length
   cumulative distribution, 130ff
   frequency distribution, 1, 10, 12, 15, 123
   mean length, 13
   mean number, 15, 128
convex body, 11

conservation laws, 106
cosmic ray, 24, 62, 174
  abundance, 25
  energy spectrum, 26
  extra galactic, 24
  flux, 27, 89
  heavy ions, 17
  neutrons, 20, 140
  protons, 26
  secondaries, 24, 34
  showers, 61, 71
CREME code, 132ff, 149
  options, 132
critical
  angle, 234
  charge, 91ff, 142, 278ff
  LET, 91ff, 129
cross coupled resistors, 307, 311
cross section, 1, 3
  coherent, 145
  interaction, 4, 5ff
  asymptotic (saturation), 147
  macroscopic, 158
  microscopic, 158
  multi-bit, 234
crow flight distance, 3
current density (ion), 193
cutoff energy, 78ff
cyclotron, 88ff, 122

depletion layer collapse, 180
destructive physical analysis (DPA), 174
device burnout, 223
displacement damage, 155, 293
DMOS, 224ff
domino parallelepiped model, 131, 235
dose rate latchup, 297
double bit error detection, 247
drift *See* charge collection

effective LET, 94, 96

electron, 26, 73
  fast, 52
  interactions, 172
electron free orbits, 218
equivalent days, 132, 133
error correction chip, 253
error detection and correction (EDAC), 249ff, 271
error rates,
  SEU
  alphas, 167
  geosynchronous, 128, 276
  ground, 169
  neutrons, 155
  protons, 140, 288
  SEL, 210, 297

fault tolerance, 247
feature size, 142, 146
feedback resistors, 271, 307, 311
figure of merit, 131, 227
  SEU error rate, 131ff, 136, 276
fission fragment
  energy distribution, 103
  mass number, 102
  products, 70, 105
  spectra, 103
  yield, 102
flares (*See* solar flares)
fluence, 2, 13
flux, 2, 4, 234
funnels, funneling, 180
  models, 182ff, 208, 227

GaAs, 276
gamma rays, 36
gamma ray bursts, 36, 57
gate rupture, 179, 288
  SEU induced (SEGR), 220ff
geomagnetic cutoff, 28, 31
geosynchronous orbit, 128, 310
    Van Allen belt correlation, 148
glossary, 322

ground level SEU, 169ff
guidelines, 275, 305

Hamming processes, 247
Heinrich
  curves, 61, 75, 131
  spectra, 76
  LET spectra, 76
high flying aircraft, 155
hot electrons, 52, 236

ion beam, 90
ionization
  dose, 23, 240ff
  energy loss, 61, 67ff
  indirect, 142
  track, 192ff
ionosphere, 48ff

junction contact potential, 199, 278
Jupiter, 54, 58

knock-on recoil particles, 155

$L_{0.1}$, 96
L shells, 30, 56, 65
lasers, 112ff
  absorption coefficient, 115, 117
  band gap narrowing, 116, 117
  beam charge density, 118
  charge deposition profile, 118
  funneling threshold, 118ff
  properties of, 112ff
latchup cross section, 297
linear energy transfer (LET), 61, 74, 83, 115, 123
  dimensions, 75
  spectra, 75
low earth orbit (LEO), 57, 145, 216, 227, 257
low pass characteristic, 307

macroscopic cross section, 158
magnetic field
  dip, 216, 218
  singularities, 218
magnetic poles, 32, 54, 90, 308
magnetopause, 27
magnetosphere, 27, 32
manufacturer's abbreviations, 323
mass number, 90
McIlwain L parameter, 30, 56
mean chord length, 13, 82
mean free path, 7, 151
mean square diffusion distance, 198
meson, 34, 173
microscopic cross section, 158
MIR space station, 219
modified figure of merit SEU rate, 283
Mott-Rutherford cross section, 157
multi-channel analyser, 104, 269, 271
multiple events
  two populations, 240
  upsets, 237
multiple
  bit SEU, 17, 232ff, 272
  bit map, 237ff, 257
  particle coherence, 270ff
muon, 34, 57, 173

neutrino, 35, 141
neutron, 155
  energy spectrum, 160
  flux spectrum, 156
  reactions, 163
  sources, 20, 296
nodes, 93
nova, 24, 25
nuclear detonations, 37, 243
nucleosynthesis, 25

orbits
  geosynchronous, 128
  inclination, 286
  low earth, 308
occultation of, 33

package
  ceramic, 162

impurities, 162
lid, 162
materials, 162
pair production, 61, 71
parallelepiped model, 14, 74, 97, 124, 136, 201
    mean projected area, 128
parity check, 247
part index tables, 311ff
pion, 34
plasma filament/track, 181, 182, 300
Poisson distribution, 255
polar orbits, 216
polysilicon resistors, 307
power MOSFETs, 223ff, 308
    burnout, 306
protons
    approximation methods, 149
    energy threshold (*See* A parameter)
    flux, 144, 287
    latchup, 301
    low earth orbits, 35, 219
    range-energy, 65
    reactions 141
    SEU cross sections 143
    computation 147ff, 149
    geosynchronous 128ff
    one parameter 144, 286, 290
    two parameter 146, 150, 292
    Van Allen belts 148
pulse height analyser (PHA) (*See* multi-channel analyser)

quasi-neutral region, 196

RADPAK, 257
rads, 243
random
    mean length, 13
    track segment, 18
range, 65
    alphas, 42, 163
    deuterons, 42
    electrons, 81
    energy, 66
    ion, 66
    particle, 139
    protons, 65
    tritons, 42
redundancy/reliability, 245, 271
    allocation of, 246ff
refined figure of merit polynomials, 282, 284, 285
renormalization, 282
rigidity, 28

safe operating area (SOA), 276
saturation cross section, 95
scaling factor, 204ff
scanning electron microscope (SEM), 276
screening factor, 183
scrubbing, 245
    single error correction, 248
    double error detection, 248ff
    fault tolerance, 247
second breakdown, 302
SEL, 179, 210, 297
    compared with dose rate, 210, 297
    cross section, 213
    latchup, 212, 214
    with temperature, 214ff
sensitive
    nodes, 93, 235
    regions, 270
sensors/detectors, 263ff
SEU
    alpha particle, 128
    charge collected, 199
    cross coupled resistors, 307
    cross section, 94, 144, 203
    databases, 321, 325
    figure of merit formulas, 127
    galactic cosmic rays, 128
    gate rupture, 220ff
    geosynchronous, 128

compared with proton, 145
ground level, 127
latchup, 297ff
latitude dependence, 216
mitigation guide, 305
multiple error rates, 295
neutron, 155, 293
proton, 140, 284
scaling, 204
sheet resistance, 278
shielding, 173, 256ff, 259, 271
submicron features, 146, 245
shield average cross section, 261
shield path, 258, 259
solar
  cells, 52
  flare, 23, 42, 57
    anomolously large, 43
    flux, 47
    minimum/maximum, 47, 146ff, 151
    models, 45, 47
    spectra, 45
    statistics, 43, 45, 48, 241
  wind, 19, 27
South Atlantic Anomaly, 216ff, 257
spallation, 68, 141, 143
SPICE code, 278
spreading resistance, 188ff
stopping power, 62ff
straggling, 82, 85
  energy, 83
  particle, 84
  range, 83
stuck bits, 212, 245
sunspots, 27
supernova, 25
surface barrier detectors (SBDs), 268
susceptibilities, 307
synchrotron radiation, 54, 68

target, 11
temperature variations, 212

test
  organizations, 324
  facilities, 325
thorium, 162
threshold LET, 96
throughput rate, 205
time of flight, 108
track
  conductance, 191
  density profile, 184, 233
  LET, 134
  segment, 18, 234
  terminal points, 18
  transport, 181, 183, 194, 233
  traversal points, 18
triple modular redundancy (TMR), 255
tunneling, 40

units
  centigrays (MKS) 1 rad
  grays (MKS) 100 rads
  rads (CGS) 100 ergs/gm, 243

Van Allen belts, 48, 284
  inner, 50, 58
  Jupiter (Jovian), 54, 58
  other planetary, 54
  outer, 51, 58
    slot region, 284
    temporary peaks, 284
Van de Graaf generator, 88, 90, 122
VELA program, 36
vendor claims, 309
VLSIC, 239

Weibull distribution, 8ff
WNR spectrum, 293
word (select) lines, 166

x-rays, 68

yttrium (YAG) laser, 116

Z (atomic number), 62, 145